THIRD EDITION

Instrumentation

for Process

Measurement

and Control

NORMAN A. ANDERSON

CHILTON COMPANY   RADNOR, PENNSYLVANIA

14 15 16 17 18   8 7 6 5 4 3 2

CONTENTS

Preface

SECTION I FEEDBACK PROCESS CONTROL 1

1 Introduction to Process Control 1

Types of Processes 5   Processes with More Than One Capacity and Resistance 6   Dead Time 6   Measurement 7   Symbols 10   The Feedback Loop 10   Feedback Control 14   Controlling the Process 14   Selecting Controller Action 16   Upsets 17   Process Characteristics and Controllability 17   Controller Responses 18   On/Off Control 19   Proportional Action 20   Integral Action (Reset) 23   Derivative Action 25   Selecting the Controller 27   Conclusion 29   Questions 30

2 Process/Pressure Measuring Instruments 33

What Is Pressure? 33   Units of Measurement 35   Pressure Measurement 35   The Pascal 36   Bar Versus Pascal 37   Gauge, Absolute, and Differential Pressure 37   Understanding the Effects of Gravity 38   Gravity-Dependent Units 38   Gravity-Independent Units 39   Pressure Standards 39   Plant Instruments That Measure Pressure Directly 44   Bell

Instrument 44   Slack or Limp-Diaphragm 44   Pressure Gauges 46
Liquid or Steam Pressure Measurement 48   Seals and Purges 48   Pulsation Dampener 48   Metallic Bellows 49   Pressure Transmitters 52   Signal Transmissions 52   Pneumatic Recorders and Indicators 53   Mechanical
Pressure Seals 55   Calibration Techniques 59   Field Standard 59   Portable Pneumatic Calibrator 59   Force-Balance Pneumatic Pressure Transmitter 60   Pneumatic Relay 62   Principle of Operation 63   Absolute
Pressure Transmitter 64   Questions 65

3 Level and Density Measurements 69

Level Measurement Methods 69   Float-and-Cable 70   Displacement
(Buoyancy) 71   Head or Pressure 73   Capacitance 78   Conductance 79
Radiation 79   Weight 79   Ultrasonic 80   Thermal 81   Density Mea-
surement Methods for Liquids and Liquid Slurries 81   Hydrostatic Head 82
Radiation 84   Vibration 84   Temperature Effects and Considerations 84
Differential Pressure Transmitter 85   Questions 87

4 Flow Measurement 90

Constriction or Differential Head Type 91   Primary Devices 94   Secondary
Devices 99   Relating Flow Rate to Differential 100   Effect of Temperature
on Flow Rate 103   Variable Area Meters (Rotameter) 109   Open-Channel
Flow Rate Measurements 109   Primary Devices 110   Installation and
Selection Considerations 116   Velocity Flowmeters 117   Magnetic Flowmeter 117   Vortex Flowmeter 121   Turbine Flowmeter 122   Other Flowmeters 123   Conclusion 124   Questions 124

5 Temperature and Humidity Measurements 126

Temperature 126   Filled Thermal Systems 128   Electrical Systems 130
Thermocouples 130   Resistance Thermal Detectors 139   Thermistors 144
Humidity Measurements 144   Questions 148

6 Analytical Measurements 151

Electrical Conductivity 151   Types of Calibration 154   Calibration in
Conductivity 154   Calibration in Terms of Concentration of
Electrolyte 155   Polarization 156   Cell Construction 156   Electrodeless
Conductivity Measurements 157   Hydrogen Ion Activity (pH) 159   Ioniza-
tion or Dissociation 159   The pH Scale 161   Measurement of pH—The
Glass Electrode 164   Reference Electrode 165   Temperature Compensa-
tion 168   Reading the Output of the pH Electrodes 169   pH Control 170
Summary 172   Oxidation-Reduction Potential 172   Ion-Selective Measure-
ment 173   Chromatography 174   Capacitance 175   References 176
Questions 176

SECTION II PNEUMATIC AND ELECTRONIC CONTROL SYSTEMS 179

7 The Feedback Control Loop 179

The Closed-Loop Control System 183   Phase Shift Through *RC* Networks 184   Oscillation 184   Stability in the Closed-Loop System 187   Nonlinearities 188   Controllers and Control Modes 189   Two-Position Control 189   Throttling Control 191   Application of Proportional Control 193   Proportional-Plus-Integral Control 195   Adding Derivative (Rate) 199   Selecting the Controller 200   Questions 202

8 Pneumatic Control Mechanisms 205

The Flapper-Nozzle Unit 205   Valve Relay or Pneumatic Amplifier 207   The Linear Aspirating Relay or Pneumatic Amplifier 208   Proportional Action 209   Control Mechanism Requirements 211   The Automatic Controller 212   Derivative and Integral 215   Manual Control Unit 216   Transfer 217   Set Point 220   The Closed-Loop Pneumatic Control System 223   The Process 225   The Controller Set Point 225   Valve and Actuator 226   Final Control Element 227   Dynamic Behavior of Closed-Loop Control Systems 228   Adjusting the Controller 230   Batch Controller 231   Principle of Operation 232   Load Bias (Preload) 233   Questions 234

9 Electronic (Analog) Control Systems 237

Feedwater Control Systems 238   Transmitters 238   The Controllers 241   Principle of Operation 243   Increase/Decrease Switch 244   Deviation Signal Generation 244   Derivative Action 244   Proportional Band Action 245   Final Summing and Switching 245   High and Low Limits 245   External Integral ($-R$) Option 246   External Summing ($-S$) Option 246   Power Supply Fault Protection Circuit 246   Automatic/Manual Switch 247   Controller Adjustments 247   General Description of the Feedwater Control System 247   Control Station 248   M/2AX + A4-R Drum Level Control Unit 249   M/2AX + A4 Feedwater Control Unit 250   M/2AP + SUM Feedwater/Steam Flow Computing Unit 250   Square Root Extractor 250   Feedwater Control 252   Closed-Loop Operation 252   Comparison of Electronic and Pneumatic Systems 254   Questions 256

SECTION III ACTUATORS AND VALVES 259

10 Actuators 259

Valve Actuator 259   Valve Positioner 261   Electrical Signals 263   Principle of Operation 265   Electric Motor Actuators 267   Questions 268

11 Control Valves 270

Capacity of a Control Valve 272   Valve Sizing 272   Determining Pressure
Drop Across the Valve 273   Cavitation and Flashing 274   Valve Range-
ability 276   Selection Factors 278   Sequencing-Control Valves 279
Viscosity Corrections 281   References 292   Questions 292

SECTION IV CONTROL LOOP ADJUSTMENT AND ANALYSIS 295

12 Controller Adjustments 295

Proportional-Only Controller 295   Proportional-Plus-Integral Controller 296
Adding Derivative Action 297   Tuning Maps 298   Adjusting a Controller
with the Proportional, Integral, and Derivative Mode 299   Closed-Loop
Cycling Method 301   Questions 303

13 Step-Analysis Method of Finding Time Constant 304

Block Diagrams 304   Step Analysis of Single-Time-Constant System 308
Step Analysis of Two-Time-Constant System 309   Percent-Incomplete
Method 310   Multicapacity System 313   Finding Control Modes by Step
Analysis 315   On/Off Control Action 316   Proportional Control Action 316
Proportional-Plus-Integral Action 318   Proportional-Plus-Derivative Con-
trol Action 319   Proportional-Plus-Derivative-Plus-Integral Control
Action 320   Using the Reaction Curve to Determine Controller Adjust-
ments 320   References 324   Questions 325

14 Frequency Response Analysis 328

Finding the Time Constant from the Bode Diagram 331   Testing a
System 331   Control Objectives 335   Adding Derivative 336   Adding
Integral 337   Adjusting the Proportional Band 339   Closed-Loop
Response 339   Conclusion 340   Questions 341

SECTION V COMBINATION CONTROL SYSTEMS 343

15 Split-Range, Auto-Selector, Ratio, and Cascade Systems 343

Duplex or Split-Range Control 343   Auto-Selector or Cutback Control 345
Nonpipeline Application of Auto-Select/Cutback Control Systems 348   Cascade Controller 352   Saturation in Cascaded Loops 355   Questions 357

16 Feedforward Control 360

Definition 360   History 360   Advantages 360   Technique 361   Application to Heat Exchanger 361   Distillation Control 369   Conclusion 370
References 370   Questions 371

SECTION VI PROCESS COMPUTERS AND SIMULATION 372

17 Computer Interface and Hardware 372

Basic Elements of the Computer 372   Input 373   Central Processing Unit (CPU) 373   Output 374   Types of Memory 374   Core 374 Drum 375   Disk 375   Semiconductor (Solid-State) 376   Operator Communication Devices 377   Process Interface Equipment 377   Conclusion 385   Questions 385

18 Computer Software and Operation 388

A Brief Introduction 388   Basic Computer Operation 390   Real-Time Clock and Power Fail/Restart Logic 394   Control Software 394   Achieving Process Control 396   Questions 400

19 Programmed Control Systems 402

Description of Circuit 403   Sequential Control Systems 405   Programmable Controllers 406   Programming Language 406   Programming 408 Logic System Automation 409   System Control Displays 410   Conclusion 411   Questions 411

20 Constructing and Instrumenting Real and Simulated Processes 413

Simulated Processes 413   Constructing an Electric Process Simulator 414 The Foxboro Electronic Process Simulator 415   Pneumatic Process Simulator 420   Operation of the Pneumatic Simulator 423   Time Base 424 Real Processes 427   Questions 433

Appendix 436

Units and Conversion Tables for Process Control 436   Specific Gravities of Common Liquids 451   Standard Pipe Dimensions 452   Properties of Saturated Steam and Saturated Water 453   Resistance Values 454 Velocity and Pressure Drop 455   Thermocouple Temperature 458   Relative Humidity (% saturation) Tables 476

Glossary 480

Answers to Questions 488

Bibliography 491

Index 492

The acceptance of previous editions of *Instrumentation for Process Measurement and Control* has affirmed the need for a basic text on the subject. In the years since the first edition of this book appeared, a number of instructors have requested that the material be treated in more detail. This edition is designed to meet that need. Questions at the conclusion of each chapter have also been expanded to enable the student to judge his progress. The answers appear at the back of the book.

A third edition was also mandated by the need to convey information about new equipment in the field of process control instrumentation. The material here attempts, in a basic way, to meet the needs of the instrumentation engineer or technician who must learn how equipment operates. Mathematics have been kept to a minimum throughout.

Digital devices are having increasing impact on the process control field. While it is beyond the scope of this text to deal with these techniques in detail, digital devices are introduced, as are some of the associated terms.

The objectives of this edition, then, are: to introduce the fields of process measurement and feedback control; to provide useful reference material to students and persons working in the field; to bridge the gap between basic technology and more sophisticated systems that offer promise for the future.

The author is indebted to the many people who assisted with the preparation of this volume. Chief among them are David Fuller, who checked its technical content; Richard Sherman, who edited the text; and Roberta Kavanaugh, who typed the manuscript. The author would be remiss if he did not also acknowledge the contributions of his fellow workers at The Foxboro Company and the support of company management. A special acknowledgment is also appropriate for the thousands of instrument students whose questions and responses over the years have turned the author's thinking toward this approach to the understanding of a most challenging and fascinating field.

Norman A. Anderson, P.E.

1

Introduction to Process Control

The technology of process instrumentation continues to grow in both application and sophistication. In 1774, James Watt employed the first control system applying feedback techniques in the form of a flyball governor to control the speed of his steam engine. Ten years later, Oliver Evans used control techniques to automate a Philadelphia flour mill.

Process instrumentation developed slowly at first because there were few process industries to be served. Such industries began to develop at the turn of the twentieth century, and the process instrumentation industry grew with them. However, only direct-connected process instruments were available until the late 1930s. In the 1940s, pneumatic transmission systems made complex networks and central control rooms possible. Electronic instrumentation became available in the 1950s, and its popularity has grown rapidly since. The most recent decade has produced digital computer techniques to improve the performance of more complex processes. However, present trends indicate that future process plants will employ combinations of analog and digital systems.

True control balances the supply of energy or material against the demands made by the process. The most basic (feedback) systems

1

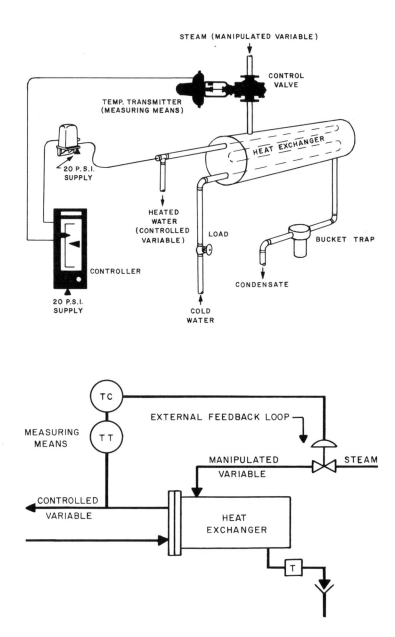

Fig. 1-1. (A) The process to be controlled occurs in a heat exchanger. All elements of the pneumatic control system are shown—transmitter, controller, valve, input water, output water, and steam. (B) Block diagram of the elements listed in A.

measure the controlled variable, compare the actual measurement with the desired value, and use the difference between them (error) to govern the required corrective action. More sophisticated (feedforward) systems measure energy and/or material inputs to a process to control the output. These will be discussed in Chapter 16.

The control loop in Figure 1-1 is shown in both actual and schematic form. The process is a shell and tube heat exchanger, and the temperature of the heated water is the controlled variable. This temperature is measured by a pneumatic temperature transmitter, which sends a pneumatic signal proportional to temperature to the pneumatic analog controller. The desired water temperature is set on the controller's set-point dial. The controller changes a pneumatic output signal according to the difference between the existing value (temperature) and the desired value (set point). The output signal is applied to the valve operator, which positions the valve according to the control signal. The required quantity of heat (steam) is admitted to the heat exchanger, causing a dynamic balance between supply and demand.

The various control equipment components that may be used to

Table 1-1. Analogy Between Characteristics of Basic Physical Systems

Variable	Electrical System	Hydraulic System	Pneumatic System	Thermal System
Quantity	Coulomb	ft³ or m³	Std. ft³ or m³	Btu or joule
Potential or effort variable	emf E (volt)	Pressure P (psi or kPa) (ft or m of head)	Pressure P (psi or kPa) (m or mm of head)	Temperature T (degrees Fahrenheit or Celsius)
Flow variable	Coulomb/s Current I (amperes)	Flow Q (ft³/s or L/s) (gal/min)	Flow Q (ft³/s or m³/s) (lb/min)	Heat flow dQ/dt (Btu/s or watts)
Resistance	R (ohm) $= \dfrac{\text{volt}}{\text{amp}}$	psi/(ft³/s) ft head/ft³/s) sec/ft²	psi/(ft³/s)	deg/(Btu/s) deg/watt
Capacitance	q(farad) $\dfrac{\text{Coulombs}}{\text{volts}}$	ft³/ft = ft²	ft²	Btu/deg
Time	Seconds	Seconds	Seconds	Seconds

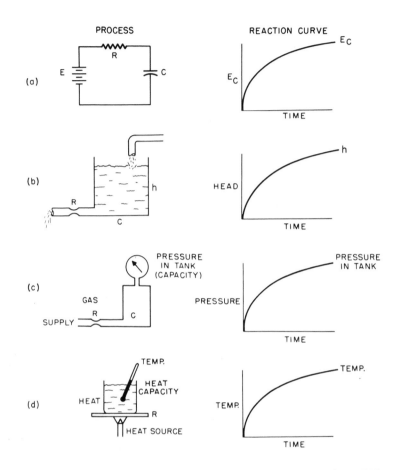

Fig. 1-2. Four types of systems: (a) electric, (b) hydraulic, (c) pneumatic, and (d) thermal. Each has a single capacity and a single resistance and all have identical response characteristics.

regulate a process and certain aspects of process behavior will be discussed in this text. Examples of some completely instrumented process systems will be given to demonstrate the practical application of the instrument components.

The physical system to be controlled may be electrical, thermal, hydraulic, pneumatic, gaseous, mechanical, or any other physical type. Figure 1-2 and Table 1-1 compare several common systems. All follow the same basic laws of physics and dynamics.

The behavior of a process with respect to time defines its *dynamic characteristics*. Behavior not involving time defines its *static characteristics*. Both static (steady) and dynamic (changing with time) responses must be considered in the operation and understanding of a process control system.

Types of Processes

The simplest process contains a single capacity and a single resistance. Figure 1-2 illustrates a single-capacity, single-resistance process in (a) electrical, (b) hydraulic, (c) pneumatic, and (d) thermal forms. To show how these behave with respect to time, we can impose a step upset (sudden change) in the input to the process and examine the output. The resulting change in process variable with respect to time is plotted in Figure 1-3. The reaction curve of all four types of systems will be identical.

This type of curve (exponential) is basic to automatic control. It can be obtained easily with an electrical capacitor and resistor arranged as in Figure 1-2a.

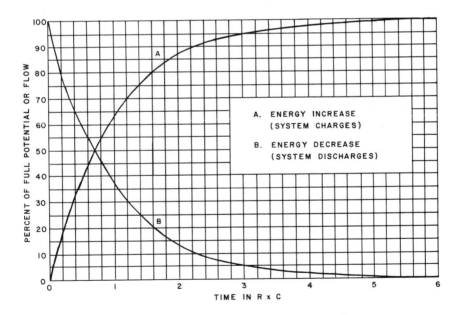

Fig. 1-3. Universal time-constant chart, showing exponential rise and decay.

Figure 1-2a shows a simple *RC* circuit—a resistor, capacitor, and battery source in series. The instant the circuit is closed, the capacitor starts to charge to the voltage of the battery. The rate at which the capacitor charges gradually decreases as the capacitor voltage approaches the battery voltage (voltage curve A in Figure 1-3). Although the rate varies, the time it takes the capacitor to charge to 63 percent of the battery voltage is a constant for any one value of *R* and *C*. Thus, no matter what the voltage of the battery, the capacitor charges to 63 percent of the battery voltage in a time interval called the time constant, or characteristic time (*T*) of the circuit. The value of *T* in seconds is the product of the resistance (in ohms) and the capacitance (in farads).

Note that the charging time increases with an increase in either *R* or *C*.

This simple *RC* circuit is often used to produce the transient waveform shown, which is called an *exponential-rise transient*.

On discharge, the circuit reacts similarly. For example, if the battery in Figure 1-2a is replaced by a solid conductor, the charged capacitor discharges 63 percent of its charge in *RC* seconds.

The simple *RC* circuit shown in Figure 1-2a symbolizes many real physical situations. It is important to examine the circuit in detail.

In *RC* seconds, the capacitor charges to 63.2 percent of the applied voltage. In the next *RC* seconds, the capacitor charges to 63.2 percent of the remaining voltage, or to 87 percent of the applied voltage. In the third interval of *RC* seconds, the capacitor charges to 95 percent of the applied voltage. Although the capacitor never charges to exactly 100 percent of the applied voltage, it does charge to 99 percent in 4.6 *RC* seconds as shown in Figure 1-3, which is a curve of capacitor voltage (or current) versus time. Note that time is plotted in *RC* (time-constant) units.

Processes with More Than One Capacity and Resistance

In practice, a process will contain many capacitance and resistance elements. Figure 1-4 illustrates a process containing two resistance elements and two capacitance elements. Figure 1-5 shows the resulting process reaction curve. Note that the additional capacitance and resistance essentially affect the initial curve shape, adding a delay to the process.

Dead Time

Dead time is a delay between two related actions. For example, assume that the temperature sensor shown in Figure 1-1 was located 10 feet

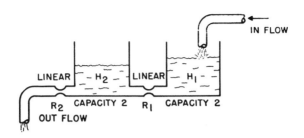

Fig. 1-4. Multicapacity system.

(3.048 m) away from the heat exchanger. If the liquid travels at a velocity of 10 feet (3.048 m) per second, a dead time of one second will occur. In some process control situations, dead time becomes the most difficult factor in the equation. Dead time may also be called pure delay, transport lag, or distance/velocity lag. Dead time is rarely found in its pure form, but occurs frequently in combination with resistance-capacitance and other types of lags. Dead time is a difficult factor to equate when applying control to the process.

Measurement

To employ feedback control, we must first measure the condition we wish to maintain at the desired standard. The condition (variable) may be temperature, pressure, flow, level, conductivity, pH, moisture content, or the like.

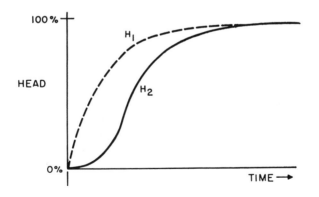

Fig. 1-5. Characteristic curve of multicapacity system.

The measuring element is connected to the control element. In many installations, the measurement is located far from the controller. This problem is solved by using a measuring transmitter (Figure 1-6). The measuring transmitter usually develops an electrical signal for an electronic controller or a pneumatic signal for a pneumatic controller.

Measuring transmitters have attained great popularity in the process industries. They perform the measurement and develop a pneumatic or electric signal proportional to the variable in one unit. This signal can be transmitted long distances. Pneumatic transmitters generally produce an air pressure change of 3 to 15 psi or 20 to 100 kPa (see p. 36 for definition of pascal unit) for measurement change of 0 to 100 per-

Fig. 1-6. Measuring transmitters convert the variable to be measured into a proportional pneumatic or electrical signal.

Table 1-2. Standard ISA and SAMA Functional Diagram Elements

(FT) FLOW TRANSMITTER	√ SQUARE ROOT EXTRACTOR	K PROPORTIONAL CONTROL ACTION
(LT) LEVEL TRANSMITTER	X MULTIPLIER	∫ INTEGRAL (RESET) CONTROL ACTION
(PT) PRESSURE TRANSMITTER	÷ DIVIDER	d/dt DERIVATIVE CONTROL ACTION
(TT) TEMPERATURE TRANSMITTER	± BIAS, ADDITION OR SUBTRACTION	f(t) TIME FUNCTION CHARACTERIZER
(ZT) POSITION TRANSMITTER	Δ COMPARATOR, DIFFERENCE	f(x) UNSPECIFIED OR NONLINEAR FUNCTION CHARACTERIZER
PANEL LIGHT	Σ ADDER, SUMMER	x.xx QUOTATION ITEM NUMBER
(I) INDICATOR	Σ/n AVERAGER	✳ MOUNTED ON THE FRONT OF PANEL
(R) RECORDER	Σ/t INTEGRATOR	REGULATED PROCESS AIR
(T) RELAY COIL	NORMALLY OPEN RELAY CONTACT	NORMALLY CLOSED RELAY CONTACT
(T) AUTO/MANUAL TRANSFER SWITCH	MANUAL SIGNAL GENERATOR	(A) ANALOG SIGNAL GENERATOR
T TRANSFER OR TRIP RELAY	S SOLENOID ACTUATOR	M ELECTRIC MOTOR
> HIGH SIGNAL SELECTOR	⊳ HIGH SIGNAL LIMITER	H/ HIGH SIGNAL MONITOR
< LOW SIGNAL SELECTOR	◁ LOW SIGNAL LIMITER	L/ LOW SIGNAL MONITOR
V⊳ VELOCITY OR RATE LIMITER	◁⊳ HIGH AND LOW LIMITER	H//L HIGH AND LOW SIGNAL MONITOR
A/D ANALOG TO DIGITAL CONV.	R/I RESISTANCE TO CURRENT CONV.	R/V RESISTANCE TO VOLTAGE CONV.
mV/V THERMOCOUPLE TO VOLTAGE CONV.	V/I VOLTAGE TO CURRENT CONV.	I/V CURRENT TO VOLTAGE CONV.
V/V VOLTAGE TO VOLTAGE CONV.	P/I PNEUMATIC TO CURRENT CONV.	P/V PNEUMATIC TO VOLTAGE CONV.
MO MOTORIZED OPERATOR	I/P CURRENT TO PNEUMATIC CONV.	V/P VOLTAGE TO PNEUMATIC CONV.
HO HYDRAULIC OPERATOR	PNEUMATIC OPERATOR	STEM ACTION (GLOBE) VALVE
f(x) UNSPECIFIED OPERATOR	THREE-WAY SELECTOR VALVE	ROTARY ACTION (BALL) VALVE

cent; that is, 0 percent of measurement yields an output pressure of 3 psi or 20 kPa, 50 percent of measurement yields 9 psi or 60 kPa and 100 percent yields 15 psi or 100 kPa output. Electronic transmitters produce either voltage or current signal outputs. For instance, the output of analog transmitters is commonly 4 to 20 mA dc.

Symbols

A set of symbols has been adopted to show instrumentation layouts and to make these layouts more uniform. Once you become familiar with these symbols, it will become easy to visualize the system.

At present, two sets of symbols are in use. One set is provided by the Scientific Apparatus Makers Association (SAMA) and the other by Instrument Society of America (ISA). In this book the ISA symbols will be used where applicable. Figure 1-7 and Tables 1-2 and 1-3 describe the symbols and identification letters often used. If you are involved in the preparation or use of instrument loop diagrams, it is suggested that you obtain the publication that defines the standards employed. A loop diagram must contain the information needed for both engineering and construction. This includes identification, description, connections and location, as well as energy sources.

The Feedback Loop

The objective of a control system is to maintain a balance between supply and demand over a period of time. As noted previously, supply and demand are defined in terms of energy or material into (the manipu-

INSTRUMENT FOR SINGLE MEASURED VARIABLE

INSTRUMENT
LOCALLY
MOUNTED

INSTRUMENT
MOUNTED ON
BOARD

INSTRUMENT
MOUNTED
BEHIND
BOARD

INSTRUMENT FOR TWO MEASURED VARIABLES

Fig. 1-7. Instrument for measured variables.

Table 1-3. Meanings of Identification Letters

	FIRST LETTER		SUCCEEDING LETTERS		
	Measured or Initiating Variable	Modifier	Readout or Passive Function	Output Function	Modifier
A	Analysis		Alarm		
B	Burner flame		User's choice	User's choice	User's choice
C	Conductivity (electrical)			Control	
D	Density (mass) or specific gravity	Differential			
E	Voltage (EMF)		Primary element		
F	Flow rate	Ratio (fraction)			
G	Gaging (dimensional)		Glass		
H	Hand (manually initiated)				High
I	Current (electrical)		Indicate		
J	Power	Scan			
K	Time or time schedule			Control station	
L	Level		Light (pilot)		Low
M	Moisture or humidity				Middle or intermediate
N	User's choice		User's choice	User's choice	User's choice
O	User's choice		Orifice (restriction)		
P	Pressure or vacuum		Point (test connection)		
Q	Quantity or event	Integrate or totalize			
R	Radioactivity		Record or print		
S	Speed or frequency	Safety		Switch	
T	Temperature			Transmit	
U	Multivariable		Multifunction	Multifunction	Multifunction
V	Viscosity			Valve, damper, or louver	
W	Weight or force		Well		
X	Unclassified		Unclassified	Unclassified	Unclassified
Y	User's choice			Relay or compute	
Z	Position			Drive, actuate or unclassified final control element	

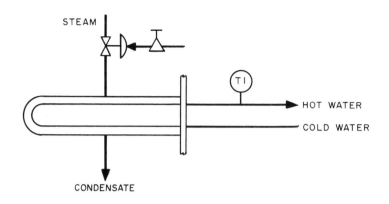

Fig. 1-8. Heat exchanger.

lated variable) and out of (the controlled variable) the process. The closed-loop control system achieves this balance by measuring the demand and regulating the supply to maintain the desired balance over time.

The basic idea of a feedback control loop is most easily understood by imagining what an operator would have to do if automatic control did not exist. Figure 1-8 shows a common application of automatic control found in many industrial plants: a heat exchanger that uses steam to heat cold water. In manual operation, the amount of steam entering the heat exchanger depends on the air pressure to the valve, which is set on the manual regulator. To control the temperature manually, the operator would watch the indicated temperature, and by comparing it with the desired temperature, would open or close the valve to admit more or less steam. When the temperature had reached the desired value, the operator would simply hold that output to the valve to keep the temperature constant. Under automatic control, the temperature controller performs the same function. The measurement signal to the controller from the temperature transmitter is continuously compared to the set-point signal entered into the controller. Based on a comparison of the signals, the automatic controller can tell whether the measurement signal is above or below the set point and move the valve accordingly until the measurement (temperature) comes to its final value.

The simple feedback control loop shown in Figure 1-9 illustrates the four major elements of any feedback control loop.

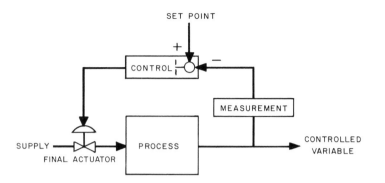

Fig. 1-9. Feedback control loop.

1. *Measurement* must be made to indicate the current value of the variable controlled by the loop. Common measurements used in industry include flow rate, pressure, level, temperature, analytical measurements such as pH, ORP and conductivity; and many others particular to specific industries.
2. For every process there must be a *final actuator* that regulates the supply of energy or material to the process and changes the measurement signal. Most often this is some kind of valve, but it might also be a belt or motor speed, louver position, and so on.
3. The kinds of *processes* found in industrial plants are as varied as the materials they produce. They range from the commonplace, such as loops to control flow rate, to the large and complex, such as distillation columns in the petrochemical industry. Whether simple or complex, they all consist of some combination of capacity resistance and dead time.
4. The last element of the loop is the *automatic controller*. Its job is to control the measurement. To ''control'' means to keep the measurement at a constant, acceptable value. In this chapter, the mechanisms inside the automatic controller will not be considered. Therefore, the principles to be discussed may be applied equally well to both pneumatic and electronic controllers and to the controllers from any manufacturer. All automatic controllers use the same general responses, although the internal mechanisms and the definitions given for these responses may differ slightly from one another.

One basic concept is that for automatic feedback control to exist, the automatic control loop must be closed. This means that information must be continuously passed around the loop. The controller must be able to move the valve, the valve must be able to affect the measurement, and the measurement signal must be reported to the controller. If this path is broken at any point, the loop is said to be open. As soon as the loop is opened—for example, when the automatic controller is placed on manual—the automatic unit in the controller is no longer able to move the valve. Thus, signals from the controller in response to changing measurement conditions do not affect the valve and automatic control does not exist.

Feedback Control

Several principles associated with feedback control can be observed by considering a familiar control situation—adjusting the temperature of water in a bathtub. This is obviously a manually controlled system. One hand feels the water in the tub while the other manipulates the inflow to reach the desired temperature. If a thermometer were used to measure the temperature, greater accuracy would result. Improved measurement generally results in improved control.

The bathtub also illustrates the important effect of process capacity. Capacity (Figure 1-2) is a measure of the amount of energy it takes to change a system a unit amount; thermal capacity is Btu/°F, or the amount of heat required to increase the temperature 1°F. Since the bathtub has a large capacity, it can be controlled in any of several ways—by partially filling the tub with cold water, for example, and then adding enough hot water to reach the desired temperature; or by mixing the hot and cold to get the same result.

Controlling the Process

In performing the control function, the automatic controller uses the difference between the set-point and the measurement signals to develop the output signal to the valve. The accuracy and responsiveness of these signals is a basic limitation on the ability of the controller to control the measurement correctly. If the transmitter does not send an accurate signal, or if there is a lag in the measurement signal, the ability of the controller to manipulate the process will be degraded. At the same time, the controller must receive an accurate set-point signal. In controllers using pneumatic or electronic set-point signals generated within the controller, miscalibration of the set-point transmitter will develop the wrong value. The ability of the controller to position the valve accurately is yet another limitation. If there is friction in the valve, the controller may not be able to move the valve to a specific stem position to produce a specific flow, and this will appear as a difference between measurement and set point. Repeated attempts to position the valve exactly may lead to hunting in the valve and in the measurement. Or, if the controller is able only to move the valve very slowly, the ability of the controller to control the process will be degraded. One way to improve the response of control valves is to use a valve positioner, which acts as a feedback controller to position the valve at the exact position corresponding to the controller output signal. However, positioners should be avoided in favor of volume boosters on fast-responding loops such as flow and liquid pressure.

For proper process control, the change in output from the controller must be in such a direction as to oppose any change in the measurement value. Figure 1-10 shows a direct-connected valve to control

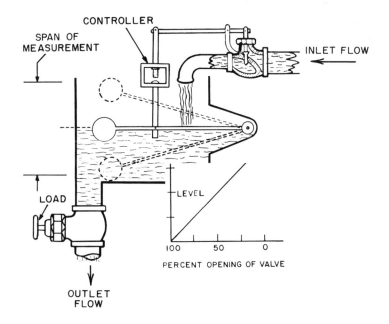

Fig. 1-10. In proportional control, the controlling valve's position is proportional to the controlled variable (level).

level in a tank at midscale. As the level in the tank rises, the float acts to reduce the flow rate coming in. Thus, the higher the liquid level, the more the flow will be reduced. In the same way, as the level falls, the float will open the valve to add more liquid to the tank. The response of this system is shown graphically. As the level moves from 0 to 100 percent, the valve moves from fully open to fully closed. The function of an automatic controller is to produce this kind of opposing response over varying ranges. In addition, other responses are available to control the process more efficiently.

Selecting Controller Action

Depending on the action of the valve, increases in measurement may require either increasing or decreasing outputs for control. All controllers can be switched between direct and reverse action. Direct action means that, when the controller sees an increasing signal from the transmitter, its output will increase. For reverse action, increasing measurement signals cause the controller output to decrease. To determine which of these responses is correct, an analysis of the loop is required. The first step is to determine the action of the valve.

In Figure 1-1, for safety reasons the valve must shut if there is a failure in the plant air supply. Therefore, this valve must be air-to-open, or fail-closed. Second, consider the effect of a change in measurement. For increasing temperature, the steam flow to the heat exchanger should be reduced; therefore, the valve must close. To close this valve, the signal from the automatic controller to the valve must decrease. Therefore, this controller requires reverse, or increase/decrease, action. If direct action is selected, increasing signals from the transmitter will result in a larger steam flow, causing the temperature to increase further. The result would be a runaway temperature. The same thing will occur on any decrease in temperature, causing a falling temperature. Incorrect selection of the action of the controller always results in an unstable control loop as soon as the controller is put into automatic.

Assuming that the proper action is selected on the controller, how does the controller know when the proper output has been reached? In Figure 1-10, for example, to keep the level constant, a controller must manipulate the flow in to equal the flow out. Any difference will cause the level to change. In other words, the flow in, or supply, must balance the flow out, or demand. The controller performs its job by maintaining this balance at a steady rate, and acting to restore this balance between supply and demand whenever it is upset.

Upsets

There are three conditions that require different flows to maintain the level in the tank. First, if the position of the output hand valve is opened slightly, more flow leaves the tank, causing the level to fall. This is a change in demand, and to restore balance, the inlet flow valve must be opened to supply a greater flow rate. A second type of unbalanced condition is a change in the set point. Maintaining any other level besides midscale in the tank causes a different flow out. This change in demand requires a different input valve position. The third type of upset is a change in the supply. If the pressure output of the pump increases, even though the inlet valve remains in the same position, the increased pressure causes a greater flow, which at first causes the level to begin to rise. Sensing the increased measurement, the level controller must close the valve on the inlet to hold the level at a constant value. In the same way, any controller applied to the heat exchanger shown in Figure 1-1 must balance the supply of heat added by the steam with the heat removed by the water. The temperature remains constant if the flow of heat in equals the flow of heat out.

Process Characteristics and Controllability

The automatic controller uses changes in the position of the final actuator to control the measurement signal, moving the actuator to oppose any change it sees in the measurement signal. The controllability of any process depends on the efficiency of the measurement signal response to these changes in the controller output. For proper control, the measurement should begin to respond quickly, but then not change too rapidly. Because of the tremendous number of applications of automatic control, characterizing a process by what it does, or by industry, is an almost hopeless task. However, all processes can be described by the relationship between their inputs and outputs. Figure 1-11 illustrates the temperature response of the heat exchanger when the control valve is opened by manually increasing the controller output signal.

At first, there is no immediate response at the temperature indication. Then the temperature begins to change, steeply at first, then approaching a final, constant level. The process can be characterized by the two elements of its response. The first element is the dead time, or the time before the measurement begins to respond. In this example, a delay arises because the heat in the steam must be conducted to the

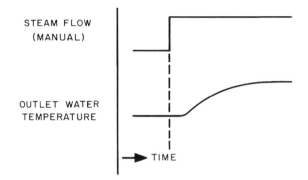

Fig. 1-11. Response of heat exchanger to step upset.

water before it can affect the temperature, and then to the transmitter before the change can be seen. Dead time is a function of the physical dimensions of a process and such things as belt speeds and mixing rates. Second, the capacity of a process is the material or energy that must enter or leave the process to change the measurements—for example, the gallons necessary to change level, the Btu's necessary to change temperature, or the standard cubic feet of gas necessary to change pressure. The measure of a capacity is its response to a step input. Specifically, the size of a capacity is measured by its time constant, which is defined as the time necessary to complete 63 percent of its total response. The time constant is a function of the size of the process and the rate of material or energy transfer. For this example, the larger the tank and the smaller the flow rate of the steam, the longer the time constant. These numbers can be as short as a few seconds, or as long as several hours. Combined with dead time, they define the time it takes the measurement signal to respond to changes in the valve position. A process will begin to respond quickly, but then not change too rapidly, if its dead time is small and its capacity is large. In short, the larger the time constant of capacity compared to the dead time, the better the controllability of the process.

Controller Responses

The first and most basic characteristic of the controller response has been shown to be either direct or reverse action. Once this distinction has been made, several types of responses are used to control a process. These are (1) on/off, two-position, control, (2) proportional action, (3) integral action (reset), and (4) derivative action.

On/off Control

On/off control is illustrated in Figure 1-12 for a reverse-acting controller and an air-to-close valve. An on/off controller has only two outputs, either full maximum or full minimum. For this system, it has been determined that, when the measurement falls below the set point, the valve must be closed to cause it to increase. Thus, whenever the signal to the automatic controller is below the set point, the controller output will be 100 percent. As the measurement crosses the set point, the controller output goes to 0 percent. This eventually causes the measurement to decrease, and as the measurement again crosses the set point, the output goes to maximum. This cycle will continue indefinitely because the controller cannot balance the supply against the load. This continuous oscillation may or may not be acceptable, depending on the amplitude and length of the cycle. Rapid cycling causes frequent upsets to the plant supply system and excessive valve wear. The time of each cycle depends on the dead time in the process because the dead time determines the time it takes for the measurement signal to reverse its direction once it crosses the set point and the output of the controller changes. The amplitude of the signal depends on how rapidly the measurement signal changes during each cycle. On large capacity processes, such as temperature vats, the large capacity causes

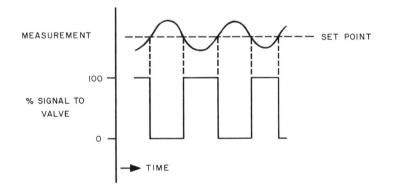

Fig. 1-12. On-off control for reverse-acting controller and air-to-close valve.

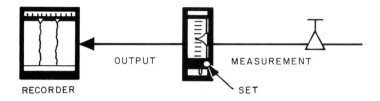

Fig. 1-13. Automatic controller with artificial signal.

a long time constant. Therefore, the measurement can change only slowly. As a result, the cycle occurs within a very narrow band around the set point, and this control may be quite acceptable, if the cycle is not too rapid. The on/off control is the one used most frequently in commercial and domestic installations. However, if the process measurement is more responsive to changes in the supply, the amplitude and frequency of the cycle begins to increase. At some point, this cycle will become unacceptable and some form of proportional control will be required.

In order to study the remaining three modes of automatic control, open-loop responses will be used. Open-loop means that only the response of the controller will be considered. Figure 1-13 shows an automatic controller with an artificial signal from a manual regulator introduced as the measurement. The set point is introduced normally and the output is recorded. With this arrangement, the specific controller responses to any desired change in measurement can be observed.

Proportional Action

Proportional response is the basis for the three-mode controller. If the other two, integral and derivative are present, they are added to the proportional response. "Proportional" means that the percent change in the output of the controller is some multiple of the percent change in the measurement.

This multiple is called the "gain" of the controller. For some controllers, proportional action is adjusted by such a gain adjustment, while for others a "proportional band" adjustment is used. Both have the same purposes and effect. (See Appendix for table showing the controller adjustments from one manufacturer to another.) Figure 1-14 illustrates the response of a proportional controller from an input/output pointer pivoting on one of three positions. With the pivot in the center between the input and the output graph, 100 percent change in

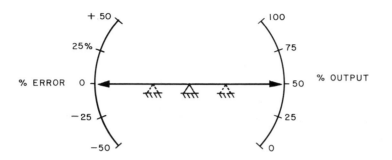

Fig. 1-14. Response of proportional controller, input to output, for three proportional band values.

measurement is required to obtain 100 percent change in output, or full valve travel. A controller adjusted to respond in this way is said to have a 100 percent proportional band. When the pivot is moved to the right-hand position, the measurement input would need to change by 200 percent in order to obtain full output change from 0 to 100 percent. This is called a 200 percent proportional band. Finally, if the pivot were in the left-hand position, and if the measurement moved over only 50 percent of the scale, the output would change over 100 percent of the scale. This is called a 50 percent proportional band. Thus, the smaller the proportional band, the smaller amount the measurement must change to cause full valve travel. In other words, the smaller the proportional band, the greater the output change for the same size measurement change. This relationship is illustrated in Figure 1-15. This graph shows how the controller output will respond as a measurement

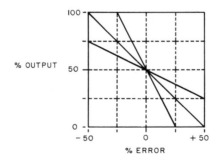

Fig. 1-15. Proportional diagram.

deviates from set point. Each line on the graph represents a particular adjustment of the proportional band. Two basic properties of proportional control can be observed from this graph:

1. For every value of proportional band, whenever the measurement equals the set point, the normal output is 50 percent.
2. Each value of the proportional band defines a unique relationship between measurement and output. For every measurement value there is a specific output value. For example, using the 100 percent proportional band line, whenever the measurement is 25 percent above the set point, the output from the controller must be 25 percent. The output from the controller can be 25 percent only if the measurement is 25 percent above the set point. In the same way, whenever the output from the controller is 25 percent, the measurement will be 25 percent above the set point. In short, there is one specific output value for every measurement value.

For any process control loop, only one value of the proportional band is the best. As the proportional band is reduced, the controller response to any change in measurement becomes increasingly greater. At some point, depending on the characteristic of each particular process, the response in the controller will be large enough to drive the measurement back so far in the opposite direction as to cause constant cycling of the measurement. This proportional band value, known as the ultimate proportional band, is a limit on the adjustment of the controller in that loop. On the other hand, if too wide a proportional band is used, the controller response to any change in measurement is too small and the measurement is not controlled as tightly as possible. The determination of the proper proportional band for any application is part of the tuning procedure for that loop. Proper adjustment of the proportional band can be observed by the response of the measurement to an upset. Figure 1-16 shows several examples of varying the proportional band for the heat exchanger.

Ideally, the proper proportional band will produce one-quarter amplitude damping, in which each half cycle is one-half the amplitude of the previous half cycle. The proportional band that will cause one-quarter wave damping will be smaller, thereby yielding tighter control over the measured variable, as the dead time in the process decreases and the capacity increases.

One consequence of the application of proportional control to the basic control loop is offset. Offset means that the controller will maintain the measurement at a value different from the set point. This is most easily seen in Figure 1-10. Note that if the load valve is opened,

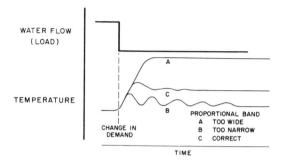

Fig. 1-16. Examples of varying proportional band for heat exchanger.

flow will increase through the valve and the valve would have to open. But note that, because of the proportional action of the linkage, the increased open position can be achieved only at a lowered level. Stated another way, in order to restore balance between the flow in and the flow out, the level must stabilize at a value below the set point. This difference, which will be maintained by the control loop, is called offset, and is characteristic of the application of proportional-only control to feedback loops. The acceptability of proportional-only control depends on whether this offset can be tolerated. Since the error necessary to produce any output decreases with the proportional band, the narrower the proportional band, the less the offset. For large capacity, small dead time applications accepting a very narrow proportional band, proportional-only control will probably be satisfactory, since the measurement will remain within a small percentage band around the set point.

If it is essential that there be no steady state difference between measurement and set point under all load conditions, an additional function must be added to the controller. This function is called integral action (an older term is reset).

Integral Action (Reset)

The open-loop response of the integral mode is shown in Figure 1-17, which indicates a step change in the artificial measurement away from the set point at some instant in time. As long as the measurement remains at the set point, there is no change in the output due to the integral mode in the controller. However, when any error exists between measurement and set point, the integral action will cause the output to begin to change and continue to change as long as the error

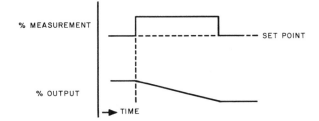

Fig. 1-17. Open-loop response of integral mode.

exists. This function, then, causes the output to change until the proper output is achieved in order to hold the measurement at the set point at various loads. This response is added to the proportional response of the controller as shown in Figure 1-17. The step change in the measurement first causes a proportional response, and then an integral response, which is added to the proportional. The more integral action there is in the controller, the more quickly the output changes due to the integral response. The integral adjustment determines how rapidly the output changes as a function of time. Among the various controllers manufactured, the amount of integral action is measured in one of two ways—either in minutes per repeat, or the number of repeats per minute. For controllers measuring integral action in minutes per repeat, the integral time is the amount of time necessary for the integral mode to repeat the open-loop response caused by proportional mode, for a step change in error. Thus, for these controllers, the smaller the integral number, the greater the action of the integral mode. On controllers that measure integral action in repeats per minute, the adjustment indicates how many repeats of the proportional action are generated by the integral mode in one minute. Table 12-1 (p. 300) relates the controller adjustments from one manufacturer to another. Thus, for these controllers, the higher the integral number, the greater the integral action. Integral time is shown in Figure 1-18. The proper amount of integral action depends on how fast the measurement can respond to the additional valve travel it causes. The controller must not drive the valve faster than the dead time in the process allows the measurement to respond, or the valve will reach its limits before the measurement can be brought back to the set point. The valve will then remain in its extreme position until the measurement crosses the set point, whereupon the controller will drive the valve to its opposite extreme, where it

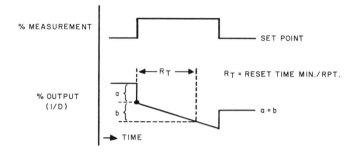

Fig. 1-18. Open-loop response of proportional plus integral modes.

will remain until the measurement crosses the set point in the opposite direction. The result will be an integral cycle in which the valve travels from one extreme to another as the measurement oscillates around the set point. When integral action is applied in controllers on batch processes, where the measurement is away from the set point for long periods between batches, the integral may drive the output to its maximum, resulting in "integral wind-up." When the next batch is started, the output will not come off its maximum until the measurement crosses the set point, causing large overshoots. This problem can be prevented by including a "batch function" in the controller, a function specifically designed to prevent "wind-up."

Derivative Action

The third response found on controllers is the derivative mode. Whereas the proportional mode responds to the size of the error and the integral mode responds to the size and time duration of the error, the derivative mode responds to how quickly the error is changing. In Figure 1-19, two derivative responses are shown. The first is a response to a step change of the measurement away from the set point. For a step, the measurement is changing infinitely fast, and the derivative mode in the controller causes a considerable change or spike in the output, which dies immediately because the measurement has stopped changing after the step. The second response shows the response of the derivative mode to a measurement that is changing at a constant rate. The derivative output is proportional to the rate of change of this error. The greater the rate of change, the greater the output due to the deriva-

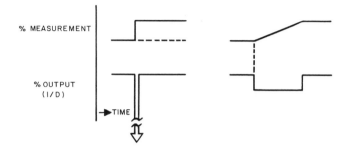

Fig. 1-19. Two derivative responses.

tive response. The derivative holds this output as long as the measurement is changing. As soon as the measurement stops changing, regardless of whether it is at the set point, above or below it, the response due to derivative action will cease. Among all brands of controllers, derivative response is commonly measured in minutes, as shown in Figure 1-20. The derivative time in minutes is the time that the open-loop, proportional-plus-derivative response, is ahead of the response due to proportional action alone. Thus, the greater the derivative number, the greater the derivative response. Changes in the error are the result of changes in either the set point or the measurement, or both. To avoid a large output spike caused by step changes in the set point, most modern controllers apply derivative action only to changes in the measurement. Derivative action in controllers helps to control processes with especially large time constants. Derivative action is unnecessary on

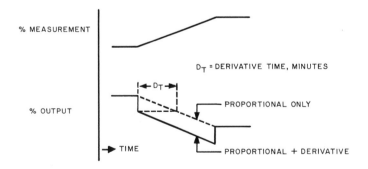

Fig. 1-20. Open-loop response of proportional plus derivative modes.

processes that respond fairly quickly to valve motion, and cannot be used at all on processes with noise in the measurement signal, such as flow, since the derivative in the controller will respond to the rapid changes in measurement it sees in the noise. This will cause large and rapid variations in the controller output, which will keep the valve constantly moving up and down, wearing the valve and causing the measurement to cycle.

As previously described, the response of an element is commonly expressed in terms of a time constant, defined as the time that will elapse until an exponential curve reaches 63.2 percent of a step change in input. Although the transmitter does not have an exact exponential response, the time constant for the pneumatic temperature transmitter and its associated thermal system is approximately 2 seconds.

The reaction curve shown in Figure 1-21 incorporates the response of the heat exchanger and measuring system, and it represents the signal that will actually reach the controller. This curve indicates that, given a sudden change inside the heat exchanger, more than 30 seconds will elapse before the controller receives a signal that is a true representation of that change. From the reaction curve, or characteristic, we can determine the type of controller required for satisfactory control under this difficult, but common, delayed response characteristic.

Selecting the Controller

The heat exchanger acts as a small-capacity process; that is, a small change in steam can cause a large change in temperature. Accurate

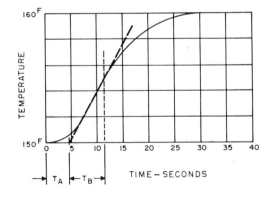

Fig. 1-21. The process reaction curve is obtained by imposing a step change at input.

regulation of processes such as this calls for proportional rather than on/off control.

Variations in water rate cause load changes that produce offset, as described previously. Thus, the integral mode should also be used.

Whether or not to include the derivative mode requires additional investigation of the process characteristic. Referring to the reaction curve (Figure 1-19), notice that the straight line tangent to the curve at the point of inflection is continued back to the 150°F or 66°C (starting) level. The time interval between the start of the upset and the intersection of the tangential line is marked T_A; the time interval from this point to the point of inflection is T_B. If T_B exceeds T_A, some derivative action will prove advantageous. If T_B is less than T_A, derivative action may lead to instability because of the lags involved.

The reaction curve of Figure 1-21 clearly indicates that some derivative will improve control action. Thus, a three-mode controller with proportional, integral, and derivative modes satisfies the needs of the heat exchanger process.

Figure 1-22 shows the combined proportional, integral, and deriva-

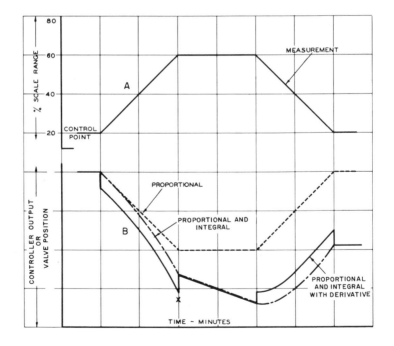

Fig. 1-22. Open-loop response of three-mode controller.

tive responses to a simulated heat exchanger temperature measurement that deviates from the set point due to a load change. When the measurement begins to deviate from the set point, the first response from the controller is a derivative response proportional to the rate of change of measurement that opposes the movement of the measurement away from the set point. This derivative response is combined with the proportional response. In addition, as the integral mode in the controller sees the error increase, it drives the valve further still. This action continues until the measurement stops changing, at which point the derivative response ceases. Since there is still an error, the measurement continues to change due to integral action, until the measurement begins to move back toward the set point. As soon as the measurement begins to move back toward the set point, there is a derivative response proportional to the rate of change in the measurement opposing the return of the measurement toward the set point. The integral response continues, because there is still error, although its contribution decreases with the error. Also, the output due to proportional is changing. Thus, the measurement comes back toward the set point. As soon as the measurement reaches the set point and stops changing, derivative response again ceases and the proportional output returns to 50 percent. With the measurement back at the set point, there is no longer any changing response due to integral action. However, the output is at a new value. This new value is the result of the integral action during the time that the measurement was away from the set point, and compensates for the load change that caused the original upset.

Conclusion

This chapter has described the responses of a three-mode controller when it is used in the feedback control of industrial measurements. The reader should have a clear understanding of the following points:

1. In order to achieve automatic control, the control loop must be closed.
2. In order to maintain a stable feedback control loop, the most important adjustment to the controller is the selection of the proper action, either reverse or direct, on the controller. Proper selection of this action will cause the controller output to change in such a way that the movement of the valve will oppose any change in the measurement seen by the controller.
3. The proper value of the settings of the proportional band, the integral mode, and derivative time depends on the characteristics of

the process. The proportional band is the basic tuning adjustment on the controller. The more narrow the proportional band, the more the controller reacts to changes in the measurement. If too narrow a proportional band is used, the measurement cycles excessively. If too wide a proportional band is used, the measurement will wander and the offset will be too large.

4. The function of the integral mode is to eliminate offset. If too much integral action is used, the result will be an oscillation of the measurement as the controller drives the valve from one extreme to the other. If too little integral action is used, the measurement will return to the set point too slowly.

5. The derivative mode opposes any change in the measurement. Too little derivative action has no significant effect. Too much derivative action causes excessive response of the controller and cycling in the measurement.

Questions

1-1. All control systems that fit into the usual pattern are:
a. Open-loop   c. Closed-loop
b. Nonself-regulating   d. On/off

1-2. If operating properly, automatic control will always:
a. Reduce manpower
b. Reduce costs
c. Make the process operate more uniformly
d. Decrease maintenance

1-3. Automatic controllers operate on the difference between set point and measurement, which is called:
a. Offset   c. Error
b. Bias   d. Feedback

1-4. A two-position controller (on/off) always:
a. Controls with a fixed offset
b. Controls around a point
c. Automatically adjusts its integral time
d. Requires precise tuning

1-5. Gain and proportional bands are:
a. Reciprocally related
b. Two different control modes
c. Adjusted independently of one another
d. Controller functions calibrated in time units

1-6. When we adjust integral time in a controller:
  a. We determine an *RC* time constant in the controller's internal feedback path
  b. We adjust the time it will take for integral to equal derivative
  c. We set the process time constant so that it will always equal 1
  d. What happens specifically depends on the type of controller, pneumatic or electronic

1-7. Match the following:
  Controller, two-position ____
  Derivative ____
  Deviation ____
  Final-controlling element ____
  Proportional band
    adjustment ____
  Regulated by control valve ____
  Reset ____
  Set point ____

  a. Gain
  b. Rate
  c. Integral
  d. Controller, on/off
  e. Valve
  f. Desired value
  g. Manipulated variable
  h. Error

1-8. A proportional controller will have an offset difference between set point and control point:
  a. At all times
  b. Equal to the proportional band setting
  c. That depends upon process load
  d. That will eventually vanish

1-9. If it were possible for a proportional controller to have a true 0 percent proportional band, the controller gain would have to be:
  a. Unity
  b. 0
  c. 100
  d. Infinite

1-10. If the proportional band of the controller is adjusted to minimum possible value, the control action is likely to be:
  a. On/off
  b. With maximum offset
  c. Excellent
  d. Inoperative

1-11. The following symbol (FC) appears in an instrument diagram. It represents a:
  a. Flow rate controller
  b. Fixed control point
  c. Frequency converter
  d. Final control element

1-12. With a proportional-only controller if measurement equals set point, the output will be:
  a. 0
  b. 100 percent
  c. 50 percent
  d. Impossible to define

1-13. If in a proportional-plus-integral controller measurement is away from the set point for a long period, the controller's output will be:

   a. 0 or 100 percent, depending on action selected
   b. Unknown
   c. 0
   d. 100 percent

1-14. In the modern controller, derivative action is applied only to the:
   a. Error   **c.** Set point
   b. Measurement   **d.** Integral circuit

1-15. The function of the integral (reset) mode is to:
   a. Oppose change in measurement
   b. Automatically adjust the controller's gain
   c. Eliminate offset
   d. Stabilize the control loop

Process/Pressure Measuring
Instruments

Pressure is a universal processing condition. It is also a condition of life on this planet: we live at the bottom of an atmospheric ocean that extends upward for many miles. This mass of air has weight, and this weight pressing downward causes atmospheric pressure. Water, a fundamental necessity of life, is supplied to most of us under pressure. In the typical process plant, pressure influences boiling point temperatures, condensing point temperatures, process efficiency, costs, and other important factors. The measurement and control of pressure, or lack of it—vacuum—in the typical process plant is critical. Instruments are available to measure a wide range of pressures. How these instruments function is the subject of this chapter.

What Is Pressure?

Pressure is force divided by the area over which it is applied. Pressure is often defined in terms of "head." For example, assume that we have a water column 1 foot square and 23 feet tall. We want to find the pressure in the bottom of the column. The weight of the column may be calculated by first finding the volume of water. This is the area of the base multiplied by height, or 1 times 23 equals 23 cubic feet. Water weighs 62.43 pounds per cubic foot. So the weight of 23 cubic feet will be 23 times 62.43, or 1,435.89 pounds. The area of the base is 1 square foot, or 12 inches times 12 inches, or 144 square inches. The pressure equals 1,434.89 divided by 144 equals 9.9715, or approximately 10 pounds per square inch. In practice, we find that only the height of the water counts. It may be present in a small pipe or beneath the surface of a pond. In any case, at a depth of 23 feet, the pressure will amount to approximately 10 pounds per square inch. If in your home the water pressure is 50 pounds per square inch and the system uses a gravity feed, the water tank, or reservoir, holds the water at a height of 50 divided by 10, or 5 times 23 equals 115 feet above the point where the pressure measurement is made. Head and pressure, then, may mean the same thing. We must be able to convert from one to the other. You may encounter reference to inches of mercury for pressure measurement. Mercury is 13.596 times as heavy as an equal volume of water. Therefore, a head of mercury exerts a pressure 13.596 times greater than an equivalent head of water. Because it is hazardous, mercury no longer is used commonly in manometers.

The head or pressure terms cited thus far are called, collectively, "gauge pressure." For example, if a tire gauge is used to check the pressure in your automobile tires, it measures gauge pressure. Gauge pressure makes no allowance for the fact that on earth we exist under a head of air, or an atmosphere. The height of this head of air varies with elevation, and also to some degree with weather conditions. If you ride an elevator from the bottom to the top floor of a tall building, you will likely feel your ears "pop." This is caused by the change in atmospheric pressure.

A simple method of measuring atmospheric pressure would be to take a length of small diameter (0.25 inches) glass tubing about 35 inches long, sealed at one end. Fill the tube entirely with mercury and temporarily seal the end. Invert this end into a deep dish of mercury and remove the seal. The result will be a column of mercury as shown in Figure 2-1 with some space remaining at the top. Atmospheric pressure on the surface of the exposed mercury will balance the height of mercury in the tube and prevent it from running out of the tube. The height of the mercury above the level in the dish is, then, a measure of atmospheric pressure. At sea level, this would amount to approximately 29.9 inches, or 14.7 pounds per square inch. When the effect of the atmosphere is included in our measurement, we then must use absolute pressure (gauge pressure plus atmospheric pressure).

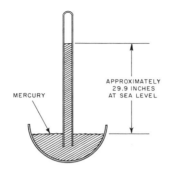

APPROXIMATELY
29.9 INCHES
AT SEA LEVEL

MERCURY

Fig. 2-1. Mercury barometer.

Units of Measurement

Every major country has adopted its own favorite units of measurement. The United States has traditionally employed the English system. However, international trade has made it necessary to standardize
units of measurement throughout the world. Fortunately, during this
standardization, there has been rationalization of the measurement system. This has led to the adoption of the *System International d'Unites*
(SI), a metric system of units. The force of common usage is so strong
that the familiar English system will undoubtedly persist for many
years, but the changeover is definitely underway. The time will soon
come when process industries will deal exclusively with SI units.

Pressure Measurement

Perhaps the area that has caused the most concern in the change to SI
units is pressure measurement. The new unit of pressure, the pascal, is
unfamiliar even to those who have worked in the older CGS (centimetre, gram, second) metric system. Once it is accepted and understood,
it will lead to a great simplification of pressure measurement from the
extremes of full vacuum to ultrahigh pressure. It will reduce the multi-
plicity of units now common in industry to one standard that is compatible with other measurements and calculations. To understand the pascal and its relationship to other units of pressure measurement, we
must return to a basic understanding of pressure.

As noted previously, pressure is force per unit area.

From Newton's laws, force is equal to mass times acceleration. In the English system, the distinction between mass and force became blurred with common usage of terms such as weight and mass. We live in an environment in which every object is subject to gravity. Every object is accelerated toward the center of the earth, unless it is restrained. The force acting on each object is proportional to its mass. In everyday terms this force is called the weight of the object.

$$W = m \cdot G$$

where:
   W = weight of the object
   m = its mass
   G = acceleration due to gravity

Because gravity on the earth's surface is roughly constant, it has been easy to talk about a weight of 1 pound and a mass of 1 pound interchangeably. However, in fact, force and mass, as quantities, are as different as apples and pears, as the astronauts have observed. A number of schemes have been devised to overcome this problem. For example, a quantity called the pound-force was invented and made equal to the force on a mass of one pound under a specified acceleration due to gravity. The very similarity between these two units led to more confusion. The pascal, by its definition, removes all these problems.

The Pascal

The SI unit of pressure is defined as the pressure or stress that arises when a force of one newton (N) is applied uniformly over an area of one square metre (m²). This pressure has been designated one pascal (Pa).

Thus, $Pa = N/m^2$. This is a small unit, but the kilopascal (KPa), 1,000 pascals, and the megapascal (MPa), one million pascals, permit easy expression of common pressures. The definition is simple, because gravity has been eliminated. The pascal is exactly the same at every point, even on the moon, despite changes in gravitational acceleration.

In SI units, the unit of force is derived from the basic unit for mass, the kilogram (kg), and the unit of acceleration (metres per second per second, m/s²). The product of mass times acceleration is force and is designated in newtons. One newton would be the force of one kilogram accelerating at one metre per second per second ($N = kg \cdot m/s^2$).

Bar versus Pascal

After the introduction of SI units, the use of the "bar" (10^5Pa) gained favor, especially in European industry, where it closely resembles the CGS unit of kg/cm² (kilograms per square centimetre). At that time, the SI unit was called the "newton per square metre." As well as being quite a mouthful, it was found to be inconveniently small (one N/m² equals 0.000145 psi). The use of the millibar in meteorology lent weight to the acceptance of the bar. However, the use of a multiple like (10^5) in such an important measurement and the resulting incompatibility of stress and pressure units led to the adoption of the N/m², giving it a new name, the pascal (Pa), in October 1971. The kilopascal (kPa), 1,000 pascals, equals 0.145 psi and most common pressures are thus expressed in kPa. The megapascal (mPa) equals 145 psi and is convenient for expressing high pressures.

Gauge, Absolute, and Differential Pressure

The pascal can be used in exactly the same way as the English or CGS metric units. The pascal may be regarded as a "measuring gauge," the size of which has been defined and is constant. This gauge can be used to measure pressure quantities relative to absolute vacuum. Used in this way, the results will be in pascal (absolute). The gauge may also be used to measure pressures relative to the prevailing atmospheric pressure, and the results will be pascal (gauge). If the gauge is used to measure the difference between pressures, it becomes pascal (differential).

The use of gauge pressure is extremely important in industry, since it is a measure of the stress within a vessel and the tendency of fluids to leak out. It is really a special case of differential pressure measurement, inside versus outside pressure. Where there is any doubt about whether a pressure is gauge, differential, or absolute, it should be specified in full. However, it is common practice to show gauge pressure without specifying, and to specify by saying "absolute" or "differential" only for absolute or differential pressures. The use of "g" as in psig is disappearing, and the use of "a" as in psia is frowned upon. Neither g nor a is recognized in SI unit symbols. However, ΔP is recognized for differential pressure in all units.

Understanding the Effects of Gravity

Before discussing the effects of gravity on pressure measurement, it is well to keep in mind the size of error that can arise if gravity is regarded as constant.

The "standard" gravitational acceleration is 9.806650 m/s². This is an arbitrary figure selected as a near average of the actual acceleration due to gravity found all over the earth. The following are typical values at different places:

Melbourne (Australia)	9.79966 m/s²
Foxboro (USA)	9.80368 m/s²
Soest (The Netherlands)	9.81276 m/s²

Hence the difference around the world is approximately ±0.1 percent from the average. This is of little practical importance in industrial applications. However, with some transmitters being sold with a rated accuracy of ±0.25 percent, it is well to consider the effect of a gravity-induced difference of more than half the tolerance that can arise if the transmitter was calibrated in Europe and tested in Australia.

Gravity-Dependent Units

Units such as psi, kg/cm², inches of water, and inches of mercury (Hg) are all gravity dependent. The English unit pounds per square inch (psi) is the pressure generated when the force of gravity acts on a mass of one pound distributed over one square inch. Consider a dead weight tester and a standard mass of one pound which is transported around the earth's surface: the pressure at each point on the earth will vary as the gravitational acceleration varies. The same applies to units such as inches of water and inches of mercury. The force at the bottom of each column is proportional to the height, density, and gravitational acceleration.

Dead weight testers are primary pressure standards. They generate pressure by applying weight to a piston that is supported by a fluid, generally oil or air. By selecting the weights and the cross-sectional area of the piston, the pressure generated in any gravity field can be calculated. Therefore, dead weight testers are gravity dependent. For accurate laboratory work, the gravity under which the tester was calibrated and that at the place of use must be taken into account. Similarly, the pressure obtained by a certain height of fluid in a manometer depends on density and gravity. These factors must be corrected for the existing conditions if precision results are to be obtained. Factors given in the conversion tables in the Appendix, it should be noted, deal with units of force, not weight. Dead weight testers will be discussed in more detail later in this chapter.

Gravity-Independent Units

While gravity plays no part in the definition of the pascal, it has the same value wherever it is measured. Units such as pounds-force per square inch and kilogram-force per square centimetre are also independent of gravity because a specific value of gravitational acceleration was selected in defining these units.

Under equal gravity conditions, the pound-mass and pound-force are numerically equal (which is the cause of considerable confusion). Under nonstandard gravity conditions (the usual case), correction factors are required to compensate for the departure from standard. It should be noted that the standard value of actual gravity acceleration is not recognized as such in the SI unit system, where only the SI unit of acceleration of one metre per second per second is used. In the future, only the measured actual gravity at the location of measurement (G) will be used when gravity plays a part in the system under investigation. The pascal is a truly gravity-independent unit and will be used to avoid the presently confusing question of whether a stated quantity is gravity dependent.

Pressure Standards

Now let us consider the calibration standards that are employed with pressure-measuring instruments and the basic instruments that are used to measure pressure. It may help to look at the ways in which the standards for pressure calibration are established. You will recall that head is the same as pressure. A measure of head, then, can be a dependable measure of pressure. Perhaps the oldest, simplest, and, in many respects, one of the most accurate and reliable ways of measuring pressure is the liquid manometer. Figure 2-2 shows a differential manometer. When only a visual indication is needed and static pressures are in a range that does not constitute a safety hazard, a transparent tube is satisfactory. When conditions for the visual manometer are unsuitable, a variety of float-type liquid manometers are often employed.

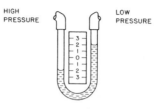

Fig. 2-2. Simple U-tube manometer.

The simplest differential gauge is the liquid-filled manometer: it is basic in its calibration and free of frictional effects. It is often used to calibrate other instruments. The most elementary type is the U-gauge, which consists of a glass tube bent in the form of a U, or two straight glass tubes with a pressure connection at the bottom. When a differential pressure is applied, the difference between the heights of the two columns of liquid is read on a scale graduated in inches, millimetres, or other units. In the more advanced designs, vertical displacement of one side of the manometer is suppressed by using a chamber of large surface area on that side. Figure 2-3 shows such a manometer. If the area ratio is in the vicinity of 1,600 to 1, the displacement in the large chamber becomes quite small and the reading on the glass tube will become extremely close to true inches or true millimetres. The large side would have to be of infinite area for the reading in the glass tube to be exact. This problem is sometimes overcome with a special calibration of the scale. However, if the glass tube becomes broken and must be replaced, the scale must be recalibrated.

A more common and quite reliable design features a zeroing gauge glass as shown in Figure 2-4. The scale may be adjusted to zero for each differential pressure change, and the reading may be taken from a scale graduated in actual units of measurement after rezeroing. The filling

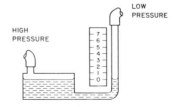

Fig. 2-3. Well or reservoir manometer.

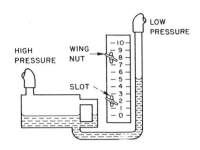

Fig. 2-4. Well manometer with zeroing adjustment.

liquid is usually water or mercury, or some other stable fluid especially compounded by the manometer manufacturers for use with their product. Incline manometer tubes, such as those shown in Figure 2-5, will give magnified readings, but must be made and mounted carefully to avoid errors due to the irregularities of the tube. It is also essential that the manometers be precisely positioned to avoid errors due to level.

Still other types of manometers for functions other than simple indication, including those used with high pressure and hazardous fluids, employ a float on one leg of the manometer. When reading a manometer, there are several potential sources of error. One is the effect of gravity, and another is the effect of temperature on the material contained within the manometer. Correction tables are available which provide the necessary correction for the conditions under which the manometer is to be read. Perhaps even more important is the meniscus correction (Figure 2-6). A meniscus surface should always be read at its center—the bottom, in the case of water, and the top, in the case of mercury. To be practical, gravity and temperature corrections are seldom made in everyday work, but the meniscus correction, or proper reading, must always be taken into account.

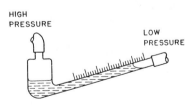

Fig. 2-5. Inclined manometer.

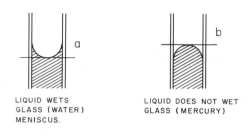

LIQUID WETS
GLASS (WATER)
MENISCUS.

LIQUID DOES NOT WET
GLASS (MERCURY)

Fig. 2-6. Reading a manometer.

The dead weight tester is shown in Figure 2-7. The principle of a dead weight is similar to that of a balance. Gravity acts on a calibrated weight, which in turn exerts a force on a known area. A known pressure then exists throughout the fluid contained in the system. This fluid is generally a suitable oil. Good accuracy is possible, but requires that several factors be well established: (1) the piston area; (2) the weight precision; (3) gravity corrections for the weight if the measurement is to be made in an elevation quite different from the original calibration elevation; (4) buoyancy, since as each weight displaces its volume of air, the air weight displaced should also be taken into consideration; (5) the absence of friction; (6) head of transmitting fluid and (7) operation technique (the weight should be spun to eliminate frictional effect); unless the instrument being calibrated and the tester are at precisely the same level, the head of material can contribute an appreciable error. Perhaps the most important part of the procedure is to keep the piston floating. This is accomplished generally by spinning the weight platform.

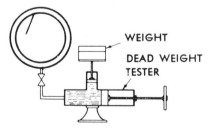

WEIGHT

DEAD WEIGHT
TESTER

Fig. 2-7. Ranges 30 psi and up: increase pressure with crank until pressure supports an accurately known weight. An accurate test gauge may be used with hydraulic pump in a similar setup.

A properly operated dead weight tester should have a pressure output accurate to a fraction of a percent (actually, 0.1 percent of its calibration or calibrated reading).

Still another type of dead weight tester is the pneumatic dead weight tester. This is a self-regulating primary pressure standard. An accurate calibrating pressure is produced by establishing equilibrium between the air pressure on the underside of the ball against weights of known mass on the top. A diagram of an Ametek pneumatic tester is shown in Figure 2-8. In this construction, a precision ceramic ball is floated within a tapered stainless steel nozzle. A flow regulator introduces pressure under the ball, lifting it toward the annulus between the ball and the nozzle. Equilibrium is achieved as soon as the ball begins to lift. The ball floats when the vented flow equals the fixed flow from the supply regulator. This pressure, which is also the output pressure, is proportional to the weight load. During operation, the ball is centered with a dynamic film of air, eliminating physical contact between the ball and the nozzle. When weights are added or removed from the weight carrier, the ball rises or drops, affecting the air flow. The regulator senses the change in flow and adjusts the pressure beneath the ball to bring the system into equilibrium, changing the output pressure accordingly. Thus, regulation of output pressure is automatic with change of weight mass on the spherical piston or ball. The pneumatic dead weight

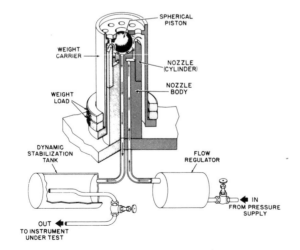

Fig. 2-8. Pneumatic tester. Courtesy of AMETEK Mfg. Co.

tester has an accuracy of ±0.1 of 1 percent of indicated reading. It is commonly used up to a maximum pressure of 30 psi or 200 kPa gauge.

Plant Instruments That Measure Pressure Directly

Thus far in this chapter we have been concerned with the definition of pressure, and some of the standards used have been described. In the plant, manometers and dead weight testers are used as standards for comparison and calibration.

The working instruments in the plant usually include simple mechanical pressure gauges, precision pressure recorders and indicators, and pneumatic and electronic pressure transmitters. A pressure transmitter makes a pressure measurement and generates either a pneumatic or electrical signal output that is proportional to the pressure being sensed. We will discuss transmitters in detail later in this chapter. Now we will deal with the basic mechanical instruments used for pressure measurement, how they operate and how they are calibrated. When the amount of pressure to be measured is very small, the following instruments might be used.

Bell Instrument

This instrument measures the pressure difference in the compartment on each side of a bell-shaped chamber. If the pressure to be measured is gauge pressure, the lower compartment is vented to atmosphere. If the lower compartment is evacuated, the pressure measured will be in absolute units. If the differential pressure is to be measured, the higher pressure is applied to the top of the chamber and the lower pressure to the bottom.

The bell chamber is shown in Figure 2-9. Pressure ranges as low as 0 to 1 inch (0 to 250 Pa) of water can be measured with this instrument. Calibration adjustments are zero and span. The difficulty in reading a manometer accurately to fractions of an inch are obvious, yet the manometer is the usual standard to which the bell differential instrument is calibrated. The bell instrument finds applications where very low pressures must be measured and recorded with reasonable accuracy.

Slack or Limp-Diaphragm

The slack or limp-diaphragm instrument is used when very small pressures are to be sensed. The most common application of this gauge

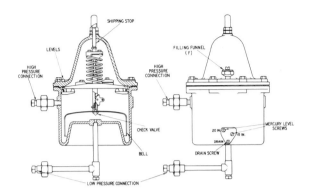

Fig. 2-9. Bell instrument.

is measurement of furnace draft. The range of this type instrument is from 0 to 0.5 inches (125 Pa) of water to 0 to 100 inches (25 kPa) of water above atmospheric pressure.

To make this instrument responsive to very small pressures, a large area diaphragm is employed. This diaphragm is made of very thin, treated nonporous leather, plastic, or rubber, and requires an extremely small force to deflect it. A spring is always used in combination with the diaphragm to produce a deflection proportional to pressure. Let us assume the instrument shown in Figure 2-10 is being used to measure pressure. The low-pressure chamber on the top is vented to the atmosphere. The pressure to be measured is applied to the high-pressure chamber on the bottom. This causes the diaphragm to move upward. It

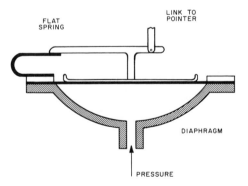

Fig. 2-10. Slack or limp-diaphragm instrument.

will move until the force developed by pressure on the diaphragm is equal to that applied by the calibrated spring. In the process of achieving balance, the lever attached to the diaphragm is tipped and this motion transmitted by the sealed link to the pointer. This instrument is calibrated by means of zero and span adjustments. The span adjustment allows the ridged connection to the spring to be varied, thus changing the spring constant. The shorter the spring the greater the spring constant; thus, a greater force is required to deflect it. As the spring is shortened, a higher pressure is required for full deflection of the pointer. The zero adjustment controls the spring's pivot point, thereby shifting the free end of the calibrated spring and its attached linkages. Turning the zero adjustment changes the pointer position on the scale, and it is normally adjusted to read zero scale with both the high and low pressure chambers vented to the atmosphere. If pressure is applied to both the high- and low-pressure chambers, the resultant reading will be in differential pressure. This instrument is extremely sensitive to overrange, and care should be taken to avoid this problem. The other long-term difficulty that can develop is damage to the diaphragm. The diaphragm may become stiff, develop leaks, or become defective, which results in error. These difficulties may be observed by periodically inspecting the diaphragm.

Pressure Gauges

In the process plant, we frequently find simple pressure gauges scattered throughout the process and used to measure and indicate existing pressures. Therefore, it is appropriate to devote some attention to the operation of a simple pressure gauge.

The most common of all the pressure gauges utilizes the Bourdon tube. The Bourdon tube was originally patented in 1840. In his patent, Eugene Bourdon stated: ''I have discovered that if a thin metallic tube be flattened out then bent or distorted from a straight line, it has the property of changing its form considerably when exposed to variations of internal or external pressure. An increase of internal pressure tends to bring the tube to a straight cylindrical form, and the degree of pressure is indicated by the amount of alteration in the form of the tube.''

Figure 2-11A shows such a tube. As pressure is applied internally, the tube straightens out and returns to a cylindrical form. The excursion of the tube tip moves linearly with internal pressure and is converted to pointer position with the mechanism shown. Once the

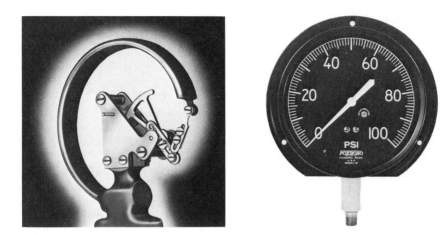

Fig. 2-11. (A) Bourdon tube. (B) Typical pressure gauge.

internal pressure is removed, the spring characteristic of the material returns the tube to its original shape.

If the Bourdon tube is overranged, that is, pressure is applied to the point where it can no longer return to its original shape, the gauge may take a new set and its calibration becomes distorted. Whether the gauge can be recalibrated depends on the extent of the overrange. A severely overranged gauge will be ruined, whereas one that has been only slightly overranged may be recalibrated and reused.

Most gauges are designed to handle approximately 35 percent of the upper range value as overrange without damage. Typically, a gauge will exhibit some amount of hysteresis, that is, a difference due to pressure moving the tube in the upscale direction versus the spring characteristic of the tube moving it downscale. A typical gauge may also exhibit some amount of drift, that is, a departure from the true reading due to changes over a long period in the physical properties of the materials involved. All of these sources of error are typically included in the manufacturer's statement of accuracy for the gauge. A typical Bourdon tube gauge, carefully made with a Bourdon tube that has been temperature-cycled or stress-relieved, will have an accuracy of ± 1 percent of its upper range value. A carefully made test gauge will have an accuracy of ± 0.25 to 0.50 percent of its upper range value. The range of the gauge is normally selected so that it operates in the upper part of the middle third of the scale.

In addition to the gauges used for visual observations of pressure throughout the plant, the typical instrument shop will have a number of test gauges which are used as calibration standards. The gauge must be vertical to read correctly. A typical pressure gauge is shown in Figure 2-11 B.

Liquid or Steam Pressure Measurement

When a liquid pressure is measured, the piping is arranged to prevent entrapped vapors which may cause measurement error. If it is impossible to avoid entrapping vapors, vents should be provided at all high points in the line.

When steam pressure is measured, the steam should be prevented from entering the Bourdon tube. Otherwise, the high temperature may damage the instrument. If the gauge is below the point of measurement, a "siphon" (a single loop or pigtail in a vertical plane) is provided in the pressure line to the gauge. A cock is installed in the line between the loop and the gauge. In operation, the loop traps condensate, preventing the steam from entering the gauge.

Seals and Purges

If the instrument is measuring a viscous, volatile, corrosive, or extremely hot or cold fluid, the use of a pressure seal or a purge is essential to keep the fluid out of the instrument. Liquid seals are used with corrosive fluids. The sealing fluid should be nonmiscible with the measured fluids. Liquid purges are used on hot, viscous, volatile, or corrosive liquids, or liquids containing solids in suspension. The purging liquid must be of such a nature that the small quantities required (in the order of 1 gph) (4 litres per hour) will not injure the product or process. Mechanical seals will be described in detail later in this chapter.

Pulsation Dampener

If the instrument is intended for use with a fluid under pressure and subject to excessive fluctuations or pulsations, a deadener or damper should be installed. This will provide a steady reading and prolong the life of the gauge.

Two other elements that use the Bourdon principle are the spiral (Figure 2-12) and helical (Figure 2-13). The spiral and helical are, in effect, multitube Bourdon tubes. Spirals are commonly used for

Fig. 2-12. Spiral.   **Fig. 2-13.** Helical.

pressure ranges up to 0 to 200 psi or 1.4 MPa, and helicals are made to measure pressures as high as 0 to 80,000 psi or 550 MPa. The higher the pressure to be measured, the thicker the walls of the tubing from which the spiral or helical is constructed. The material used in the construction may be bronze, beryllium copper, stainless steel, or a special Ni-Span C alloy. Spirals and helicals are designed to provide a lever motion of approximately 45 degrees with full pressure applied. If this motion is to be translated into pen or pointer position, it is common practice to utilize a four-bar linkage, and this necessitates a special calibration technique. If, instead of measuring gauge pressure, it is necessary to read absolute pressure, the reading must make allowances for the pressure of the atmosphere. This may be done by utilizing an absolute double spiral element. In this element, two spirals are used. One is evacuated and sealed; the second has the measured pressure applied.

The evacuated sealed element makes a correction for atmospheric pressure as read on the second element. Thus, the reading can be in terms of absolute pressure, which is gauge plus the pressure of the atmosphere, rather than gauge.

An absolute double spiral element of this type may be used to measure pressures up to 100 psi or 700 Kpa absolute. This element is shown in Figure 2-14.

Metallic Bellows

A bellows is an expandable element made up of a series of folds or corrugations called convolutions (Figure 2-15). When internal pressure is applied to the bellows, it expands. Because a sizable area is involved,

Fig. 2-14. Absolute double spiral.

the applied pressure develops a sizable force to actuate an indicating or recording mechanism. (In some instruments, the bellows is placed in a sealed can and the pressure is applied externally to the bellows. The bellows will then compress in a fashion similar to the expansion just described. Such an arrangement is shown in Figure 2-16).

A variety of materials are used to fabricate bellows, including brass, beryllium copper, copper nickel alloys, phospher bronze, Monel, steel, and stainless steel. Brass and stainless steel are the most commonly used. Metallic bellows are used from pressure ranges of a few

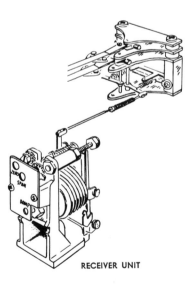

RECEIVER UNIT

Fig. 2-15. Bellows receiver unit.

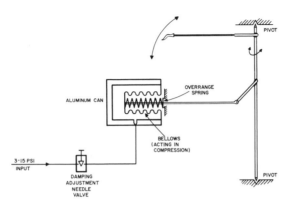

Fig. 2-16. Bellows in sealed can.

ounces to many pounds per square inch. A bellows will develop many times the power available from a helical, spiral, or Bourdon tube.

A bellows is typically rated in terms of its equivalent square inch area. To create a linear relationship between the excursion of the bellows and the applied pressure, it is common practice to have the bellows work in conjunction with a spring, rather than with the spring characteristic of the metal within the bellows itself. Each bellows and spring combination has what is called a spring rate. The springs used with the bellows are usually either helical or spiral. Typically, the spring rate of the helical spring is ten times or more that of the bellows material itself.

Using a spring with a bellows has several advantages over relying on the spring characteristics of the bellows alone. The calibration procedure is simplified, since adjustments are made only on the spring. Initial tension becomes zero adjustment and the number of active turns becomes span adjustment. A spring constructed of stable material will exhibit long-term stability that is essential in any component.

When a measurement of absolute pressure is to be made, a special mechanism employing two separate bellows may be used. It consists of a measuring bellows and a compensating bellows, a mounting support, and an output lever assembly (Figure 2-17). The measuring and compensating bellows are fastened to opposite ends of the fixed mounting support: the free ends of both bellows are attached to a movable plate mounted between them. The motion of this movable plate is a measure of the difference in pressure between the two bellows.

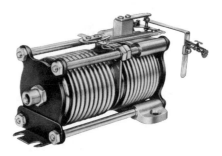

Fig. 2-17. Absolute pressure bellows.

Since the compensating bellows is completely evacuated and sealed, this motion is a measure of pressure above vacuum, or absolute pressure applied to the measuring bellows. This movable plate is attached to the output lever assembly which, in turn, is linked to the instrument pen, or pointer.

In many applications, the bellows expands very little, and the force it exerts becomes its significant output. This technique is frequently employed in force-balance mechanisms which will be discussed in some detail in Chapter 8.

Pressure Transmitters

Signal Transmissions

In the process plant, it is impractical to locate the control instruments out in the plant near the process. It is also true that most measurements are not easily transmitted from some remote location. Pressure measurement is an exception, but if a high pressure measurement of some dangerous chemical is to be indicated or recorded several hundred feet from the point of measurement, a hazard will be created. The hazard may be from the pressure and/or from the chemical carried in the line.

To eliminate this problem, a signal transmission system was developed. In process instrumentation, this system is usually either pneumatic (air pressure) or electrical. Because this chapter deals with pressure, the pneumatic, or air pressure system, will be discussed first. Later it will become evident that the electrical transmitters perform a similar function.

Using the transmission system, it will be possible to install most of the indicating, recording, and control instruments in one location. This makes it practical for a minimum number of operators to run the plant efficiently.

When a transmission system is employed, the measurement is converted to a pneumatic signal by the transmitter scaled from 0 to 100 percent of the measured value. This transmitter is mounted close to the point of measurement in the process. The transmitter output—air pressure for a pneumatic transmitter—is piped to the recording or control instrument. The standard output range for a pneumatic transmitter is 3 to 15 psi, 20 to 100 kPa; or 0.2 to 1.0 bar or kg/cm². These are the standard signals that are almost universally used.

Let us take a closer look at what this signal means. Suppose we have a field-mounted pressure transmitter that has been calibrated to a pressure range of 100 psi to 500 psi (689.5 to 3447.5 kPa).

When the pressure being sensed is 100 psi (69 kPa), the transmitter is designed to produce an output of 3 psi air pressure (or the approximate SI equivalent 20 kPa). When the pressure sensed rises to 300 psi (2068.5 kPa), or midscale, the output will climb to 9 psi, or 60 kPa, and at top scale, 500 psi (3447.5 kPa), the signal output will be 15 psi, or 100 kPa.

This signal is carried by tubing, usually ¼-inch copper or plastic, to the control room, where it is either indicated, recorded, or fed into a controller.

The receiving instrument typically employs a bellows element to convert this signal into pen or pointer position. Or, if the signal is fed to a controller, a bellows is also used to convert the signal for the use of the controller. The live zero makes it possible to distinguish between true zero and a dead instrument.

The top scale signal is high enough to be useful without the possibility of creating hazards.

Pneumatic Recorders and Indicators

Pressure recorders and indicators were described earlier in this chapter. Pneumatic recorders and indicators differ only in that they always operate from a standard 3 to 15 psi or 20 to 100 kPa signal. The indicator scale, or recorder chart, may be labeled 0 to 1,000 psi. This would represent the pressure sensed by the measuring transmitter and converted into the standard signal that is transmitted to the receiver.

The receiver converts the signal into a suitable pen or pointer position. Because the scale is labeled in proper units, it is possible to read the measured pressure.

A typical pneumatic indicator is shown in Figure 2-18 (top) and its operation may be visualized by studying Figure 2-18 (bottom, right).

The input signal passes through an adjustable needle valve to provide damping, then continues to the receiver bellows. This bellows,

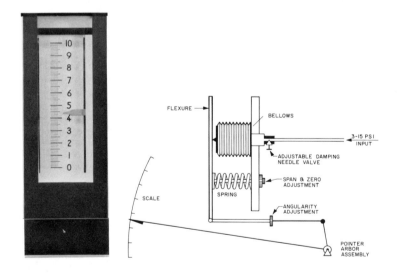

Fig. 2-18. Pneumatic indicator.

acting in expansion, moves a force plate. A spring opposing the bellows provides zero and span adjustments. Regulating the amount of spring used (its effective length) provides a span adjustment. Setting the initial tension on the spring provides a zero adjustment.

The force plate is connected to a link that drives the pointer arbor assembly. Changing the length of the link by turning the nut on the link provides an angularity adjustment. The arm connecting the pointer and the arbor is a rugged crushed tube designed to reduce torsional effects. A takeup spring is also provided to reduce mechanical hysteresis. Overrange protection is provided to prevent damage to the bellows or pointer movement assembly for pressures up to approximately 30 psi or about 200 kPa.

A pneumatic recorder also uses a receiver bellows assembly as shown in Figure 2-19 (top). This receiver provides an unusually high torque due to an effective bellows area of 1.1 square inches.

The input signal first passes through an adjustable damping restrictor. The output of this signal is the input to the receiver bellows assembly. The receiver consists of a heavy duty impact extruded aluminum can containing a large brass bellows working in compression. The large effective area of the bellows assures an extremely linear pressure-to-pen position relationship. The motion, created by the input signal, is picked up by a conventional link and transferred to an arbor. Calibration is easily accomplished by turning zero, span, and angularity adjustments on this simple linkage system. All calibration adjustments are accessible through a door (Figure 2-19) on the side of the instrument and may be made while instrument is in operation. The arm interconnecting the pen and the arbor incorporates a rugged tubular member to reduce torsional effects and takeup spring to reduce mechanical hysteresis. Overrange protection is provided to prevent damage to the bellows or pen movement assembly for pressures up to 30 psi or about 200 kPa.

Mechanical Pressure Seals

Application
A sealed pressure system is used with a pressure measuring instrument to isolate corrosive or viscous products, or products that tend to solidify, from the measuring element and its connective tubing.

Definition
A sealed pressure system consists of a conventional pressure measuring element or a force-balance pressure transmitter capsule assembly

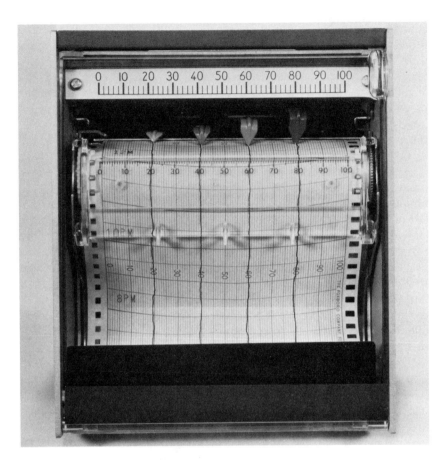

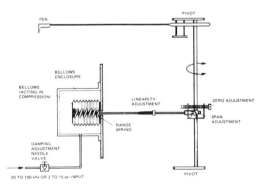

Fig. 2-19. Pneumatic recorder.

connected, either directly or by capillary tubing, to a pressure seal as seen in Figure 2-20. The system is solidly filled with a suitable liquid transmission medium.

The seal itself may take many forms, depending on process conditions, but consists of a pressure sensitive flexible member, the diaphragm, functioning as an isolating membrane, with a suitable method of attachment to a process vessel or line.

Principle of Operation
Process pressure applied to the flexible member of the seal assembly forces some of the filling fluid out of the seal cavity into the capillary tubing and pressure measuring element, causing the element to expand in proportion to the applied process pressure, thereby actuating a pen, pointer, or transmitter mechanism. A sealed pressure system offers high resolution and rapid response to pressure changes at the diaphragm. The spring rate of the flexible member must be low when compared with the spring rate of the measuring element to ensure that the fill volume displacement will full stroke the measuring element for the required pressure range. A low-diaphragm spring rate, coupled with maximum fill volume displacement, is characteristic of the ideal system.

Fig. 2-20. Seal connected to 6-inch pressure gauge.

A sealed pressure system is somewhat similar to a liquid-filled thermometer, the primary differences being a flexible rather than rigid member at the process, and no initial filling pressure. The flexible member of the seal should ideally accommodate any thermal expansion of the filling medium without perceptible motion of the measuring element. A very stiff seal member, such as a Bourdon tube, combined with a low-pressure (high-volume change) element, will produce marked temperature effects from both varying ambient or elevated process temperatures.

The Foxboro 13DMP Series pneumatic d/p Cell transmitters with pressure seals (Figure 2-21) measure differential pressures in ranges of 0 to 20 to 0 to 850 inches of water or 0 to 5 to 0 to 205 kPa at static pressures from full vacuum up to flange rating. They transmit a proportional 3 to 15 psi or 20 to 100 kPa or 4 to 20 mA dc signal to receivers located up to several hundred yards or meters from the point of measurement.

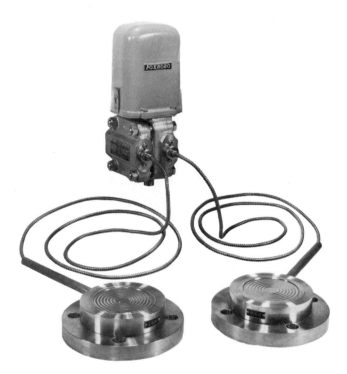

Fig. 2-21. Seals connected to differential pressure transmitter.

Filling Fluids

Ideally, the filling fluid used in a sealed pressure system should be noncompressible, have a high-boiling point, a low-freezing point, a low coefficient of thermal expansion, and low viscosity. It should be noninjurious to the diaphragm and containing parts, and should not cause spoilage in the event of leakage. Silicone-based liquid is the most popular filling fluid.

The system is evacuated before the filling fluid is introduced. The system must be completely filled with fluid and free from any air pockets that would contract or expand during operation, resulting in erroneous indications at the pen or pointer or in an output signal. The degree of accuracy of any filled pressure system depends on the perfection of the filling operation.

Calibration Techniques

The procedure for calibration of a pressure instrument consists of comparing the reading of the instrument being calibrated with a standard. The instrument under calibration is then adjusted or manipulated to make it agree with the standard. Success in calibration depends not only on one's ability to adjust the instrument, but on the quality of the standard as well.

Field Standards

Field standards must be reasonably convenient to use and must satisfy the accuracy requirements for the instrument under calibration. A 100-inch water column, for example, is extremely accurate but not practical to set up out in the plant. For practical reasons, we find that most field standards are test gauges. The test gauge is quite similar, in most cases, to the regular Bourdon gauges. However, more care has gone into its design, construction, and calibration, making it very accurate. A good quality test gauge will be accurate to within ±0.25 percent of its span. This is adequate for most field use.

Under some conditions, a manometer may be used in the field. This usually occurs when a low-pressure range is to be calibrated and no other suitable standard is readily available.

Portable Pneumatic Calibrator

All of the ingredients required to perform a calibration have been combined into a single unit called a portable pneumatic calibrator. This unit

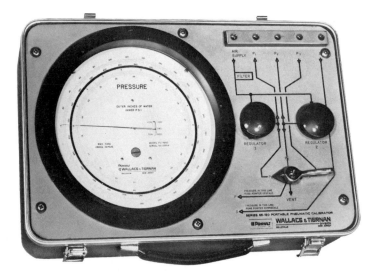

Fig. 2-22. Portable pneumatic calibration. Courtesy of Wallace & Tiernan.

contains an accurate pressure gauge or standard, along with a pressure regulator and suitable manifold connections. The portable pneumatic calibrator will accurately apply, hold, regulate, and measure gauge pressure, differential pressure, or vacuum. The gauge case may be evacuated to make an absolute pressure gauge. The pointer in the operation of the gauge in this calibrator differs in that it makes almost two full revolutions in registering full scale, providing a scale length of 45 inches. This expanded scale makes the gauge very easy to read.

The calibrator is available in seven pressure ranges (SI unit ranges). Of these, three are sized for checking 3 to 15 psi or 20 to 100 kPa pneumatic transmission instrumentation. The calibrator is shown in both picture and schematic form in Figures 2-22 and 2-23. The air switching arrangement makes it possible to use the calibrator as both a source or signal and a precise gauge for readout. This portable pneumatic calibrator is manufactured by Wallace and Tiernan, Inc., Belleville, New Jersey.

Force-Balance Pneumatic Pressure Transmitter

A pneumatic pressure transmitter senses a pressure and converts that pressure into a pneumatic signal that may be transmitted over a reason-

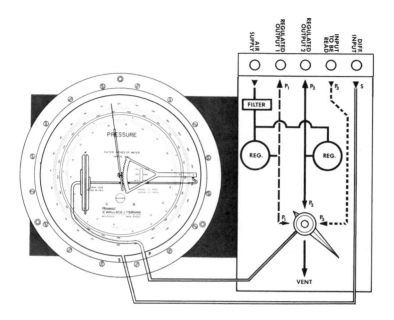

Fig. 2-23. Connections for different pressure readouts (courtesy of Wallace & Tiernan): For Gauge Pressure: test pressure is applied to the capsule through the appropriate P connection; the case is open to atmosphere through S.
For Differential Pressure: high-test pressure is applied to the capsule through the appropriate P connection. Low-test pressure is applied to the case through S.
For Absolute Pressure: test pressure is applied to the capsule through the appropriate P connection and the case is continuously subjected to full vacuum through S.
For Vacuum: the capsule is open to atmosphere through connection P; the case is connected to test vacuum at S.
For Positive and Negative Pressures: test pressure is applied to the capsule through the appropriate P connection and the case is open to atmosphere through S.

able distance. A receiving instrument is then employed to convert the signal into a pen or pointer position or a measurement input signal to a controller. The Foxboro Model 11GL force-balance pneumatic pressure transmitter is an example of a simple, straightforward device that performs this service (Figure 2-24). Before its operation is discussed, the operation of two vital components must be described: the flapper and the nozzle, or detector; and the pneumatic amplifier, or relay. These two mechanisms are found in nearly every pneumatic instrument.

The flapper and nozzle unit converts a small motion (position) or

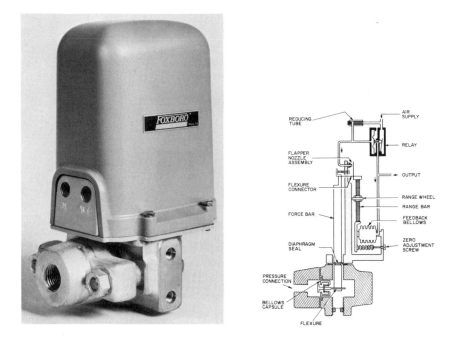

Fig. 2-24. Model 11GM pressure transmitter is a force-balance instrument that measures pressure and transmits it as a proportional 3 to 15 psi or 20 to 100 kPa pneumatic signal.

force into an equivalent (proportional) pneumatic signal. Flapper movement of only six ten-thousandths of an inch (0.0015 cm) will change the nozzle pressure by 0.75 psi or 5.2 kPa. This small pressure change applied to the pneumatic amplifier or relay becomes an amplified change of 3 to 15 psi or 20 to 100 kPa in the amplifier output.

Let us take a closer look at the operation of the amplifier or relay. It is shown in a cross-sectional view in Figure 2-25.

Pneumatic Relay

A relay is a pneumatic amplifier. Like its electronic counterpart, the function of the relay is to convert a small change in the input signal (an air pressure signal) to a large change in the output signal. Typically, a 1 psi or 7 kPa change in input will produce approximately a 12 psi or 80 kPa change in output.

The supply enters the relay through a port on the surface of the

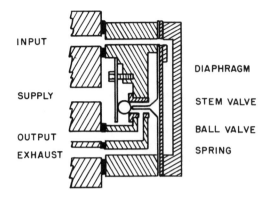

INPUT

SUPPLY

OUTPUT

EXHAUST

DIAPHRAGM

STEM VALVE

BALL VALVE

SPRING

Fig. 2-25. Cross-sectional view of amplifier.

instrument on which the relay is mounted. The input signal (nozzle pressure) enters the relay through another port and acts on the diaphragm. Since the diaphragm is in contact with a stem valve, the two move in unison.

As the input signal increases, the stem pushes against a ball valve which in turn moves a flat spring, allowing the supply of air to enter the relay body. Further motion of the stem valve causes it to close off the exhaust port. Thus, when the input pressure increases, the stem (exhaust) valve closes and the supply valve opens; when the input decreases, the stem valve opens and the supply valve closes. This varies the pressure to the output.

Principle of Operation

The 11GM Pneumatic Transmitter (Figure 2-24) is a force-balance instrument that measures pressure and transmits it as a proportional 3 to 15 psi pneumatic signal (20 to 100 kPa).

The pressure is applied to a bellows, causing the end of the bellows to exert a force (through a connecting bracket) on the lower end of the force bar. The metal diaphragm is a fulcrum for the force bar. The force is transmitted through the flexure connector to the range rod, which pivots on the range adjustment wheel.

Any movement of the range rod causes a minute change in the clearance between the flapper and nozzle. This produces a change in the output pressure from the relay to the feedback bellows until the force of the feedback bellows balances the pressure on the measurement bellows. The output pressure which is established by this balance is the transmitted signal and is proportional to the pressure applied to the measurement bellows.

If the pressure to be measured is high, such as 5,000 psi or 35 MPa, a different sensing element is employed.

The pressure being measured is applied to a Bourdon tube. This pressure tends to straighten the tube and causes a horizontal force to be applied to the lower end of the force bar. The diaphragm seal serves as both a fulcrum for the force bar and as a seal for the pressure chamber. The force is transmitted through a flexure connector to the range rod, which pivots on the range adjustment wheel.

Any movement of the range rod causes a minute change in the clearance between the flapper and nozzle. This produces a change in the output pressure from the relay to the feedback bellows until the force in the bellows balances the force created by the Bourdon tube.

The output pressure, which establishes the force-balancing, is the transmitted pneumatic signal that is proportional to the pressure being measured. This signal is transmitted to a pneumatic receiver to record, indicate, or control.

The calibration procedure for this transmitter is the same as for the low pressure transmitter, but the calibration pressure is developed with a dead weight tester. For safety, oil or liquid, never air, should be used for high pressure calibrations.

Pressure measurements found in some applications require that absolute, rather than gauge pressure, be determined. To handle these applications, the absolute pressure transmitter may be used.

Absolute Pressure Transmitter

The pressure being measured is applied to one side of a diaphragm in a capsule. The space on the other side of the diaphragm is evacuated, thus providing a zero absolute pressure reference (Figure 2-26).

The pressure exerts a force on the diaphragm that is applied to the lower end of the force bar. The diaphragm seal serves both as a fulcrum for the force bar and as a seal for the pressure chamber. The force is transmitted through the flexure connector to the range bar, which pivots on the range adjustment wheel.

Any movement of the range bar causes a minute change in the clearance between the flapper and nozzle. This produces a change in the output pressure from the relay (Figure 2-26) to the feedback bellows

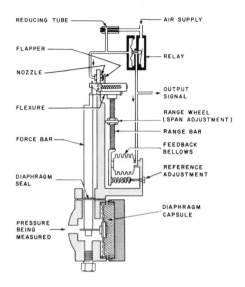

REDUCING TUBE — AIR SUPPLY

FLAPPER —

RELAY

NOZZLE —

OUTPUT
SIGNAL

FLEXURE —

RANGE WHEEL
(SPAN ADJUSTMENT)

RANGE BAR

FORCE BAR —

FEEDBACK
BELLOWS

REFERENCE
ADJUSTMENT

DIAPHRAGM
SEAL

DIAPHRAGM
CAPSULE

PRESSURE
BEING
MEASURED

Fig. 2-26. Pneumatic absolute pressure transmitter.

until the force in the bellows balances the force on the diaphragm capsule.

The output pressure, which establishes the force-balance, is the transmitted pneumatic signal, which is proportional to the absolute pressure being measured. This signal is transmitted to a pneumatic receiver to record, indicate, or control.

Questions

2-1. An ordinary commercial Bourdon gauge has a scale of 0 psi to 250 psi, and an accuracy of ± 1 percent of span. If the gauge reads 175 psi, within what maximum and minimum values will the correct pressure fall?
- **a.** 174 to 176 psi
- **b.** 172.5 to 177.5 psi
- **c.** 176.5 to 178.5 psi
- **d.** 179 to 180 psi

2-2. A manometer, read carefully:
- **a.** Always has zero error
- **b.** Has an error caused by the liquid's impurity unless corrected
- **c.** May have an error caused by temperature and gravity effects on the various components and so on.
- **d.** Has an error that varies only with altitude

2-3. Absolute pressure is:
  a. Gauge pressure plus atmospheric pressure
  b. Gauge pressure less atmospheric pressure
  c. Gauge pressure plus atmospheric pressure divided by two
  d. Always referenced to a point at the peak of Mt. Washington, NH

2-4. The pressure at the bottom of a pond where it is exactly 46 feet deep will be:
  a. 100 psi   **c.** 20 psi
  b. 46 psi   **d.** 20 psi absolute

2-5. The advantages of making absolute pressure measurements rather than gauge are:
  a. Greater accuracy
  b. Eliminates errors introduced by barometric variations
  c. Is more indicative of true process conditions
  d. Is more related to safety than gauge pressure

2-6. A bellows element is used as a receiver for a 3 to 15 psi pneumatic signal and by error a 30-psi pressure is applied to it. The result will be:
  a. A damaged bellows
  b. An instrument in need of recalibration
  c. No damage
  d. A severe zero shift

2-7. A pressure instrument is calibrated from 100 to 600 psi. The span of this instrument is:
  a. 600   **c.** 500
  b. 100   **d.** 400

2-8. Instruments that measure pressure are generally classified as:
  a. nonlinear   **c.** free of hysteresis
  b. linear   **d.** none of the above

2-9. When reading a manometer, it is good practice to:
  a. Read at the bottom of the meniscus
  b. Read at the top of the meniscus
  c. Read at the center of the meniscus
  d. Carefully estimate the average of the meniscus

2-10. If a pressure range of 0 to 1 inch of water is to be measured and recorded, an instrument capable of doing this would be:
  a. A bellows   **c.** A Bourdon tube
  b. A bell with a liquid seal **d.** A mercury manometer

2-11. When measuring a pressure that fluctuates severely:
  a. A large-capacity tank should be installed
  b. A pulsation dampener should be employed

c. No problem is created

d. Read the peak and minimum and divide by two to obtain the true pressure

2-12. A measurement of absolute pressure is to be made using a mechanism employing two separate bellows. Measurement is applied to one bellows and the other:

a. Is sealed at an atmospheric pressure of 14.7 psi

b. Is completely evacuated and sealed

c. Contains alcohol for temperature compensation

d. Has an active area twice that of the measuring bellows and is sealed at a pressure twice atmospheric

2-13. The danger of having a high-pressure line carrying a dangerous chemical rupture in the control room is:

a. Eliminated by using special duty piping

b. Ignored

c. Eliminated through the use of a transmission system

d. Minimized by placing the line within a protective barrier

2-14. The standard pneumatic transmission signal most generally used in the United States is:

a. 3 to 27 psi   c. 3 to 15 psi

b. 10 to 50 psi   d. 2 to 12 psi

2-15. A sealed pressure system:

a. Is similar in some ways to a liquid-filled thermometer

b. Seals the process material in the instrument

c. Must be used at a fixed temperature

d. Is always used with manual temperature compensation

2-16. A pneumatic relay:

a. Is a set of electrical contacts pneumatically actuated

b. Is a signal booster

c. Is a pneumatic amplifier

d. Contains a regulator actuated by a bellows

2-17. When the clearance between flapper and nozzle changes by 0.0006 inches the output of the transmitter will change by:

a. 3 psi   c. 12 psi

b. 15 psi   d. 6 psi

2-18. An instrument is to be calibrated to measure a range of 0 to 6,000 psi. For such a calibration:

a. oil or liquid should be used

b. Air must be used

c. An inert gas such as nitrogen is required

d. Any source of high pressure is acceptable

2-19. The kilopascal (kPa), an SI unit of pressure:
   a. Is always equivalent to psi $\times$ 6.895
   b. Is gravity dependent
   c. Is a force of 1.000 newtons (N) applied uniformly over an area of one square metre
   d. Changes substantially from place to place

2-20. A pneumatic pressure transmitter is calibrated to a pressure range of 100 to 500 psi. The signal output is 10.2 psi. What is the measured pressure in psi?
   a. 272 psi   **c.** 267 psi
   b. 340 psi   **d.** 333 psi

2-21. A viscose line is located 100 feet from the control room where a record of line pressure must be made. The maximum pressure is approximately 85 psi. Select the instrumentation that can handle this job:
   a. A sealed system with 100 feet of capillary connected to a special element in a recording instrument.
   b. A sealed system connected to a nearby pneumatic transmitter sending a signal to a pneumatic receiving recorder
   c. A direct connected spiral element recorder
   d. A pneumatic transmitter directly connected to the line

2-22. A pressure recorder reads 38 psi and the barometer reads 30.12 inches of mercury. The absolute pressure is
   a. 23.21 psi absolute   **c.** 52.79 psi absolute
   b. 68.12 psi absolute   **d.** 38 psi absolute

2-23. The lowest ranges of pressure may be measured by
   a. a water column   **c.** a bell meter
   b. a bellows   **d.** a slack or limp-diaphragm

2-24. A slack diaphragm indicator is to be calibrated 0 to 1 inch of water. The standard would likely be:
   a. a water column
   b. a column of kerosene
   c. a mercury-filled inclined manometer
   d. a water-filled inclined manometer

Level and Density Measurements

Level Measurement Methods

The typical process plant contains many tanks, vessels, and reservoirs. Their function is to store or process materials. Accurate measurement of the contents of these containers is vital. The material in the tanks is usually liquid, but occasionally it may consist of solids.

Initially, level measurement appears to present a simple problem. However, a closer look soon reveals a variety of problems that must be resolved. The material may be very corrosive; it may tend to solidify; it may tend to vaporize; it may contain solids; or it may create other difficulties.

The common methods employed for automatic continuous liquid level measurements are as follows (see also Table 3-1):

1. Float-and-cable
2. Displacement (buoyancy)
3. Head (pressure)
4. Capacitance
5. Conductance
6. Radiation (nucleonic)
7. Weight
8. Ultrasonic
9. Thermal

69

Table 3-1. Liquid Level Measurement

Method	Available Upper Range Values	Closed or Open Tank	Condition of Liquid
Bubble tube	10 in to 250 ft 0.25 to 75 m	Open	Any type including corrosive or dirty
Diaphragm box	4 in to 250 ft 0.1 to 75 m	Open	Clean
Head Pressure	2 in to many ft. 50 mm to many metres	Open	Any type with proper selection
Diff. Pressure	5 in to many ft. 0.25 to many metres	Both	Any type with proper selection
Displacement (buoyancy)	6 in to 12 ft 0.15 to 3.6 m	Both	Any type with proper selection
Float-and-Cable	3 in to 50 ft. 75 mm to 15 m	Open	Any type
Weight	Inches to Feet mm to m Depends on Tank Dim.	Both	Any type
Radiation (nucleonic)	Wide	Both	Any type
Capacitance	Wide	Both	Nonconductive
Ultrasonic	Wide	Both	Any type
Conductance	One or more Points	Both	Conductive

Float-and-Cable

A float-and-cable or float-and-tape instrument (Figure 3-1) measures liquid level by transmitting to a mechanism the rise and fall of a float that rides on the surface of the liquid. Mechanisms are available to accommodate level variations ranging from a few inches to many feet. Float-and-cable devices are used primarily in open tanks, whereas float level switches may be designed to operate in a pressurized tank.

Float devices have the advantage of simplicity and are insensitive to density changes. Their major disadvantage is their limitation to reasonably clean liquids. Turbulence may also create measurement problems. The float and cable technique does not lend itself to the transmitter concept as well as do some of the following techniques.

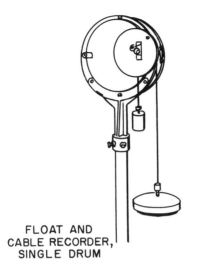

FLOAT AND
CABLE RECORDER,
SINGLE DRUM

Fig. 3-1. Float-and-cable recorder.

Displacement (Buoyancy)

The displacement, or buoyancy, technique is a type of force-balance transmitter (Figure 3-2). It may be used to measure liquid level, interface, or density by sensing the buoyant force exerted on a displacer by the liquid in which it is immersed. The buoyant force is converted by a force-balance pneumatic or electronic mechanism to a proportional 3 to 15 psi, 20 to 100 kPa, 4 to 20 mA/dc or 10 to 50 mA/dc signal.

Fig. 3-2. Buoyancy type level measuring transmitter.

Archimedes' principle states that a body immersed in a liquid will be buoyed upward by a force equal to the weight of the liquid displaced. The displacer element is a cylinder of constant cross-sectional area and heavier than the liquid displaced. It must be slightly longer than the level change to be measured (Figure 3-3). Since the displacer must accurately sense the buoyant force throughout the full range of measurement, displacer lengths and diameters will vary according to the particular process requirements.

The formula to determine the buoyant force span for liquid level applications is as follows:

$$f = V \left(\frac{Lw}{L} \right) (B)(SG) \tag{3-1}$$

where:

f = buoyant force span (lbf or N)
V = total displacer volume (cubic inches or cubic centimeters)
Lw = working length of displacer (inches or millimeters)
L = total displacer length (inches or millimeters)
B = a constant (weight of unit volume of water) (0.036 lb^f/in^3 or 9.8×10^{-3} N/cm^3
SG = specific gravity of the liquid

The minimum buoyant force span for the buoyancy level transmitters is 1.47 pounds-force or 6.7 newtons. The maximum mass of the displacer-plus-hanger cable must not exceed 12 pounds or 5.4 kilograms (or buoyant force span × 6, whichever is smaller).

For a cylindrical displacer, the formula for the volume is:

$$V = \left(\frac{\pi}{4} d^2 \right) L \tag{3-2}$$

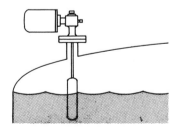

Fig. 3-3. Buoyancy type level transmitter installation.

In the above formula, as well as the sizing of displacers, the working volume considered is limited to the cylindrical volume of the displacer and does not include the volume of the dished end pieces or hanger connection.

Unless certain precautions are taken, buoyancy transmitters are not recommended for extremely turbulent process conditions. The turbulence may cause the displacer to swing erratically, resulting in unpredictable measurement effects or physical damage to the displacer, transmitter, or vessel. In such cases, some form of displacer containment should be used.

Buoyancy transmitters may be applied readily to glass-lined vessels, vessels in which a lower connection is not permissible or possible, density applications with fluctuating pressures or levels, and high-temperature service. Interface level measurement may be accomplished by permitting the interface level to vary over the length of the displacer.

$$f = V \text{ (B) } (SG \text{ diff})$$

SG diff $= SG$ lower liquid minus the SG upper liquid

Head or Pressure

Measurement of head, or pressure, to determine level is the most common approach. The ways in which level may be determined by measuring pressure are many and varied. Some of the more common techniques are presented here.

Bubble Tube Method

In the air purge, or bubble tube system (Figure 3-4), liquid level is determined by measuring the pressure required to force a gas into the liquid at a point beneath the surface. In this way, the level may be obtained without the liquid entering the piping or instrument.

A source of clean air or gas is connected through a restriction to a bubble tube immersed a fixed depth in the tank. The restriction reduces the air flow to a minute amount, which builds up pressure in the bubble tube until it just balances the fluid pressure at the end of the bubble tube. Thereafter, pressure is kept at this value by air bubbles escaping through the liquid. Changes in the measured level cause the air pressure in the bubble tube to build up or drop. A pressure instrument connected at this point can be made to register the level or volume of liquid. A small V-notch is filed in the bottom of the tube so that air

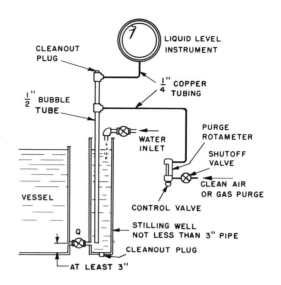

Fig. 3-4. Using bubble pipe.

emerges in a steady stream of small bubbles rather than in intermittent large bubbles (Figure 3-5).

An advantage of the bubble tube method is that corrosive or solids bearing liquids can damage only an inexpensive, easily replaced pipe.

Diaphragm Box

The diaphragm box is shown in Figure 3-6. It is similar to the diaphragm seals used for pressure gauges (Figure 2-20, p. 57) except that the fill is air and the diaphragm is very slack, thin, and flexible. The

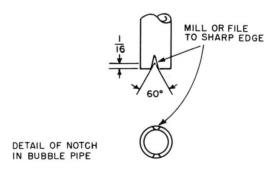

Fig. 3-5. Detail of notch in bubble pipe.

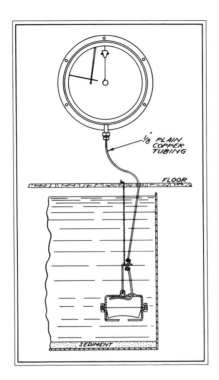

Fig. 3-6. Diaphragm box.

diaphragm box is usually suspended from a chain. The instrument that senses the pressure changes and relates them to level may be mounted above or below the vessel. The diaphragm box is used primarily for water level measurement in open vessels.

Pressure and Differential Pressure Methods Using Differential Pressure Transmitters

These are the most popular methods of measuring liquid level and are shown in Figure 3-7. With an open tank, the pressure at the high pressure side of the cell is a measure of the liquid level. With a closed tank, the effect of tank pressure on the measurement is nullified by piping this pressure to the opposite side of the cell.

When a sealing fluid is used, it must possess a specific gravity higher than that of the liquid in the vessel. The sealing fluid must also be immiscible with the liquid in the vessel.

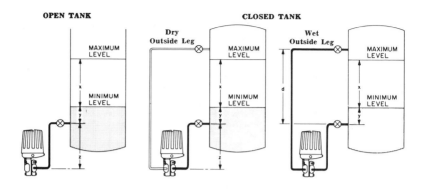

Fig. 3-7.

| Span = xG_L | Span = xG_L | Span = xG_L |
| Suppression = $yG_L + zG_s$ | Suppression = $yG_L + zG_s$ | Elevation = $dG_s - yG_L$ |

Where G_L = specific gravity of liquid in tank

G_s = specific gravity of liquid in outside filled line or lines

If transmitter is at level of lower tank tap, or if air purge is used, $z = 0$.

(Note: The density of the gas in the tank has been disregarded in these calculations.)

EXAMPLE: Assume an open tank with X = 80 inches, y = 5 inches, and z = 10 inches. The specific gravity of the tank liquid is 0.8; the specific gravity of the liquid in the connecting leg is 0.9.

Span = 80 (0.8) = 64 inches head of water

Suppression = 5 (0.8) + 10 (0.9)

= 4 + 9 = 13 inches head of water

Range = 13 to 77 inches head of water

EXAMPLE: Assume a closed tank with X = 70 inches, y = 20 inches, and d = 100 inches. The specific gravity of the tank liquid is 0.8; a sealing liquid with a specific gravity of 0.9 is used.

Span = 70 (0.8) = 56 inches head of water

Elevation = 100 (0.9) − 20 (0.8)

= 90 − 16 = 74 inches head of water

Range = −74 to −18 inches head of water

Fig. 3-8. The repeater is a flanged, force-balance, pneumatic instrument that measures pressure and delivers an output equal to the measured pressure.

(minus sign indicates that the higher pressure is applied to the low side of the cell.)

At times the liquid to be measured possesses characteristics that create special problems. Assume, for example, that the liquid being measured will solidify if it is applied to a wet leg and it is virtually impossible to keep the leg dry. This type of application could use a pressure repeater. The repeater (Figure 3-8) is mounted above the maximum level of the liquid, and the liquid level transmitter is mounted near the bottom of the tank. The pressure in the vapor section is duplicated by the repeater and transmitted to the instrument below, (Figure 3-9). Thus, the complications of a wet leg are avoided, and a varying pressure in the tank will not affect the liquid level measurement.

The pressure range that may be repeated is 0 to 100 psi or 0 to 700 kPa. The supply pressure must exceed the pressure to be repeated by

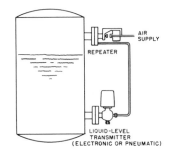

Fig. 3-9. Liquid-level transmitter (electronic or pneumatic) used with a repeater.

20 psi or 140 kPa. A somewhat similar configuration may be arranged by using a sealed system as described in Chapter 2 and shown in Figure 2-21, p. 58.

Capacitance

If a probe is inserted into a tank and the capacitance measured between it and the tank, a sizable change in capacitance will occur with liquid level. This phenomenon is due primarily to the substantial difference between the dielectric constant of air and that of the liquid in the tank. This technique is best applied to nonconductive liquids, since it is best to avoid the problems generated by conducting materials like acids (Figure 3-10).

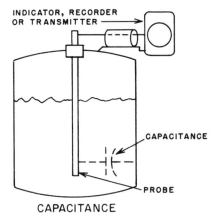

Fig. 3-10. Capacitance.

Conductance

Conductivity level sensors consist of two electrodes inserted into the vessel or tank to be measured. When the level rises high enough to provide a conductive path from one electrode to the other, a relay (solid state or coil) is energized. The relay may be used for either alarm or control purposes. Conductivity then becomes either point control or an alarm point. The liquid involved must be a conductor and must not be hazardous if a spark is created. Level by conductivity finds occasional applications in process plants (Figure 3-11).

Radiation

A radiation level measurement generally consists of a radioactive source on one side of the tank and a suitable detector on the other. As the radiation passes through the tank, its intensity varies with the amount of material in the tank and can be related to level. One advantage is that nothing comes in contact with the liquid. Among the disadvantages are the high cost and the difficulties associated with radioactive materials. Radioactive techniques do have the ability to solve difficult level-measuring problems (Figure 3-12).

Weight

Occasionally the measurement of the contents of a tank is so difficult that none of the usual schemes will work. When this occurs it may be advantageous to consider a weighing system. Weight cells, either hy-

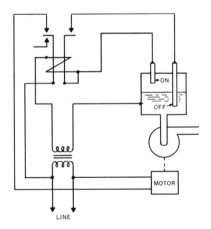

Fig. 3-11. Conductivity.

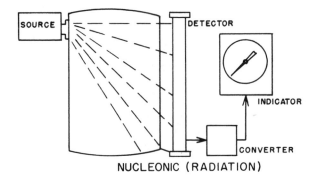

NUCLEONIC (RADIATION)

Fig. 3-12. Radiation technique.

draulic or strain gauge, are used to weigh the vessel and its contents. The tare weight of the tank is zero adjusted out of the reading, which will result in a signal proportional to tank contents (Figure 3-13).

One advantage of the weight system is that there is no direct contact with the contents of the tank and the sensor. However, the system is not economical, and varying densities may confuse the relationship between signal and true level.

Ultrasonic

The ultrasonic level sensor (Figure 3-14) consists of an ultrasonic generator or oscillator operating at a frequency of approximately

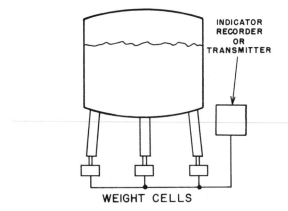

Fig. 3-13. Level measurement using weight method.

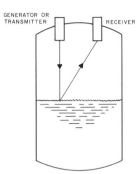

Fig. 3-14. Ultrasonic technique.

20,000 Hz and a receiver. The time required for the sound waves to travel to the liquid and back to the receiver is carefully measured. The time is a measure of level. This technique has excellent reliability and good accuracy. Furthermore, nothing comes in contact with liquid in the tank, which minimizes corrosion and contamination. The only general limitation is economic.

Thermal

One approach employing a thermal sensor to determine level relies on a difference in temperature between the liquid and air above it. When the liquid contacts the sensor, a point determination of level is made.

Another approach depends on a difference in thermal conductivity between the liquid level sensed and the air or vapor above it. Thermal techniques are inexpensive, but have achieved only modest popularity in process applications.

Density Measurement Methods for Liquids and Liquid Slurries

Density is defined as mass per unit volume. Specific gravity is the unitless ratio of the density of a substance to the density of water at some standard temperature. In Europe, this is known as relative density.

The measurement and control of liquid density are critical to a great number of industrial processes. Although density can be of interest, it is usually more important as an inference of composition, of concentration of chemicals in solution, or of solids in suspension.

Many methods are available to determine concentration, including measurement of changes in electrolytic conductivity and of rise in boiling point. This section deals with common industrial methods of measuring density continuously (Table 3-2).

Agitation in a process tank where density is being measured must be sufficient to ensure uniformity of the liquid. Velocity effects must be avoided; agitation should be sufficient to maintain a uniform mixture without affecting head pressure at measurement points.

Nearly all industrial liquid density instruments utilize the effect of density changes on weight as a measurement. Such instrumentation falls into three major categories, defined by the principles employed. These include two other categories, radiation and vibration methods.

For density measurement, a cylindrical displacer is located so as to be fully submerged. The buoyant force on the displacer changes only as a function of changing liquid density, and is independent of changing liquid level in the vessel. Buoyancy transmitters based upon the Archimedes principle may be readily adapted to the measurement of liquid density or specific gravity.

Hydrostatic Head

Proved suitable for many industrial processes is the method that continuously measures the pressure variations produced by a fixed height of liquid (Figure 3-15). Briefly, the principle is as follows: The difference in pressure between any two elevations below the surface (A and B) is equal to the difference in liquid head pressure between these elevations. This is true regardless of variation in level above elevation

TABLE 3-2. Liquid Density Measurement

Method	Minimum Span	Condition of Liquid	Accuracy as % Span
Hydrometer	0.1	Clean	±1%
Displacer	0.005	Clean	±1%
Hydrostatic head	0.05	Any	½ to 1%
Radiation	0.05	Any	1%
Weight of fixed volume	0.05	Clean	1%
Vibrating U-Tube	0.05	Clean	1–3%

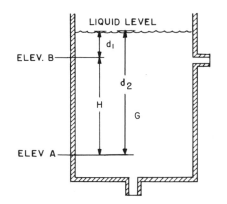

Fig. 3-15. Fixed height of liquid for density measurement.

A. This difference in elevation is represented by dimension H. Dimension (H) must be multiplied by the specific gravity (G) of the liquid to obtain the difference in head in inches or metres of water, which are the standard units for instrument calibration.

To measure the change in head resulting from a change in specific gravity from G_1 to G_2, it is necessary only to multiply H by the difference between G_2 and G_1. Expressed mathematically:

$$P = H \ (G_2 - G_1) \tag{3-3}$$

The change in head (P) is differential pressure in inches or metres of water. G_1 is minimum specific gravity and G_2 is maximum specific gravity.

It is common practice to measure only the span of actual density changes. Therefore, the measurement zero is suppressed; that is, the instrument "zero," or lower range value, is elevated to the minimum head pressure to be encountered. This allows the entire instrument measurement span to be devoted to the differential caused by density changes. For example, if G_1 is 1.0 and H is 100 inches, the span of the measuring device must be elevated 100 inches of water ($H \times G_1$). For a second example, if G_1 is 0.6 and H is 3 metres, the zero suppression should be 1.8 metres of water.

The two principal relationships that must be considered in selecting a measuring device are:

$$\text{Span} = H \times (G_2 - G_1)$$

$$\text{Zero suppression} = H \times G_1$$

(3-4)

For a given instrument span, a low gravity span requires a greater H dimension (a deeper tank).

Radiation

Density measurements by this method are based on the principle that absorption of gamma radiation increases with the increasing specific gravity of the material measured.

The principal instrumentation includes a constant gamma source (usually radium), a detector, and an indicating or recording instrument. Variations in radiation passing through a fixed volume of flowing process liquids are converted into a proportional electrical signal by the detector.

Vibration

Damping a vibrating object in contact with the process fluid increases as the density of the process fluid increases. This principle is applied to industrial density measurement. The object that vibrates (from externally applied energy) is usually an immersed reed or plate. A tube or cylinder that conducts, or is filled with, the process fluid can also be used. Density can be inferred from one of two measurements: changes in the natural frequency of vibration when energy is applied constantly, or changes in the amplitude of vibration when the object is "rung" periodically, in the "bell" mode.

Temperature Effects and Considerations

The density of a liquid varies with expansion due to rising temperature, but not all liquids expand at the same rate. Although a specific gravity measurement must be corrected for temperature effects to be completely accurate in terms of standard reference conditions for density and concentration, in most cases this is not a practical necessity.

For applications in which specific gravity measurement is extremely critical, it may be necessary to control temperature at a constant value. The necessary correction to the desired base temperature can then be included in the density instrument calibration.

Differential Pressure Transmitter

There are a variety of system arrangements for hydrostatic head density measurements with d/p Cell transmitters. Although flange-mounted d/p Cell transmitters are often preferred, pipe-connected transmitters can be used on liquids where crystallization or precipitation in stagnant pockets will not occur.

These d/p Cell transmitter methods are usually superior to those using bubble tubes. They can be applied wherever the vessel is high enough to satisfy the minimum transmitter span. They are also well suited for pressure and vacuum applications.

Constant level overflow tanks permit the simplest instrumentation as shown in Figure 3-16. Only one d/p Cell transmitter is required. With H as the height of liquid above the transmitter, the equations are still:

$$\text{Span} = H \times (G_2 - G_1), \text{ zero suppression} = H \times G_1$$

Applications with level and/or static pressure variations require compensation. There are three basic arrangements for density measurement under these conditions. First, when a seal fluid can be chosen that is always heavier than the process fluid and will not mix with it, the method shown in Figure 3-17 is adequate. This method is used extensively on hydrocarbons with water in the wet leg.

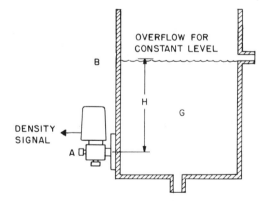

Fig. 3-16. Constant level, open overflow tanks require only one d/p Cell transmitter for density measurement.

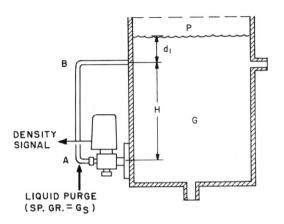

Fig. 3-17. In an open or closed tank with varying level and/or pressure, a wet leg can be filled with seal fluid heavier than the process liquid.

For a wet leg fluid of specific gravity G_s, an elevated zero transmitter must be used. The equations become:

$$\text{Span} = H \times (G_2 - G_1), \text{ zero elevation} = H \times (G_s - G_1)$$

When no seal or purge liquid can be tolerated, there are ways to provide a "mechanical seal" for the low-pressure leg, or for both legs, if needed. Figure 3-18 shows the use of a pressure repeater for the upper connection.

The repeater transmits the total pressure at elevation B to the low pressure side of the d/p Cell transmitter. In this way, the pressure at elevation B is subtracted from the pressure at elevation A. Therefore, the lower transmitter measures density (or $H \times G$, where G is the specific gravity of the liquid). The equations for the lower transmitter are:

$$\text{Span} = H \times (G_2 - G_1)$$

$$\text{Zero suppression} = H \times G_1$$

The equation for the upper repeater is:

$$\text{Output (maximum)} = d_1 \text{ (max)} \times G_2 + P \text{ (max)},$$

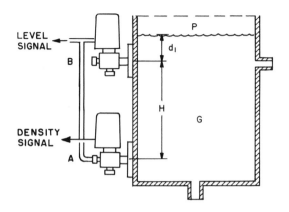

Fig. 3-18. In an open or closed tank with varying level and/or pressure where seal fluid or purge is not suitable, a pressure repeater can be used.

where d_1 is the distance from elevation B to the liquid surface, and P is the static pressure on the tank, if any. When there is no pressure on the tank, the upper repeater output is a measurement of level.

Transmitter locations should be as high as possible above the bottom of the tank. The process liquid then can be drained below this level for removal and maintenance of the transmitter.

Questions

3-1. An open tank contains a liquid of varying density and the level within the tank must be accurately measured. The best choice of measuring system would be:
   a. Bubble tube
   b. Diaphragm box
   c. Float and cable
   d. Head type with differential pressure transmitter

3-2. A chain-suspended diaphragm box and pressure instrument are used to measure liquid level in a tank which is 10 feet in diameter and 15 feet deep. The tank contains a liquid that has a specific gravity of 0.9 at ambient temperature. The span of the pressure instrument required is approximately:
   a. 10 psi   **c.** 5 psi
   b. 25 psi   **d.** 15 psi

3-3. A pressurized tank contains a liquid, SG = 1.0, and the level measuring pressure taps are 100 inches apart. A pneumatic differential pressure transmitter is used to measure level. The leg that connects the top of the tank to the transmitter is filled with tank liquid. The top tap should be connected to:

    a. The low-pressure tap on the transmitter.
    b. The high-pressure tap on the transmitter.
    c. Either the high- or the low-pressure tap.
    d. The high-pressure tap but through a seal chamber.

3-4. In Question 3-3 the bottom connection is made to the low-pressure tap on the transmitter and the top is made to the high-pressure tap. The signal output, if the transmitter is calibrated 0 to 100 inches, when the tank is full will be:

    a. Top scale value   **c.** 0
    b. Bottom scale value   **d.** A measure of density

3-5. A plant has a water tower mounted on top of an 80-foot platform. The tank is 30 feet high. What is the height of water in the tank if a pressure gauge on the second floor, height 15 feet, reads 40 psi.

    a. Full   **c.** 4.74 feet
    b. 12.42 feet   **d.** 27.42 feet

3-6. A displacer is 5 inches in diameter and 30 inches long. If it is submerged to a depth of 20 inches in liquid SG = 0.8, what force will it exert on the top works?

    a. 11.3 pounds   **c.** 35.5 pounds
    b. 426.3 pounds   **d.** 1.47 pounds

3-7. A closed tank level (SG = 0.83) is measured with a differential pressure transmitter. The level may vary from 10 to 100 inches. The high-pressure tap is 10 inches above the transmitter and a water seal fluid is used. A pressure repeater is used for the top tank pressure. The differential pressure transmitter should be calibrated:

    a. 10 to 100 inches
    b. 10 to 84.7 inches
    c. 8.3 to 74.7 inches
    d. 8.3 to 98.3 inches

3-8. To resist the corrosive effects of a very unusual, highly explosive chemical, a storage tank is lead-lined. Level measurement is difficult because the material also solidifies once it enters a measuring tap. To measure the level with accuracy the best choice would be:

    a. Weigh the tank and its contents and zero out the tare weight of the tank.
    b. Use a radioactive level measurement.
    c. Install a conductivity level measurement.
    d. Use a thermal conductivity level detector.

3-9. A displacer 48 inches long and 1⅝ inches in diameter is used to measure density on a 4 to 20 mA dc transmitter. Its total submerged weight changes from 1.42 pounds to 5.7 pounds. If the density is checked and determined to be 0.4 when the output is 4 mA dc, the density at an output of 12 mA dc is:

    a. 0.6   **c.** 1.0

    b. 0.8   **d.** 1.6

3-10. A density measuring system is set up as shown in Figure 3-18. The H distance is 100 inches and the lower transmitter is calibrated to a 120-inch span. If the transmitter is pneumatic and delivering a 12-psi signal, the specific gravity of the tank's contents is:

    a. 1.0   **c.** 0.9

    b. 0.83   **d.** 1.2

4

Flow Measurement

The process industries by their very nature deal constantly with flowing fluids, and measurement of these flows is essential to the operation of the plant. These measurements are indicated, recorded, totalized, and used for control. Flow measurement is generally the most common measurement found in the process plant.

Fluid flow measurement is accomplished by:

A. Displacement
  1. Positive displacement meters
  2. Metering pumps
B. Constriction Type, Differential Head
  1. Closed conduit or pipe
    a. Orifice plate
    b. Venturi tube
    c. Flow nozzle
    d. Pitot tube
    e. Elbow
    f. Target (drag force)
    g. Variable area (rotameter)
  2. Open channel
    a. Weir
    b. Flume

C. Velocity Flowmeters
   1. Magnetic
   2. Turbine
   3. Vortex or swirl
   4. Ultrasonic
   5. Thermal
D. Mass Flowmeters
   1. Weight types
   2. Head and magnetic types compensated for temperature, pressure, and density
   3. Gyroscope precision types
   4. Centrifugal force (torque) types

Positive displacement meters and metering pumps measure discrete quantities of flowing fluid. This flow is indicated in terms of an integrated or totalized flow volume (gallons, cubic feet, litres, cubic metres, and the like). A typical application of this type of flow measurement is custody transfer, and familiar examples are domestic water metres and gasoline pumps. The other types of meters listed above measure flow rate.

It is flow rate—quantity per time, such as gallons per minute or liters per second—which is most generally used for measurement and related control applications in the process plant. The most common rate meter is the constriction or head type.

Constriction or Differential Head Type

In this type of flow measurement a primary device or restriction in the flow line creates a change in fluid velocity that is sensed as a differential pressure (head). The instrument used to sense this differential pressure is called the secondary device, and this measurement is related to flow rate. The flow-measuring system, often called a flowmeter, consists of primary and secondary devices properly connected.

The operation of the head-type flow-measuring system depends on a theorem first proposed by Daniel Bernoulli (1700–1782). According to this theorem the total energy at a point in a pipeline is equal to the total energy at a second point if friction between the points is neglected.

The energy balance between Points 1 and 2 in a pipeline can be expressed as follows:

$$\frac{P_2}{\rho} + \frac{(Vm_2)^2}{2G} + Z_2 = \frac{P_1}{\rho} + \frac{(Vm_1)^2}{2G} + Z_1$$

where:

  P = Static pressure, absolute
  Vm = Fluid stream velocity
  Z = Elevation of center line of the pipe
  ρ = Fluid density
  G = Acceleration due to gravity

If the energy is divided into two forms, static head and dynamic or velocity head, some of the inferential flow devices can be more easily studied.

For example, with an orifice plate, the change in cross-sectional area between the pipe and the orifice produces a change in flow velocity (Figure 4-1). The flow increases to pass through the orifice. Since total energy at the inlet to, and at the throat of, the orifice remains the same (neglecting losses), the velocity head at the throat must increase, causing a corresponding decrease in static head. Therefore, there is a head difference between a point immediately ahead of the restriction and a point within the restriction or downstream from it. The resulting differential head or pressure is a function of velocity that can then be related to flow. Mechanical flow-measuring instruments use some device or restriction in a flow line that results in such a differential head.

The following should clarify these relationships. Assume a tank as shown in Figure 4-2. A flow line enters the tank and replaces the out-

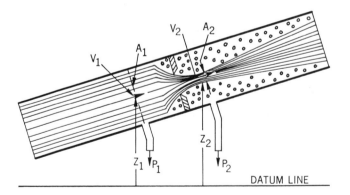

Fig. 4-1. Orifice plate differential producer. The difference in head (pressure) at P_1 and P_2 is a function of velocity which can be related to flow.

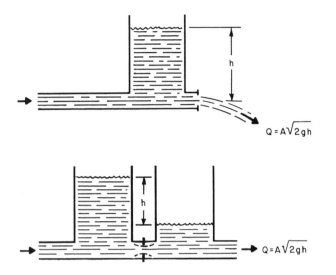

Fig. 4-2. Free and tank-to-tank flow, used to simplify concept of flow formula (4-1).

flow through the orifice located near the bottom of the tank. If the level in the tank is H, the velocity of outflow will be:

$$V^2 = 2GH$$

or

$$V = \sqrt{2GH}$$

This is the law of falling bodies that is developed in most physics books. The volume of liquid discharged per time unit through the orifice can be calculated by geometry.

$$Q = AV$$

Substituting the expression for velocity in this equation:

$$Q = A\sqrt{2GH} \tag{4-1}$$

This expression may also be used to calculate the rate of flow past a point in a pipe. The actual flow rate will be less than this equation will calculate. The reduction is caused by several factors, including friction and contraction of the stream within the pipe. Fortunately all common primary devices have been extensively tested and their coefficients determined. These data become part of the flow calculations.

Primary Devices

The orifice plate is the most popular primary device found in most process plants. Orifice plates are applicable to all clean fluids, but are not generally applicable to fluids containing solids in suspension (dirty fluids).

A conventional orifice plate consists of a thin circular plate containing a concentric hole (Figure 4-3). The plate is usually made of stainless steel but other materials, such as monel, nickel, hastelloy, steel, glass, and plastics, are occasionally used. The most popular orifice plate is the sharp-edged type. The upstream face of the plate usually is polished and the downstream side is often counterbored to prevent any interference with the flowing fluid. The bore in the plate is held to a tolerance of a few ten-thousandths of an inch in small sizes and to within a few thousandths in sizes above 5 inches in diameter.

In addition to the conventional, sharp-edged, concentric plate, there are others that have been designed to handle special situations. These special types constitute only a small percentage of the total, but they do occasionally solve a difficult problem.

There are two plates (Figure 4-3) designed to accommodate limited amounts of suspended solids. The eccentric plate has a hole that is bored off-center, usually tangent to the bottom of the flow line (inside periphery of the pipe). The segmental orifice plate has a segment removed from the lower half of the orifice plate. In addition, there are

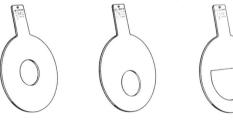

CONCENTRIC   ECCENTRIC   SEGMENTAL

Fig. 4-3. Orifice plate types.

quadrant-edged or special-purpose orifice plates with rounded edges. This design minimizes the effect of low Reynolds number (see below).

Orifice plates along with other primary devices operate most satisfactorily at high flow rates. The reason is the change in flow coefficient that occurs when the flow rate changes from high to very low. When this happens the flow velocity distribution changes from uniform to parabolic. This may be related to a ratio of inertia forces to viscous forces called Reynolds number.

$$\text{Reynolds Number} = \frac{\text{Average Fluid Velocity} \times \text{Orifice Diameter}}{\text{Absolute Viscosity}}$$

In most process plants the Reynolds numbers encountered will be high—between 10,000 and 1,000,000. Under these conditions the flow coefficients stay fixed and head-type flow measurements are reasonably accurate. To achieve the ultimate in accuracy, a Reynolds number correction is made and this is described in the text *Principles and Practices of Flowmeter Engineering* by L. K. Spink. At low Reynolds number, the flow changes from turbulent to parabolic or laminar and the coefficients undergo a change. The changeover cannot be precisely defined, which makes the application of head-type flow measurements difficult at low-flow velocities. The quadrant-edged orifice has the ability to perform well at lower velocities than the square-edged plate. However, it does not perform as well at high Reynolds numbers (velocities).

Orifice plates that are to handle liquids containing small amounts of dissolved air should contain a small vent hole bored at the top to permit the air to pass through the plate. Specifications have been established for the size of this vent hole in relation to the size of the orifice. Plates for use with steam should have a similar hole drilled at the bottom to drain any condensed steam. In certain flow measurements, the orifice plate should also have a bottom drain hole to permit the condensation to drain.

Table 4-1 summarizes application factors to help in deciding whether the orifice plate can do a particular job.

Another primary device that is frequently found in the process plant is the Venturi tube shown in Figure 4-4. The Venturi tube produces a large differential with a minimum permanent pressure loss. It has the added advantage of being able to measure flows containing suspended solids. The most significant disadvantage is its cost which, when compared with other primary devices, is high.

Table 4-1. Primary Flow Measurement Devices

	Turbine Flowmeter	*Target Meter*	*Concentric*	*Segmental (Eccentric)*	*Quadrant*	*Venturi*	*Nozzle*	*Pitot*	*Elbow*	*Lo-loss Tube*	*Magnetic Flowmeter*	*Vortex, or Swirl*
						ORIFICE						
Accuracy (empirical data)	E	F	E	F	G	G	G	*	P	G	E	E
Differential produced for given flow and size	None	G	E	E	E	G	G	F	P	E	None	G
Pressure loss #	P	P	P	P	P	G	P	None	None	E	E	G
Use on dirty service	P	E	P	F	P	E	G	VP	P	G	E	U
For liquids containing vapors	E	E	†	E	†	E	G	F	F	G	E	U
For vapors containing condensate	P	E	‡	E	‡	E	G	P	F	G	None	U
For viscous flows	F	G	F	U	E	G	G	§	U	F	E	VP
First cost	P	G	E	G	G	P	F	G	E	P	P	P
Ease of changing capacity	E	G	E	G	E	P	F	VP	VP	P	E	G
Ease of installation	F	G	G	G	G	F	F	E	G	F	F	F

E	Excellent	
G	Good	
F	Fair	
P	Poor	
VP	Very poor	
U	Unknown	
#	E indicates lowest loss, P highest, etc.	

* For measuring velocity at one point in conduit the well-designed Pitot tube is reliable. For measuring total flow, accuracy depends on velocity traverse.
† Excellent in vertical line if flow is upward.
‡ Excellent in vertical line if flow is downward.
§ Requires a velocity traverse.

Several other primary devices may be classified as modifications of the Venturi tube. The Lo-Loss and Dall tubes are similar in many ways to the Venturi. The flow nozzle (Figure 4-5) has some similarity but does not have a diffuser cone, and this limits its ability to minimize permanent pressure loss. The Venturi family of primary devices is often

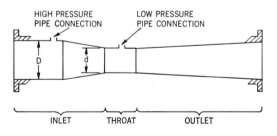

Fig. 4-4. The Herschel Venturi tube, developed in 1888, to produce a large differential pressure with a small head loss.

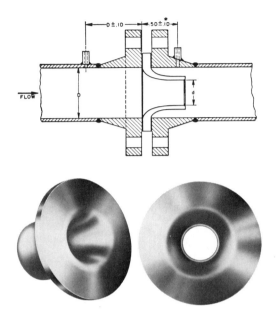

Fig. 4-5. Flow nozzle assembly.

chosen to minimize permanent pressure loss or to cope with solids suspended in the flowing material. Table 4-1 provides aid in making a selection.

The Pitot tube is another primary device. It has the advantage of practically no pressure drop. Its limitations are its inability to handle solids carried by the flowing material and somewhat limited accuracy. Solids tend to plug the openings in the tube, and the classical Pitot tube senses impact pressure at only one point, thus decreasing accuracy. A multi-impact opening type (annubar tube) is available which improves the potential accuracy. In the process plant Pitot tubes are used more for testing than for continuous use. It is possible to install a Pitot tube into a flow line under pressure if the required equipment is available.

Occasionally a pipe elbow (Figure 4-6) may be used as a primary device. Elbow taps have an advantage in that most piping configurations contain elbows that can be used. If an existing elbow is used, no additional pressure drop is created and the expense involved is minimal. The disadvantages are that accuracy will be lacking and dirty flows may tend to plug the taps. Repeatability should be reasonably good. Two taps locations are used at either 45 or 22½ degrees from the

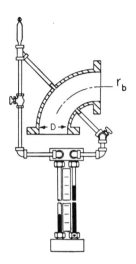

Fig. 4-6. Elbow (used as primary device).

inlet face of the elbow. At low-flow velocities, the differential produced is inadequate for good measurement, thus the elbow is a usable choice only for high top-scale velocities.

The target or drag body flow device (Figure 4-7) is in many ways a variation of the orifice plate. Instead of a plate containing a bore through which the flow passes, this device contains a solid circular disc and the flow passes around it. An essential part of the measurement is the force-balance transducer that converts this force into a signal. This signal is related to flow squared, just as it would be with any other primary device connected to a differential pressure transmitter. The

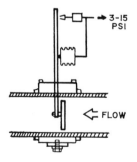

Fig. 4-7. Target flowmeter.

target meter may be generally applied to the measurement of liquids, vapors, and gases. Condensates pass freely along the bottom of the pipe, while gases and vapors pass easily along at the top. The target flow device has demonstrated the ability to handle many difficult assignments not possible with other primary elements. Both pneumatic and electronic force-balance transducers can be used, making either type signal, pneumatic or electric, available.

Secondary Devices

Secondary instruments measure the differential produced at the primary devices and convert it into a signal for transmission or into a motion for indication, recording, or totalization.

Secondary instruments include the mercury manometer, the so-called mercuryless or diaphragm (bellows) meter and various types of force-balance and motion-balance pneumatic and electronic transmitters.

In the past the mercury manometer was an extremely popular secondary device, but the high price and danger of mercury have all but eliminated it from everyday use. However, a few mercury manometers are still employed for flow rate measurements in gas pipelines. The bellows or diaphragm secondary devices are popular when a direct indication or record is desired. The most popular secondary device is the force-balance transmitter. The reasons for this popularity are virtually unlimited adjustability, no harm from extensive overrange, and the availability of a standard signal that can be fed into a recorder, a controller, or other instruments that may be combined to form a system.

Still another detail is the mounting of the secondary device (locations are shown in Figure 4-8). With liquid flow, the secondary device is

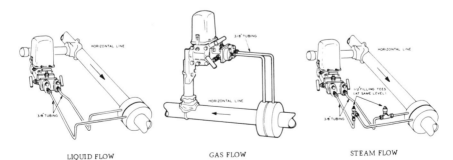

LIQUID FLOW   GAS FLOW   STEAM FLOW

Fig. 4-8. Flow.

below the pipeline being measured to ensure that the connecting lines attached to the side of the pipe remain liquid filled. With steam, the lines should always remain filled with condensate; but with gas, the secondary device is mounted above the flow line to drain away any liquid that may be present.

When any primary device is installed in a pipeline, the accuracy will be improved if as much straight run as possible precedes it. Straight run beyond the primary device is of far less concern.

Still another important consideration is the location of the pressure taps. With nozzles, Venturi, Dall and Pitot tubes, and elbows, the pressure tap locations are established. For orifice plates, however, a variety of tap locations are used. These are flange, corner, vena contracta, or D and D/2, along with full flow or pipe taps. Figure 4-9 describes these tap locations, and Figure 4-10 shows the importance of tap location. In general, flange taps are preferred except when physical limitations make pipe taps advantageous. Corner taps require a special flange and vena contracta tap locations relate to orifice openings. Corner taps are advantageous for pipe sizes under 2 inches.

Figure 4-11 describes the permanent head loss caused by the primary device selected.

Relating Flow Rate to Differential

The two basic formulas have already been introduced. In the English system, $G = 32$ feet per second squared, and in the metric system, $G = 980$ centimetres per second squared. The height (h) would be expressed in feet in the English system and centimetres in the metric

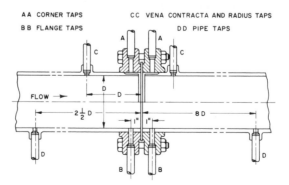

Fig. 4-9. Tap locations for orifices.

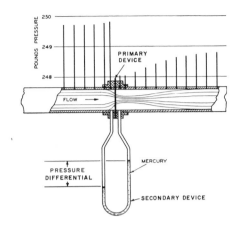

Fig. 4-10. Differential pressure profile with orifice plate.

system. In Equation 4-1, area (*A*) would be expressed in square feet and in the metric system in square centimetres yielding cubic feet per second or cubic centimetres per second. In the United States, the English system is the most common. This system is in the process of being changed to the SI metric, but it appears that the total conversion

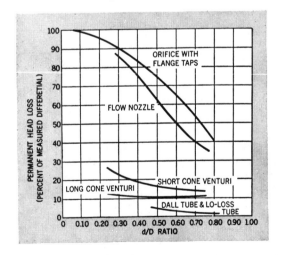

Fig. 4-11. Permanent head loss in a differential producer may be plotted as a percent of measured differential for the several types of primary devices. These values must be interpreted according to the acceptable head loss limit for any particular application.

will take a long time to complete. For this reason the sample calculations that follow will use the English system.

In Equation 4-1, V is in feet per second; the acceleration due to gravity (G) is in feet per second squared; H is the height in feet of a column of the fluid caused by the differential pressure across a primary device. To express this equation in terms of an equivalent differential (h), in inches of water, H is replaced by $(h/12)G_f)$ where G_f is the specific gravity of the fluid at flowing temperature.

Substituting V into Equation 4-1:

$$Q = A \sqrt{2G \frac{h}{12 \times G_f}} \qquad (4\text{-}2)$$

where:

Q = Volumetric flow (cubic feet per second)
A = Cross-sectional area of the orifice or throat of the primary device (square feet)
h = Differential across the primary device (inches of water)
G_f = Specific gravity of the fluid (dimensionless)
G = Acceleration due to gravity (a constant: 32.17 feet per second^2)

For liquids, it is more useful to express Q in gallons per minute. Also, it is convenient to express the area of the orifice or throat in terms of its diameter (d in inches). Substituting in Equation 4-2:

$$Q \text{ (gpm)} = 60 \times 7.4805 \times \frac{\pi d^2}{4 \times 144} \sqrt{\frac{2 \times 32.17}{12}} \times \sqrt{\frac{h}{G_f}}$$

$$Q \text{ (gpm)} = 5.667 \times d^2 \times \sqrt{\frac{h}{G_f}} \qquad (4\text{-}3)$$

Equation 4-3 must be modified to take into account several factors, such as the contraction of the jet, frictional losses, viscosity, and the velocity of approach. This modification is accomplished by applying a discharge coefficient (K) to the equation. K is defined as the actual flow rate divided by the theoretical flow rate through the primary device. Basic texts carry K factors for most primary devices. For instance, K is listed according to various pressure tap locations, and d/D ratios. Where extremely accurate results are required, however, the K factor must be determined in the laboratory by actual flow calibration of the primary device.

Applying K to Equation 4-3 gives:

$$Q \text{ (gpm)} = 5.667 \, Kd^2 \sqrt{\frac{h}{G_f}} \qquad\qquad (4\text{-}4)$$

Since both K and d are unknown, another value is defined. The quantity (S) is set equal to $K \, (d/D)^2$. Since Kd^2 then equals SD^2, substitution produces:

$$Q \text{ (gpm)} = 5.667 \, SD^2 \sqrt{\frac{h}{G_f}} \qquad\qquad (4\text{-}5)$$

S values (sizing factor) are tabulated against beta ratios (d/D) for various differential devices in Table 4-2.

Effect of Temperature on Flow Rate

The above equations give the flow through the primary device at conditions that exist at the time the measurement is made. In most industries, it is more desirable to know the equivalent volume flow at a stated reference temperature, usually 60°F (15.6°C).

For liquids, this correction can be applied by multiplying the equation by the ratio of the liquid specific gravity at the flowing temperature (G_f) divided by the liquid specific gravity at the reference temperature (G_1). Thus Q gallons per minute at the reference temperature:

$$Q = 5.667 \, SD^2 \sqrt{\frac{h}{G_f}} \times \frac{G_f}{G_1} \qquad\qquad (4\text{-}6)$$

which simplifies to:

$$Q = 5.667 \, \frac{SD^2 \sqrt{G_f h}}{G_1} \qquad\qquad (4\text{-}7)$$

Similar equations can be developed for steam or vapor flow in weight units (such as pounds per hour) or for gas flow in volume units such as standard cubic feet per hour (scfh).

$$W \text{ (pounds per hour)} = 359 \, SD^2 \sqrt{h\gamma_f} \qquad\qquad (4\text{-}8)$$

$$Q \text{ (scfh)} = 218.4 \, SD^2 \frac{T_b}{p_b} \sqrt{\frac{hp_f}{T_f G}} \qquad\qquad (4\text{-}9)$$

Table 4-2. Sizing Factors

Beta or d/D Ratio	Square-Edged Orifice, Flange Corner or Radius Taps	Full-Flow (Pipe) 2½ & 8D Taps	Nozzle and Venturi	Lo-Loss Tube	Dall Tube	Quadrant-Edged Orifice
0.100	0.005990	0.006100				
0.125	0.009364	0.009591				
0.150	0.01349	0.01389				
0.175	0.01839	0.01902				
0.200	0.02402	0.02499				0.0305
0.225	0.03044	0.03183				0.0390
0.250	0.03760	0.03957				0.0484
0.275	0.04558	0.04826				0.0587
0.300	0.05432	0.05796	0.08858			0.0700
0.325	0.06390	0.06874	0.1041			0.0824
0.350	0.07429	0.08068	0.1210	0.1048		0.0959
0.375	0.08559	0.09390	0.1392	0.1198		0.1106
0.400	0.09776	0.1085	0.1588	0.1356	0.1170	0.1267
0.425	0.1109	0.1247	0.1800	0.1527	0.1335	0.1443
0.450	0.1251	0.1426	0.2026	0.1705	0.1500	0.1635
0.475	0.1404	0.1625	0.2270	0.1900	0.1665	0.1844
0.500	0.1568	0.1845	0.2530	0.2098	0.1830	0.207
0.525	0.1745	0.2090	0.2810	0.2312	0.2044	0.232
0.550	0.1937	0.2362	0.3110	0.2539	0.2258	0.260
0.575	0.2144	0.2664	0.3433	0.2783	0.2472	0.292
0.600	0.2369	0.3002	0.3781	0.3041	0.2685	0.326
0.625	0.2614	0.3377	0.4159	0.3318	0.2956	0.364
0.650	0.2879	0.3796	0.4568	0.3617	0.3228	
0.675	0.3171	0.4262	0.5016	0.3939	0.3499	
0.700	0.3488	0.4782	0.5509	0.4289	0.3770	
0.725	0.3838		0.6054	0.4846	0.4100	
0.750	0.4222		0.6667	0.5111	0.4430	
0.775	0.4646			0.5598	0.4840	
0.800	0.5113			0.6153	0.5250	
0.820				0.6666	0.5635	

FOR AN ELBOW WITH 45° TAPS. $S = .68\sqrt{\dfrac{r_b}{D}}$

In Equations 4-8 and 4-9:

γ_f = Specific weight of the steam or vapor at operating conditions in pounds per cubic foot

T_b = Reference temperature (absolute); (i.e., 460 plus the reference temperature in °F.)

p_b = Reference pressure (psi absolute)

T_f = Operating temperature at the primary device (absolute); (i.e., 460 plus operating temperature in °F.)

p_f = Operating pressure (psi absolute)

G = Specific gravity of the gas (molecular weight of the gas divided by the molecular weight of air or the weight of a volume of the gas at a given temperature and pressure divided by the weight of an equal volume of air at the same temperature and pressure).

The reference temperature is very often 60°F and the reference pressure atmospheric 14.7 psi absolute. If these are the standard conditions, Equation 4-9 can be simplified to:

$$Q(\text{scfh}) = 7{,}727 \, SD^2 \, \sqrt{\frac{hP_f}{T_f G}} \qquad \text{(4-10)}$$

Equation 4-10 is applicable to gas flow only when the pressure differential is small enough so that gas density does not change significantly. A simple rule of thumb is that the maximum differential in inches of water should not exceed the absolute operating pressure in psi absolute. For example, if the gas operating pressure is 22 psi absolute, at a particular installation, and the maximum differential is 20 inches of water, Equation 4-10 can be used.

Flow rate measurements for gas and steam are more difficult to make with accuracy than those for liquid. The reason is changes in specific gravity, weight, temperature, pressure, and so on, that may occur under operating conditions.

These changes will have an effect on measurement accuracy and under certain conditions may be difficult to predict. An abbreviated set of tables for the formulas given are included in this book. If more accuracy is required, more exact equations, along with detailed tables, such as those found in *Principles and Practice of Flow Meter Engineering* by Spink, should be used.

Another method of performing these flow calculations is to use a flow slide rule. The flow slide rule has the table values incorporated into its scales. If the tables given are used, the resulting accuracy should be as good as the flow slide rule.

Now several sample problems are given to demonstrate the procedure followed for each type of calculation. Additional problems are given at the end of this chapter.

SAMPLE PROBLEM 1 A 4-inch schedule 40 pipe carries water that is measured by a concentric, sharp-edged, orifice plate, d = 2.000 inches, with flange taps. The differential is measured with an electronic differential pressure transmitter. The transmitter is calibrated 0 to 100 inches of water pressure and has an output of 4 to 20 mA dc. If the signal from the transmitter is 18.4 mA dc, find the flow rate.

Step 1. Convert the electrical signal to differential pressure.

$$\frac{18.4 - 4}{20 - 4} \times 100 = 90 \text{ inches of water}$$

Step 2. Determine *ID* of 4-inch schedule 40 pipe (see Appendix, table A-4) = 4.026 inches.

Step 3. Calculate $\dfrac{d}{D} = \dfrac{2.000}{4.026} = 0.4968$

Step 4. From Table 4-2 determine S. This will require interpolation.

0.475	0.1404
0.4968	S
0.5000	0.1568

$$S = \left(\frac{0.4968 - 0.475}{0.5000 - 0.475}\right)(0.1568 - 0.1404) + 0.1404$$

$$S = 0.1547$$

Step 5. Substitute in Equation 4-7:

$$Q \text{ (gpm)} = 5.667 \times 0.1547 \times 4.026 \times 4.026 \sqrt{\frac{90}{1}}$$

$$= 5.667 \times 0.1547 \times 16.21 \times 9.487$$

$$= 134.83 \text{ or } 135 \text{ gpm}$$

SAMPLE PROBLEM 2 An elbow is used as a primary device. The taps are made at 45 degrees. The line is a 6-inch schedule 40 pipe. What is the water flow rate if the effective radius of curvature is 9 inches and a differential pressure of 35 inches of water pressure is produced.

$$S = 0.68 \sqrt{\frac{r_b}{D}} \; 0.68 \sqrt{\frac{9}{6.065}} = 0.8283$$

$$Q \text{ (gpm)} = 5.667 \, SD^2 \sqrt{\frac{h}{G_f}}$$

$$= 5.667 \times 0.8283 \times 6.065 \times 6.065 \sqrt{\frac{35}{1}}$$

$$= 172.66 \sqrt{35} = 1{,}021.47 \text{ gpm}$$

SAMPLE PROBLEM 3 Dry-saturated steam is measured with a flow nozzle. The d/D is 0.45 and the line size is a 8-inch schedule 80 pipe. The static pressure is 335 psi. Calculate the flow rate at a differential pressure of 200 inches of water in pounds per hour.

$$W\text{(pounds per hr.)} = 359 \, SD^2\sqrt{h\gamma_f} \qquad \textbf{(4-8)}$$

from Table 4-2

$S = 0.2026$

$D = 7.625$ inches (Table A-4)

$\gamma_f = 0.754$ pounds per cubic feet

$$W\text{(pounds per hour)} = 359 \times 0.2026 \times 7.625^2 \sqrt{200 \times 0.754}$$

$$= 4{,}228.77 \times 12.28$$

$$W = 51{,}929 \text{ pounds per hour}$$

SAMPLE PROBLEM 4 A 6-inch schedule 40 pipe (ID-6.065) carries fuel gas with a specific gravity of 0.88. The line pressure is 25 psi. Flowing temperature is 60°F and the flow is measured with an orifice plate with flange taps. The maximum flow rate is 2,000,000 standard cubic feet per day. Find the diameter of the hole to be bored in the concentric orifice plate if 20 inches of water pressure is full-scale differential.

$$Q \text{ (scfh)} = 7,727 \, SD^2 \sqrt{\frac{hP_f}{T_fG}} \tag{4-10}$$

$$\frac{2,000,000}{24} = 7,727 \times S \times 6.065^2 \sqrt{\frac{20(25 + 15)}{(460 + 60)0.88}}$$

$$8,3333.33 = 7,727 \times S \times 36.78 \times 1.322$$

$$S = \frac{83,333.33}{7,727 \times 36.78 \times 1.322} = 0.2218$$

From Table 4-2

.575	.2144
$\dfrac{d}{D}$	.2218
.600	.2369

$$\frac{d}{D} = \left(\frac{.2218 - .2144}{.2369 - .2144} \right) .025 + .575 = .5832$$

$$d = (.583)(6.065) = 3.536 \text{ inches}$$

SAMPLE PROBLEM 5 Gasoline is carried in a 2-inch schedule 40 pipe (*ID* = 2.067). A concentric sharp-edged orifice plate with corner taps is used to measure flow rate. The orifice bore is 1 inch in diameter, and at full flow, 50 inches of water differential is produced. The specific gravity of the gasoline is 0.75. What is the flow rate?

$$Q \text{ (gpm)} = 5.667 \, SD^2 \sqrt{\frac{h}{G_f}} \tag{4-7}$$

$$\frac{d}{D} = \frac{1.000}{2.067} = 0.484$$

$\dfrac{d}{D}$	S
0.4	0.0978
0.484	
0.5	0.1568

Interpolation

$$\left[\frac{(0.484 - 0.4)}{(0.5 - 0.4)} \times (0.1568 - 0.0978) \right] + 0.0978$$

$$S = \left[\frac{0.084}{0.1} \times 0.059 \right] + 0.0978 = 0.1474$$

$$Q = 5.667 \times 0.1474 \times 2.067 \times 2.067 \times \sqrt{\frac{50}{0.75}}$$

$$= 5.667 \times 0.1474 \times 2.067 \times 2.067 \times 8.165$$

$$= 29.14 \text{ gpm}$$

Variable Area Meters (Rotameter)

The variable area (Figure 4-12) meter is a form of head meter. In this flowmeter the area of the flow restriction varies so as to maintain a constant differential pressure. The variable area meter, which is often called a rotameter, consists of a vertical tapered tube through which the fluid flow being measured passes in an upward direction. A float or rotor is contained within the tapered tube. This rotor or float is made of some material—usually metal—more dense than the fluid being measured. As the flow moves up through the tapered tube, it elevates the float until balance between gravity acting on the float and the upward force created by the flow is achieved. In achieving this balance of forces, the area through which the fluid passes has automatically been adjusted to accommodate that flow rate.

The tapered tube is often made of transparent material so the float position can be observed and related to a scale calibrated in units of flow rate. The rotameter is often used to measure low-flow rate. When very large flow rates are to be measured, a bypass rotameter may be used. This consists of an orifice in the main flow line with the variable area meter connected in parallel. An industrial rotameter is useful over the upper 90 percent of its scale. The accuracy of the rotameter, like any head type, depends on many factors. Typical errors may vary between 1 and 10 percent of full scale value.

Open-Channel Flow Rate Measurements

Flow rate measurements in open channels are fundamental to handling water and waste. Environmental considerations have made these mea-

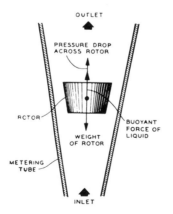

Fig. 4-12. Variable area (rotameter) flowmeter.

surements common in the typical process plant. Open-channel measurements utilize head meter techniques.

Primary Devices

The primary devices used in open-channel flow rate measurements are weirs and flumes. There are two basic weirs—the rectangular and the V-notch. The rectangular weir is made in three varieties (Figure 4-13). The first has contractions or extensions into the channel that produce a boxlike opening. The second modification completely suppresses these contractions, extending the weir across the entire width of the channel. The third type, the so-called Cippoletti weir, has end contractions set at a 4:1 angle, rather than being perpendicular to the edge of the weir. Rectangular weirs are primarily designed for larger flow; their limit on range is dictated by the design of the associated channels. Figure 4-14 shows the ranges of weir capacity for the different types and sizes of weirs.

V-notch weirs are essentially plates (usually metal) that contain a V-shaped notch (Figure 4-13). The angle of the V can vary, but the formulas given are for the most common angles 30, 60, and 90 degrees. V-notch weirs are employed for lower flow rates than those that would be measured by a rectangular weir.

In weir measurement (Figure 4-15) the nappe, or profile of water over the weir, must be completely aerated if good accuracy is to be obtained. All weirs then produce some head loss as the liquid falls free. If head loss is a problem, a flume might be a better choice.

Flumes, a further development of the basic weir concept, are designed primarily to reduce the head loss that is experienced with the

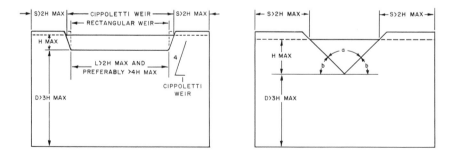

Fig. 4-13. Rectangular, Cippoletti, and V-notch weirs.

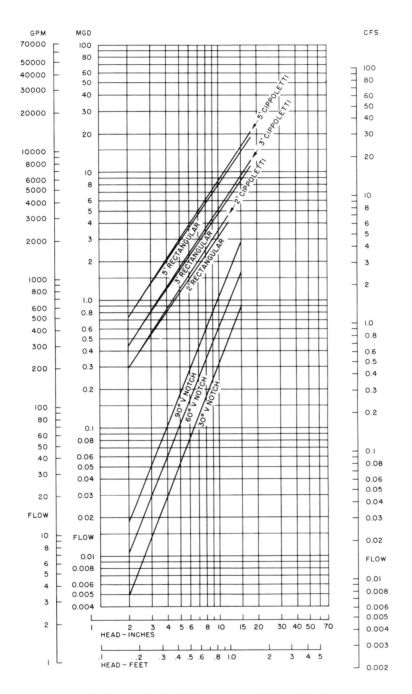

Fig. 4-14. Range of weir capacity for different types and sizes of weirs.

111

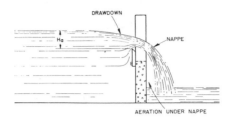

Fig. 4-15. Aeration under nappe of weir.

weir. Flumes are also able to handle solids suspended in the flowing liquid.

A very popular flume is the Parshall flume. It is somewhat similar to one-half of a rectangular Venturi tube. In flume measurement, head losses are reduced because it is not necessary to create the nappe. The Parshall flume is shown in Figure 4-16 and Table 4-3 gives the dimen-

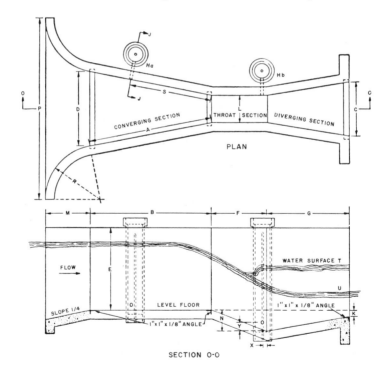

Fig. 4-16. Diagram and dimensions of Parshall flume.

Table 4-3. Dimensions and Maximum and Minimum Flows for Parshall Flumes

L Ft-In.	A Ft-In.	S Ft-In.	B Ft-In.	C Ft-In.	D Ft-In.	E Ft-In.	F Ft-In.	G Ft-In.	K In.	N In.	R Ft-In.	M Ft-In.	P Ft-In.	X In.	Y In.	FREE-FLOW CAPACITY MIN. Sec-Ft	MAX. Sec-Ft
0-3	1-6³/₈	1-¹/₄	1-6	0-7	0-10³/₁₆	2-0	0-6	1-0	1	2¹/₄		1-0	2-6¹/₄	1	1¹/₂	0.03	1.9
0-6	2-⁷/₁₆	1-4⁵/₁₆	2-0	1-3¹/₂	1-3³/₈	2-0	1-0	2-0	3	4¹/₂		1-0	2-11¹/₂	2	3	0.05	3.9
0-9	2-10⁵/₈	1-11¹/₈	2-10	1-3	1-10⁵/₈	2-6	1-0	1-6	3	4¹/₂		1-0	3-6¹/₂	2	3	0.09	8.9
1-0	4-6	3-0	4-4⁷/₈	2-0	2-9¹/₄	3-0	2-0	3-0	3	9	Smooth Curve Radius Not Critical	1-3	4-10³/₄	2	3	0.11	16.1
1-6	4-9	3-2	4-7⁷/₈	2-6	3-4³/₈	3-0	2-0	3-0	3	9		1-3	5-6	2	3	0.15	24.6
2-0	5-0	3-4	4-10⁷/₈	3-0	3-11¹/₂	3-0	2-0	3-0	3	9		1-3	6-1	2	3	0.42	33.1
3-0	5-6	3-8	5-4³/₄	4-0	5-1⁷/₈	3-0	2-0	3-0	3	9		1-3	7-3¹/₂	2	3	0.61	50.4
4-0	6-0	4-0	5-10⁵/₈	5-0	6-4¹/₄	3-0	2-0	3-0	3	9		1-6	8-10³/₄	2	3	1.3	67.9
5-0	6-6	4-4	6-4¹/₂	6-0	7-6⁵/₈	3-0	2-0	3-0	3	9		1-6	10-1¹/₄	2	3	1.6	85.6
6-0	7-0	4-8	6-18³/₈	7-0	8-9	3-0	2-0	3-0	3	9		1-6	11-3¹/₂	2	3	2.6	103.5
7-0	7-6	5-0	7-4¹/₄	8-0	9-11³/₈	3-0	2-0	3-0	3	9		1-6	12-6	2	3	3.0	121.4
8-0	8-0	5-4	7-10¹/₈	9-0	11-1³/₄	3-0	2-0	3-0	3	9		1-6	13-8¹/₄	2	3	3.5	139.5

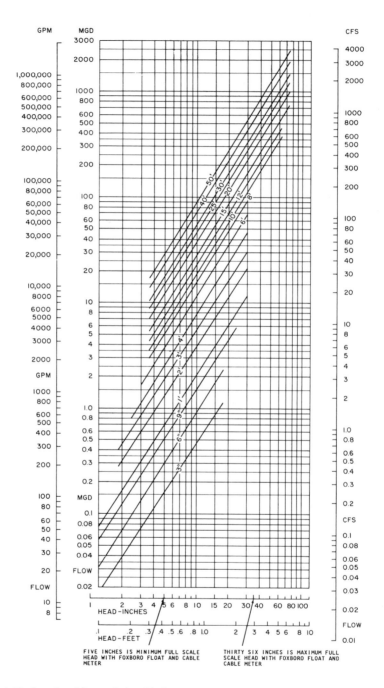

Fig. 4-17. Level and flow relationship for Parshall flume.

114

sions. By referring to Figure 4-17 the dimensions of a flume capable of measuring a prescribed flow rate may be selected. Small flumes may be purchased and installed while large flumes are generally fabricated on site. Several companies specialize in the construction and calibration of flumes.

Under ideal conditions, measurements by weirs are very accurate. In actual plant measurement, however, this may not be true because it is not practical to minimize the velocity of approach factor. The correlation between level and flow typically has an error of 3 to 5 percent. When the error of making the level measurement and relating it to flow rate is added, the total overall error may be as high as ± 5 percent. Equations for the more common open-channel flow devices are given in Table 4-4.

The equations for most open-channel primary devices are nonlinear. To convert head into flow rate it would be necessary to perform an extensive calculation. This step is generally eliminated by having the instrument perform the calculation. This is accomplished either with a

Table 4-4. Equations for Common Open-Channel Flow Devices

V-Notch Weir
30° $Q_{cfs} = .6650\ H^{5/2}$
60° $Q_{cfs} = 1.432\ H^{5/2}$
90° $Q_{cfs} = 2.481\ H^{5/2}$

Rectangular Weir (with end contractions)
$Q = 3.33\ (L\text{-}0.2H)\ H^{3/2}$

Rectangular Weir (without end contractions)
$Q = 3.33\ LH^{3/2}$

Cippoletti Weir
$Q = 3.367\ LH^{3/2}$

Parshall flume

Flume size (in.)		Flume size (in.)	
3	$Q = 0.992\ H^{1.547}$	18	$Q = 6.00\ H^{1.538}$
6	$Q = 2.06\ H^{1.58}$	24	$Q = 8.00\ H^{1.550}$
9	$Q = 3.07\ H^{1.53}$	36	$Q = 12.00\ H^{1.566}$
12	$Q = 4.00\ H^{1.522}$	48	$Q = 16.00\ H^{1.578}$

Q = cfs; L = crest length (ft); H = head (ft)

cam or by means of a special chart or scale. The head is measured with any one of several instruments. Several popular types are: float-and-cable, ball float, bubble, and pressure-sensing. A typical installation is shown in Figure 4-18.

Installation and Selection Considerations

To determine the type of primary device to be used for open-channel flow measurement, the following factors must be considered:

Probable flow range

Acceptable head loss

Required accuracy

Condition of liquid

Probably all of the devices will work, but usually one type will be the most suited to the installation. It must be remembered that a weir will not function properly if the weir nappe is not aerated, and the flume must be maintained at critical depth; otherwise these devices do not function properly. Nappe aeration is largely a question of level in the receiving channel. If the liquid rises so that clearance is insufficient for aeration, accurate flow measurement is impossible. When the flow drops so that the nappe is not aerated but is drawn to the weir edge by capillary attraction, the device does not operate properly.

Minimum weir flows are difficult to calculate because they depend to some degree on the nature of the weir. If the weir is properly sized, minimum flow will produce a substantial, measurable amount of head over the weir. Figure 4-17 shows the relationship between level and flow for the Parshall flume (Figure 4-16). Dimensions and maximum and minimum flows for the flume are tabulated in Table 4-3.

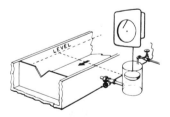

Fig. 4-18. V-Notch weir with float-and-cable.

Velocity Flowmeters

Magnetic Flowmeter

The principle of the magnetic flowmeter was first stated by Faraday in 1832, but did not appear as a practical measurement for the process plant until the 1950s. Its advantages are no obstruction to flow, hence no head loss; it can accommodate solids in suspension; and it has no pressure connections to plug up. It is very accurate and has a linear flow rate to output relationship. Its disadvantages are that measured material must be liquid; the liquid must have some electrical conductivity; and it is expensive.

Operation
Operation of the magnetic flowmeter is based on Faraday's well-known law of electromagnetic induction: The voltage (E) induced in a conductor of length (D) moving through a magnetic field (H) is proportional to the velocity (V) of the conductor. The voltage is generated in a plane that is mutually perpendicular to both the velocity of the conductor and the magnetic field. Stated in mathematical form:

$$E = CHDV \tag{4-11}$$

where C is a dimensional constant. Industrial power generators and tachometer generators used for speed measurement operate on the same principle, electrically conductive process liquid acts in a similar manner to a rotor in a generator. That is, the liquid passes through a magnetic field (Figure 4-19) induced by coils (Figure 4-20) built around a section of pipe. The process liquid is electrically insulated from the pipe, or flow tube, by the use of a suitable lining when a metal tube is used, so that the generated voltage is not dissipated through the pipeline. Two metallic electrodes are mounted in the flow tube and insulated from it; a voltage is developed across these electrodes that is directly proportional to the average velocity of the liquid passing through the magnetic field. Because the coils are energized by alternating current power, the magnetic field and resultant induced voltage is alternating. This generated voltage is protected from interference, amplified, and transformed into a standard dc current signal by a transmitter or converter. Line voltage variations are canceled by the circuits employed.

The magnetic flowmeter measures volume rate at the flowing temperature independent of the effects of viscosity, density, turbulence or

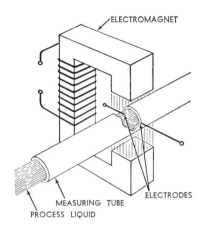

Fig. 4-19. Principle of the electromagnetic meter.

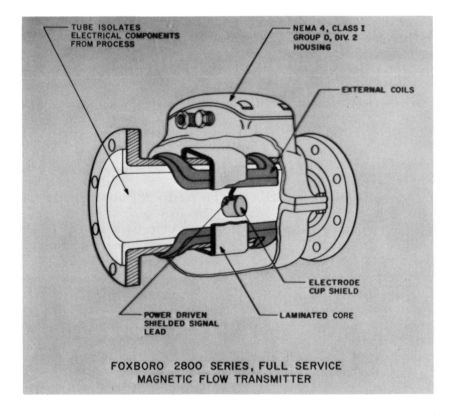

Fig. 4-20. Cut-away view.

suspended material. Most industrial liquids can be measured by magnetic flowmeters. Exceptions are some organic chemicals and most refinery products. Water, acids, bases, slurries, liquids with suspended solids, and industrial wastes are common applications. The limitation is the electrical conductivity of the liquid. The magnetic flowmeter offers no more restriction to flow than an equivalent length of pipe, Figure 4-21.

Structurally, a magnetic flowmeter consists of either a lined metal tube, usually stainless steel because of its magnetic properties, or an unlined nonmetallic tube. Linings for the metal tubes can be polytetrafluoroethylene (Teflon), polyurethane, or some other nonmagnetic, nonconducting material.

The electrodes are suitably insulated from the metal tube. Nonmetallic fiberglass flow tubes do not require any lining. The electrodes must be insulated so that the voltage generated can be measured across the electrodes. The insulation has no bearing on the actual voltage generation, but without the insulator, voltage would bleed off through the metallic walls of the tube.

The coils are similar in design to the deflection coils used on a television picture tube. Two coils are used and work together to create a uniform magnetic field. The coils are generally series-connected, but may be parallel-connected if the measured flow velocity is low.

The signal output from the flow tube's electrodes is an alternating

Fig. 4-21. Unobstructed flow.

voltage at supply frequency. This low-level voltage is generally between 1 mV and 30 mV at full flow rate.

This low-level alternating voltage must be measured and converted either into a record or display or into a dc common denominator transmission signal. This signal, typically 4 mA at zero and 20 mA at full scale, can be fed into a recorder or controller. The device that can accomplish this is a special type of transmitter. This transmitter is located either directly on the flow tube (Figure 4-22), near it, or within the control room. The preferred location is as near to the flow tube as possible; temperature and corrosive conditions are the constraints that dictate location. In some applications a digital or pulse rate signal output from the transmitter may be required, and this option is available.

Accuracy
The accuracy of most magnetic flowmeter systems is 1 percent of full-scale measurement. This includes the accuracy of both the meter itself and its secondary instrument. Because this type of meter is inherently linear, its accuracy at low-flow rates exceeds the practical accuracy of such inferential devices as the Venturi tube. Accuracy of the Venturi is ±0.75 percent, according to ASME Fluid Meters Report, and that of the secondary instrument is about ±0.5 percent. At the low end of the measurement scale, secondary instrument readability decreases owing to the square root relationship. The magnetic flowmeter can be labora-

Fig. 4-22. Foxboro magnetic flow tube and transmitter.

tory calibrated to an accuracy of 0.5 percent of full scale and is linear throughout.

Vortex Flowmeter

The Foxboro Vortex flowmeter, as shown in Figure 4-23, measures liquid, gas, or steam flow rates using the principle of vortex shedding. The transmitter produces either an electronic analog or pulse rate signal linearly proportional to volumetric flow rate.

Principle of Operation
The phenomenon of vortex shedding can occur whenever a nonstreamlined obstruction is placed in a flowing stream. As the liquid passes around the obstruction, the stream cannot follow the sharp contours and becomes separated from the body. High-velocity liquid particles flow past the lower-velocity (or still) particles in the vicinity of the body to form a shear layer. There is a large velocity gradient within this shear layer, making it inherently unstable. The shear layer breaks down after some length of travel into well-defined vortices as shown in Figure 4-24.

These vortices are rotational flow zones that form alternately on each side of the element with a frequency proportional to the liquid flow

Fig. 4-23. Vortex flowmeter.

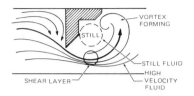

Fig. 4-24. Vortex shedding phenomenon.

rate. Differential pressure changes occur as the vortices are formed and shed. This pressure variation is used to actuate the sealed sensor at a frequency proportional to vortex shedding.

Thus, a train of vortices generates an alternating voltage output with a frequency identical to the frequency of vortex shedding. This frequency is proportional to the flow velocity.

The voltage signal from the detector (Figure 4-25) is conditioned and amplified for transmission by electronics located in a housing mounted integral with the flowmeter body. The final output signal is available either in pulse form with each pulse representing a discrete quantity of fluid for totalizing or, optionally, as a 4 to 20 mA dc analog signal for flow rate recording or control.

Turbine Flowmeter

The turbine meter derives its name from its operating principle. A turbine wheel (rotor) is set in the path of the flowing fluid. As the fluid enters the open volume between the blades of the rotor, it is deflected by the angle of the blades and imparts a force causing the rotor to turn. The speed at which the rotor turns is related, over a specified range, linearly to flow rate.

Several methods are employed to transmit this motion to a readout device outside of the conduit. In some applications a mechanical device conveys the rotor motion directly to a register. In process applications,

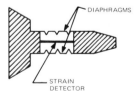

Fig. 4-25. Geometry of vortex generator.

however, the usual method is to use an electrical method. A coil containing a permanent magnet is mounted on the meter body. The turbine flowmeter (Figure 4-26) consists of a section of metal pipe, a multi-bladed rotor mounted in the center of the straight-through passage, and a magnetic pickup coil mounted outside the fluid passage. A shaft held in place by fixed radian vanes supports the rotor assembly. As each blade tip of the rotor passes the coil it changes the flux and produces a pulse. The total number of pulses indicates the volume of fluid which has passed through the meter and the rate of the pulses generated becomes a measure of flow rate.

Turbine flowmeters are frequently employed as sensors for inline blending systems.

Turbine flowmeters have excellent accuracy and good rangeability. They are limited to clean fluids. They are expensive, but do have unique features.

Other Flowmeters

Other flowmeters that may be found in some process plants include the ultrasonic, the thermal, a variety of positive displacement types, metering pumps, and others. The flowmeters described account for the majority of the flow rate measuring devices in everyday use.

Fig. 4-26. Turbine flowmeter.

Conclusion

Unfortunately, there is no best flowmeter. Factors such as accuracy, pressure loss, material to be measured, ease of changing capacity, and ease of installation along with cost must be carefully considered. Once the details of the problem have been gathered and the possible alternatives considered, the best selection will usually be clear. Some of the systems described read correctly to within a few percent. Some are considerably better. All should have good repeatability—the all important criterion for control.

Questions

4-1. Match the head meter primary devices with the application (select a single best answer for each):

Orifice plate ____	**a.** High-pressure recovery
Flow nozzle ____	**b.** Air ducts
Venturi tube ____	**c.** Sediment in liquid
Pitot tube ____	**d.** Economy and accuracy are important
Elbow taps ____	**e.** No straight run available

4-2. A differential pressure transmitter is calibrated 0 to 80 inches of water and transmits a 4 to 20 mA dc signal. This transmitter is placed across an orifice plate which is sized to create 80 inches of water differential at 6 gallons per minute.
What is the flow rate when the signal is 13 mA dc?

    a. 4.5 gpm   **c.** 3.4 gpm
    b. 3.9 gpm   **d.** 4.9 gpm

4-3. For the equipment described in the preceding question, what will the signal be if the flow rate is 4 gpm?

    a. 13.3 mA dc   **c.** 8.9 mA dc
    b. 11.2 mA dc   **d.** 6.7 mA dc

4-4. Assume the line used in Questions 2 and 3 is a ½-inch schedule 40 pipe (ID = 0.622 inches). Find the orifice bore used to satisfy the conditions given if the measured fluid is water at 60°F.

    a. 0.190 inches   **c.** 0.414 inches
    b. 0.556 inches   **d.** 0.352 inches

4-5. The best choice of orifice taps in the preceding problem would be:

    a. flange taps   **c.** pipe taps
    b. vena contracta taps   **d.** corner taps

4-6. A 2-inch schedule 40 line (ID = 2.067) is used to carry gasoline (SP GR = 0.75). The flow rate is measured with an orifice plate (d = 1.034) and pipe taps are used. At full flow rate a differential pressure of 50 inches of water is produced. What is the approximate full flow rate in gpm?

    a. 30 gpm   **c.** 250 gpm
    b. 53 gpm   **d.** 36 gpm

4-7. Assume an 8-inch schedule 160 pipe (ID = 6.813 inches) carries a full flow of 40,000 pounds per hour of dry saturated steam. The static pressure is 335 psi. Flange taps are used and the differential pressure across the orifice plate at full-flow rate is 100 inches of water. What is the size of the bore in the orifice plate?

    a. 4.357 inches   **c.** 2.271 inches
    b. 3.645 inches   **d.** 5.109 inches

4-8. Fuel gas is carried in a 6-inch schedule 40 pipe (ID= 6.065 inches). Flow rate is measured with an orifice plate using flange taps and the bore in the orifice is 3.457 inches. The specific gravity is 0.88, the flowing temperature is 60°F, and the static pressure is 25 psi. Maximum flow rate creates a differential of 20 inches of water. What is the approximate flow rate in cubic feet per day?

    a. 1,500,000 SCFD   **c.** 2,500,000 SCFD
    b. 2,000,000 SCFD   **d.** 3,000,000 SCFD

4-9. The output of a target flowmeter is:

    a. Proportional to volumetric flow rate
    b. Proportional to the square of volumetric flow rate
    c. Proportional to the square root of volumetric flow rate

4-10. The output signal or reading of a magnetic flowmeter is:

    a. Proportional to volumetric flow rate
    b. Proportional to the square of volumetric flow rate
    c. Proportional to the square root of volumetric flow rate
    d. Inversely proportional to volumetric flow rate

4-11. With suspended solids and/or entrapped gas in a flowing liquid, the magnetic flowmeter will:

    a. Read high
    b. Read low
    c. Read the liquid flow only
    d. Read the correct total volume of the mixture

4-12. A turbine flowmeter produces an output in the form of pulses. The total number of pulses is:

    a. Inversely proportional to flow
    b. Directly proportional to total flow
    c. Proportional to the square root of flow
    d. Proportional to the square of flow

Temperature and Humidity Measurements

Temperature and moisture measurements are important in process control, both as direct indications of system or product state and as indirect indications of such factors as reaction rates, energy flow, turbine efficiency, and lubricant quality.

Temperature

Present temperature scales have been in use for about 200 years. The earliest instruments—variations of the common stem thermometer—were based on the thermal expansion of gases and liquids. Such filled systems are still employed frequently for direct temperature measurement, although many other types of instruments are available. Representative temperature sensors include:

Filled thermal systems

Liquid-in-glass thermometers

Thermocouples

Resistance temperature detectors (RTDs)

Thermistors

Bimetallic devices

Optical and radiation pyrometers

Temperature-sensitive paints

Various types of measurement control systems are compared in Table 5-1.

Instrument selection must anticipate overall control requirements. Low cost often justifies consideration of filled systems for measurements below 1,200°F or 650°C. Other advantages of mechanically or pneumatically transmitted temperature measurements include low-explosion hazard, simple maintenance requirements, high reliability, and independence from external power. Advantages of electrical systems include higher accuracy and sensitivity, practicality of switching or scanning several measurement points, larger distances possible between measuring elements and controllers, replacement of components

Table 5-1. Comparison of Temperature Measuring Systems

Comparison Factors	Most Favorable	Intermediate	Least Favorable
Purchase cost	F	P	R
Long distance transmission	E	P	F
Change or replacement of components	T	E and P	F
Installation costs	E°	F	P
Maintenance	F	P	E
Averaging measurement	F	T	R
Surface measurement	R	T	F
Time constant (bare bulb and no well)	T	F	R
Temperature difference	R	F	T
Sensitivity	E	P	F
Accuracy	R	F	P and T
Operating costs	F	E	P

Key: F Filled system.
P Pneumatically transmitted filled system.
T Electrical thermocouple system.
R Electrical RTD system.
E Electrical thermocouple and RTD systems equally rated.
°General purpose wiring only. Explosion-proof electrical systems cost considerably more than other types.

(rather than complete systems) in the event of failure, fast response, and ability to measure higher temperatures.

Filled Thermal Systems

Filled thermal systems, which traditionally have been used most in the food, paper, and textile industries, consist of sensors (bulbs) connected through capillary tubing to pressure or volume sensitive elements (Figure 5-1). These systems are simple and inexpensive and generally have fast dynamic responses. Their use with pneumatic and electronic transmitters has removed inherent distance limitations of filled systems and has minimized the danger of capillary damage. Moreover, transmitter amplification has made narrow spans practical and has improved linearity and response.

Application specifications of several types of filled systems are listed in Table 5-2. These include the Scientific Apparatus Makers Association (SAMA) Classes I (liquid-expansion), II (vapor-pressure), III (gas-pressure), and V (mercury-expansion). The SAMA Class II classifications also include alphabetical designations, by which A and B indicate sensor above and below case ambient, respectively; C indicates a system in which the sensor can cross ambient; and D denotes a system which can operate at ambient conditions.

Liquid-expansion systems are characterized by narrow spans, small sensors, uniform scales, high accuracy, and capability for differential measurements. Class IA devices have an auxiliary capillary and

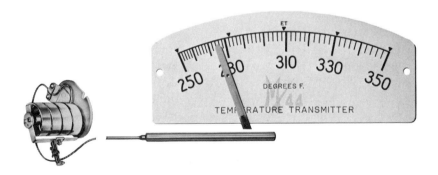

Fig. 5-1. Filled measurement systems generally are the most inexpensive way of measuring and controlling temperature.

Table 5-2. Instruments for Filled Thermal System Sensors

Type	Liquid	Vapor (a)	Gas
Principle	Volume change	Pressure change	Pressure change
SAMA Class	I	II	III
Fluids	Organic liquids (Hydro-carbons)	Organic liquids (Hydro-carbons), water	Pure gases
Lower range limit	−200°F (−130°C)	−425°F (−255°C)	−455°F (−270°C)
Upper range limit	+600°F (+315°C)	+600°F (+315°C)	+1,400°F (+760°C)
Narrowest span (b)	40°F (25°C)	70°F (40°C) (c)	120°F (70°C)
Widest span	600°F (330°C)	400°F (215°C)	1,000°F (550°C)
Ambient temperature Compensation	IA Full IB Case	Not required	IIIB case
Sensor size	Smallest	Medium	Largest
Typical sensor size for 100°C span	9.5mm (0.375 in.) OD × 48mm (1.9 in.) long	9.5mm (0.375 in.) OD × 50mm (2 in.) long	22mm (⅞ in.) OD × 70mm (6 in.) long
Overrange capability	Medium	Least	Greatest
Sensor elevation effect	None	Class IIA, Yes Class IIB, None	None
Barometric pressure effect (altitude)	None	Slightly (greatest on small spans)	Slightly (greatest on small spans)
Scale uniformity	Uniform	Non-Uniform	Uniform
Accuracy	±0.5 to ±1.0% of span	±0.5 to ±1.0% of span upper ⅔ of scale	±0.5 to ±1.0% of span
Response (d) #1 Fastest #4 Slowest	#4	#1—Class IIA #3—Class IIB	#2
Cost	Highest	Lowest	Medium
Maximum standard Capillary length	Class IA 30m or 100 ft. Class IB 6m or 20 ft.	30m or 100 ft.	30m or 100 ft.

(a) Class II systems are supplied as either SAMA Class IIA or IIB. In Class IIA, sensor is always hotter than tubing or instrument case. In Class IIB, sensor is always cooler than tubing or case.
(b) Narrowest spans vary at elevated temperatures.
(c) Smaller spans available in cryogenic regions.
(d) Actual values depend on range, capillary length, sensor dimensions, and type of instrument used.

element to provide ambient temperature compensation and IB systems often utilize bimetallic techniques. However, fully compensated liquid-expansion systems are complex and, therefore, expensive.

Vapor-pressure systems are reliable, inherently accurate, and require no compensation for ambient temperature effects. Instruments follow the vapor-pressure curves of the filled fluid, and associated dials and charts are thus nonuniform, featuring more widely spaced increments at high temperatures. Measurements occur at the interface between liquid and vapor phases of the filling medium. If the temperature in the sensor exceeds that in the capillary and indicating element, the sensor is filled with vapor while the capillary and indicator contain liquid.

The converse is true when the relative temperature polarity is reversed. A transition between liquid and vapor can cause erratic operation, so that vapor systems may be unsuitable for ranges that cross the element and capillary temperatures. Systems may also be unacceptable if uniform recording or indicating scales are desired.

Gas-pressure systems, which rank second to the vapor-pressure devices in simplicity and cost, offer the widest range of all filled systems. Conventional designs use large-volume sensors, which may be shaped to suit particular applications. For example, in duct-temperature averaging, the sensor may be constructed of a long length of tubing of small cross section. Conventional recorders are not recommended for temperature spans of less than 200°F or 110°C, but transmitters operating on force-balance principles can be utilized with spans as narrow as 50°F or 28°C. With gas-filled systems, it is difficult to compensate for ambient temperature errors, but a sufficiently large sensor size may reduce effects to acceptable limits.

Mercury-expansion systems are classified separately from other liquid-filled systems because of the unique properties of the fluid. For example, mercury is toxic and harmful to some industrial processes and products, and high-liquid density places limitations on sensor-to-instrument elevation differences. Sensors used in mercury-expansion systems are generally larger in diameter and more expensive than those used in either liquid or vapor systems. For these reasons, mercury is frequently bypassed in favor of other filling media.

Electrical Systems

Electrical temperature sensors have long been popular in the metal and paper industries, but increasing use of electronic devices has stimulated application in a variety of other areas. Thermocouples and resistance temperature detectors (RIDs) are most widely used, and representative application data are given in Table 5-3.

Thermocouples

Thermoelectricity was discovered by Seebeck in 1821. He observed an electromotive force (emf) generated in a closed circuit of two dissimilar metals when their junctions were at different temperatures. This electricity, produced by the direct action of heat, is used today to measure temperatures from subzero to high ranges.

Table 5-3. Quick-Reference Selector Chart for Standard
Thermocouples & RTDs

		TEMPERATURE LIMITS	
Sensor	*Calibration*	*°C*	*°F*
RTDs	Nickel SAMA Type II	−200 to 315	−320 to 600
	Platinum SAMA 100 ohm, or		
	DIN 43760	−200 to 650	−320 to 1,200
	Copper SAMA	0 to 150	32 to 300
Thermocouples	Iron-Constantan, ISA Type J	−210 to 760	−350 to 1,400
	Copper-Constantan, ISA Type T	−270 to 370	−455 to 700
	Chromel-Alumel, ISA Type K	−270 to 1,260	−455 to 2,000
	Chromel-Constantan, ISA Type E	−270 to 870	−455 to 1,600
	Platinum-Platinum Rhodium,	−50 to 1,480	−55 to 2,700
	ISA Types R & S		
	ISA Type B	0 to 1,700	0 to 3,100

A thermocouple consists basically of two dissimiliar metals, such
as iron and constantan wires, joined to produce a thermal electromotive
force when the junctions are at different temperatures (Figure 5-2). The
measuring, or hot, junction is inserted into the medium where the temperature is to be measured. The reference, or cold, junction is the open
end that is normally connected to the measuring instrument terminals.

The emf of a thermocouple increases as the difference in junction
temperatures increases. Therefore, a sensitive instrument, capable of
measuring emf, can be calibrated and used to read temperature directly.

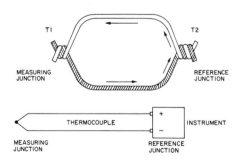

Fig. 5-2. Thermocouple.

Figure 5-3 indicates the approximate emf-millivolt output versus temperature relationship for the most popular thermocouple types. Deviation from the essentially linear relationship is generally less than 1 percent.

Introduction of intermediate metals into a thermocouple circuit will not affect the emf of the circuit, provided the new junctions remain at the same temperature as the original junction. The algebraic sum of the emf's in a circuit consisting of any number of dissimilar metals is zero, if all of the circuit is at a uniform temperature. Repeating the law, if in any circuit of solid conductors the temperature is uniform from Point 1 through all the conducting material to Point 2, the algebraic sum of the emf's in the entire circuit is totally independent of the intermediate material and is the same as if Points 1 and 2 were put in contact. If the individual metals between junctions of the circuit are homogeneous, the

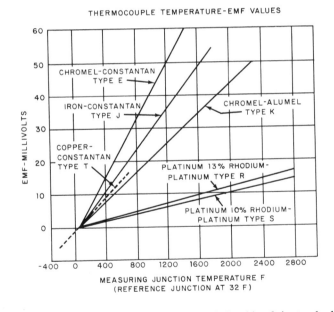

THERMOCOUPLE TEMPERATURE-EMF VALUES

Fig. 5-3. Temperature-emf values approximate the relationship of six standard thermocouples. The conditions of a particular installation may require lowering the maximum temperature to a value below that listed. Temperature/millivolt curves are available in the appendix.

sum of the thermal emf's will be zero, provided only that the junctions of the metals are all at the same temperature.

The emf in a thermoelectric circuit is independent of the method employed in forming the functions as long as all the junction is at a uniform temperature and the two wires make good electrical contact. The junction may be made directly, by welding, or by soldering. Furthermore, an instrument for measuring the emf may be introduced into a circuit at any point without altering the resultant emf, provided the junctions that are added to the circuit by introducing the instrument are all at the same temperature. If the temperatures of the new junctions are not uniform, the effect is that of introducing additional thermocouples into the circuit.

Reference Junction

To make accurate temperature measurements with thermocouples, the reference junction temperature must remain constant; if it varies, suitable compensation for these variations must be provided.

Should there be an uncompensated variation in the reference junction temperature, there will be a corresponding change in the millivoltage with a resultant error in temperature measurement (Table 5-4).

When used in the laboratory and for other checking and testing purposes, the thermocouple reference junction can be placed in a vacuum bottle filled with shaved ice saturated with water. This method provides close temperature control (within a fraction of a degree) and permits accurate reading.

Table 5-4. Standard Limits of Error

Couple or Wire	Range	Limits
Copper-constantan	−300 to −75°F	±2% of reading
	−75 to +200°F	±1½°F
	200 to 700°F	±¾% of reading
Iron-constantan	−100 to +530°F	±4°F
	530 to 1,400°F	±¾% of reading
Chromel-alumel	0 to 530°F	±4°F
	530 to 2,300°F	±¾% of reading
Platinum rhodium—	0 to 1,000°F	±3°F
platinum	1,000 to 2,700°F	±0.3% of reading

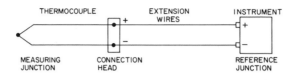

Fig. 5-4. Typical thermocouple and extension leads.

To ensure accurate readings, most thermocouples are now installed with instruments that provide automatic reference junction compensation. In most instruments, this is accomplished by passing current through a temperature-responsive resistor, which measures the variations in reference temperature and automatically provides the necessary compensating emf by means of the voltage drop produced across it.

An industrial installation generally consists of a thermocouple with its connection head, the necessary length of extension wire, and an indicating, recording, or controlling instrument with internal and automatic reference junction compensation. The extension wires generally consist of the same materials as the thermocouple elements, or may be composed of other materials and alloy wires that generate essentially the same millivoltage as the thermocouple for application temperatures up to approximately 400°F or 200°C (Figure 5-4). The ISA symbols and color codes for thermocouple lead wires are given in Table 5-5.

Table 5-5. Extension Wire Type Designations and Standard Limits of Error

ISA Type	Extension Wire	Color Code	Temperature Range F	Limits of Error	Couple Used With
TX	Copper-constantan	+ (blue) − (red)	−75 to +200	±1½ F	Copper-constantan
JX	Iron-constantan	+ (white) − (red)	0 to 400	±4 F	Iron-constantan
WX	Iron-cupronel	+ (green) − (red)	75 to 400	±6 F	Chromel-alumel
KX	Chromel-alumel	+ (yellow) − (red)	75 to 400	±4 F	Chromel-alumel
SX	Copper-CuNi alloy	+ (black) − (red)	75 to 400	±10 F	Platinum Rhodium-platinum

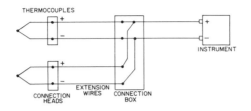

Fig. 5-5. Thermocouples in parallel.

Average Temperatures

To measure the average temperature across a large duct or vessel, or around a retort, any number of thermocouples may be used in parallel connections. The voltage at the instrument, or at the point of parallel connection, is the average of that developed by the number of thermocouples used. This voltage is equal to the sum of the individual voltages divided by the number of thermocouples. For accurate measurement, the resistances of all thermocouples and extension wire circuits should be identical. Since the resistance of the actual thermocouple will vary with temperature, and since the lengths of extension wires may also vary, the effect of these variations can be minimized by using swamping resistors. The values of the swamping resistors should be high in comparison with the change or difference in resistances encountered. A resistor value of 1,500 ohms generally works well.

In order to prevent the flow of current through a ground loop, the thermocouples should not be grounded. All thermocouples must be of the same type and must be connected by the correct extension wires (Figures 5-5 and 5-6).

Parallel connection of thermocouples for average temperature measurement is advantageous because the instrument construction and

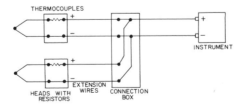

Fig. 5-6. Parallel thermocouples with swamping resistors.

calibration can be the same as that for a single thermocouple. Two or more thermocouples may be connected in series to measure average temperatures; however, this circuit requires that the instrument have special reference junction compensation to provide the increased compensating emf for the specific number of thermocouples in the circuit. The instrument also must be calibrated for the total millivoltage output of the number of thermocouples used in series. Extension wires from each thermocouple must be extended back to the actual reference junction at the instrument.

Differential Temperatures

Two thermocouples may be used for measuring differential temperatures between two points. Two similar thermocouples are connected with extension wire of the same material as is used in the thermocouples. The connections are made in such a way that the emf's developed oppose each other. Thus, if the temperatures of both thermocouples are equal, regardless of the magnitude, the net emf will be zero. When different temperatures exist, the voltage generated will correspond to this temperature difference, and the actual difference may be determined from the proper calibration curve. An instrument calibrated either for 0 millivolts at midscale or 0 millivolts at the end of the scale, depending upon whether thermocouples are operated at high or low temperatures with respect to each other, may be furnished and used to read temperature difference directly.

Copper leads may be used between the instrument and the connection box that links the instrument to the thermocouple or extension wires (Figure 5-7). The instrument should not have the reference junction compensation normally furnished when measuring temperatures with thermocouples. As in the case of parallel thermocouple connections, the thermocouples should not be grounded and the resistance of both thermocouple circuits should be the same.

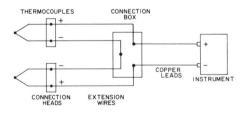

Fig. 5-7. Two thermocouples used to measure temperature difference.

Thermocouple Calibration Curves

Thermocouples develop an extremely small dc voltage, measured in thousandths of a volt. The millivolt output falls within a range of −11 to +75 millivolts, depending on the type of thermocouple and its temperature working range. Calibration curves for thermocouples are included in the Appendix. Now let us consider some typical thermocouple instruments and investigate their functions.

Potentiometric Recorder

Figure 5-8 shows a simplified circuit diagram of the Foxboro potentiometric recorder (Figure 5-9). An emf input (E_x) is derived from a thermocouple or other measuring element. A constant emf (E_z) is supplied by a Zener diode regulated power supply. As E_x varies, an error signal is developed, and the circuit becomes unbalanced.

The error signal is converted to an ac voltage by a field effect transistor chopper and amplified by the integrated circuit amplifier. The amplified output drives a two-phase balancing motor. The direction of rotation depends on whether E_x has become greater or smaller. The motor moves the wiper contact on the slidewire R_s to a position where the circuit is rebalanced. The slidewire contact is mechanically connected to the recorder pen, and both are positioned simultaneously.

With a thermocouple input, a temperature sensitive resistor (R_c) automatically compensates for reference junction ambient temperature variations. This resistor changes the balance of the circuit to cancel

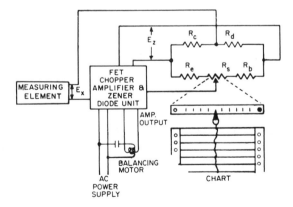

Fig. 5-8. Simplified circuit of potentiometric recorder.

Fig. 5-9. Foxboro strip chart recorder.

changes in E_x with reference junction temperature variations. For straight millivolt inputs, R_c is a temperature-stable resistor.

The slidewire is characterized to allow the use of uniform charts and scales with thermocouple and certain other nonlinear measurements. This is accomplished by varying the resistance per unit length along the slidewire.

If the temperature measurement is used in conjunction with an electrical control system, a thermocouple-to-current transmitter may be used. A typical transmitter is shown in Figure 5-10. The transmitter provides a two-wire output; the same wiring is used for both power and output. The load resistance is connected in series with a dc power supply (approximately 25 volts), and the current drawn from the supply is the 4 to 20 mA dc output signal.

Field mounting the transmitter at or near the actual measurement point eliminates the installation of special thermocouple extension wires to the receiver. In some cases, this procedure can also improve the performance of the measurement loop, since the possibility of extension wire error is avoided.

As shown in Figure 5-10, the thermocouple/millivolt bridge circuit consists of a precision-constant current source, a reference-junction compensator, a burn-out detection current source, and associated resistors. The amplifier is a very high gain, chopper-stabilized difference

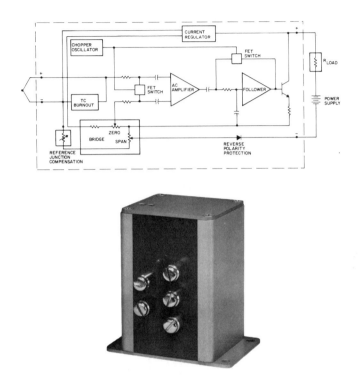

Fig. 5-10. Simplified circuit diagram of the Foxboro thermocouple transmitter. (*Bottom*) Foxboro thermocouple transmitter.

amplifier. It functions as a null detector by controlling the 4 to 20 mA dc output current to maintain a null between the compensated thermocouple and feedback voltages. The transmitter is protected from reverse polarity connection of the power supply by means of a diode in the negative lead of the output circuit. The maximum error for this transmitter is approximately ±0.25 percent of span.

Many control instruments, such as this transmitter, feature thermocouple burn-out protection. In the event of an open thermocouple, a small voltage applied to the instrument causes it to read full-scale output. This causes the control system to shut off the heat and prevent damage.

Resistance Thermal Detectors

Resistance thermometry is based on the change of electrical conductivity with temperature. Therefore, a coil of wire can act as a temperature sensor, with a direct relationship established between resistance and temperature. Standard curves are available, with certified accuracies within 0.1°F or °C. Platinum RTDs used as laboratory standards can be obtained with tolerances well within this limit, and are capable of precise temperature measurement up to 1,650°F or 900°C. If an RTD is adjusted to conform to its curve, it may be interchanged with other RTDs calibrated according to the same curve.

Resistance-thermal detector temperature measurement may be read out on many different types of instruments. If the measurement of temperature were to be used in conjunction with a pneumatic control system, the RTD could be attached to a resistance-to-pneumatic convertor, such as the Foxboro Model 34B (Figure 5-11).

The resistance-to-pneumatic convertor converts the temperature measured by a resistance-temperature element into a proportional 3 to 15 psi or 20 to 100 kPa pneumatic output signal. This signal is suitable for use with many types of pneumatic receiving instruments, such as controllers, recorders, and various computing instruments.

This transmitter has several features. For example, when used with

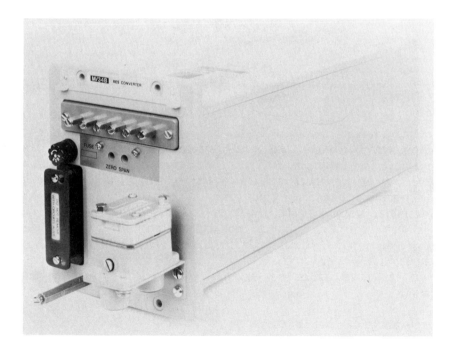

Fig. 5-11. Foxboro 34B Series resistance pneumatic transmitter.

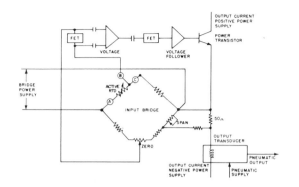

Fig. 5-12. Simplified circuit diagram of 34B Series nickel RTD transmitter.

the appropriate nickel bulb, the converter has an output that is linear with temperature. Spans as low as 5°F or 3°C can easily be achieved. An adjustable range allows a simple field calibration procedure to change the temperature input range required for a 3 to 15 psi or 20 to 100 kPa output. It also operates on all normal supply voltages.

This transmitter is an electromechanical device consisting of a solid-state, integrated circuit amplifier and an output transducer. Figure 5-12 is a simplified circuit diagram of the nickel RTD version of the transmitter. The transducer is shown in Figure 5-13.

The RTD is wired into a measurement bridge and excited from a regulated direct current power supply. The change in resistance of the RTD causes a bridge output change that is proportional to temperature. Negative feedback is obtained from the output current and applied to the opposite side of the measurement bridge. A change in feedback by the span adjustment changes the gain of the amplifier and thereby changes the span of measurement.

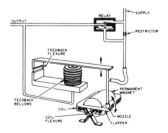

Fig. 5-13. Output transducer.

The platinum RTD transmitter incorporates a negative impedance, integrated circuit amplifier that shunts the platinum RTD to linearize the temperature versus the RTD characteristic.

The amplifier section provides the high input impedance and gain required to amplify the low-input signal. The input signal is first converted to ac by an FET (used as a chopper) and then applied to a high-gain ac amplifier. The signal is next demodulated and applied to a voltage follower. The output of this amplifier is applied to a power transistor to supply the 10 to 50 mA dc required by the transducer coil.

A functional diagram applying to the output transducer is shown in Figure 5-13. The direct current input signal is converted into a proportional air output signal as follows: A coil is positioned in the field of a fixed, permanent magnet. When a direct current flows through the coil, the electromagnet and permanent magnet forces are in opposition. Any increase in the current, which causes a proportional increase in the force on the coil, tends to close the flapper against the nozzle. The air output pressure from the pneumatic relay is thereby increased. This increased pressure is fed to the feedback bellows to exert a force on the force beam to rebalance the force from the coil. All of the forces which act upon the force beam are balanced in this manner, and the pneumatic output tracks the current input. The capacity tank (C_2) and the associated restrictor provide dynamic compensation to the circuit. During rapid input changes, the restrictor will effectively keep the capacity tank out of the circuit for faster feedback response through the relay. For slow changes, the capacity tank acts to damp pressure fluctuations and to provide a smoother output. The small capacity tank (C_1) in the output circuit provides necessary constant damping.

During alignment or calibration of the output transducer, zero adjustment is made by turning a screw to advance or retract a spring which bears on the force beam. Coarse span adjustments are made by repositioning the feedback bellows with respect to the pivot.

Wheatstone Bridge Recorder

One of the most popular instruments used with RTDs is the slidewire type Wheatstone bridge recorder. This instrument is almost identical to the potentiometric recorder previously described and shown in Figure 5-9.

Figure 5-14 shows a simplified circuit diagram of the Wheatstone bridge recorder. The resistance temperature detector (RTD) is one arm of a Wheatstone bridge excited by a Zener diode regulated dc power supply. Point A (the slidewire contact) and point B form the input to the amplifier. When the temperature changes, the resistance of the RTD

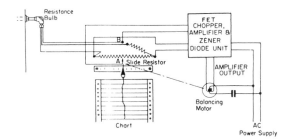

Fig. 5-14. Simplified Wheatstone bridge recorder circuit.

changes. This unbalances the bridge and creates an error signal between Points A and B.

As in the potentiometric recorder, the error signal is converted to an ac voltage by a field-effect transistor chopper and amplified by the transistorized amplifier. The output drives a two-phase balancing motor. The direction of rotation depends on the polarity of the error signal. The motor moves the wiper contact on the slidewire until the bridge is rebalanced and no error signal exists. The pen and slidewire contact are mechanically connected and therefore positioned simultaneously.

With all resistance-temperature measurements, the use of three-conductor RTD cable is recommended. The effect of ambient temperature variations on the cable is thereby minimized.

If the cable connecting the RTD to the instrument has only two conductors, these conductors become part of the resistance being measured. The result then is an error that will vary with ambient temperature. Remember, with all resistance temperature measurements, three-conductor RTD cable is recommended. The purpose of the three-conductor cable is to stretch out the measuring bridge. Note the three-conductor cable in Figure 5-14. One of the conductors is common to both sides of the bridge while the other two connect one to each side of the bridge. Any change in cable temperature will be canceled as both sides of the bridge are changed a like amount.

Occasionally RTD sensors use a four-wire cable. This is generally in conjunction with a Kelvin double bridge. The four-wire method does an excellent job of reducing temperature effects on the cable. The improvement over the three-wire method, however, is minimal. In practice, an RTD may be used with as much as 500 feet of three-conductor cable without the cable creating a perceptible error.

Thermistors

Thermistors are made of heat-treated metallic oxides, and most thermistors differ from ordinary resistors by having a negative coefficient of resistance. Thermistors are available with a nearly linear temperature resistance relationship, and other types are available with a sharp change in slope at some characteristic temperature.

A thermistor can replace an RTD as a temperature sensor. The difficulty lies in obtaining units that fit the desired characteristic curve within acceptable limits of accuracy. When this is accomplished and the thermistor is mounted so as to stand up under process conditions, it performs the same function as the conventional RTD.

One advantage of the thermistor is that it has a greater resistance change for a given temperature change than that of the conventional wire RTD. A disadvantage is that the accuracy available, although good, is slightly inferior to that of the conventional RTD. This presumably accounts for the thermistor's limited application in the process instrumentation field.

Radiation pyrometers utilize an optical system to focus energy radiated from a body onto a sensing system. Manual devices are often used, in which energy at infrared or visible wavelengths is focused on a target and compared with the light output of a calibrated optical filament. In automated devices, the energy (usually in the infrared band) is focused on a series array of thermocouples. This thermopile produces a millivolt output related to the temperature of the source. Pyrometers are used where high temperatures are to be measured or where contact with the object is impossible. Accuracy is influenced by such factors as reflections, gases present in the radiation path, and surface emissivity of the body under measurement.

Humidity Measurements

Humidity, or the amount of moisture in gases, is expressed in several different ways, including:
  1. Relative humidity—the actual quantity of water vapor present in a given space expressed as a percentage of the quantity of water vapor that would be present in the same space under saturated conditions at the existing temperature.
  2. Absolute humidity—mass of vapor per unit mass of dry gas.
  3. Dew point—the saturation temperature of the mixture at the corresponding vapor pressure. If the gas is cooled at constant pressure to the dew point, condensation of vapor will begin.

A list of the methods used to measure humidity would include the following:

Relative Humidity Measurement
  Mechanical (hair, wood, skin)
  Wet and dry bulb thermometers
  Surface resistivity devices (including Dunmore)
  Crystal frequency change

Absolute Moisture Measurement
  Gravimetry
  Electrolysis
  Infrared
  Conductivity
  Capacitance
  Color change
  Karl Fischer titration
  RF power absorption
  Neutron reflection
  Heat of absorption or desorption
  Nuclear magnetic resonance

Dew Point Measurement
  Chilled mirror
  Lithium chloride (including DEWCEL)
  Wet bulb thermometer

Direct measurement of relative humidity may be accomplished by equating the change in dimension of a hygroscopic substance with the variation in the moisture content of the air. The hair element (Figure 5-15) consists of a band composed of a number of selected and treated human hairs. The membrane element consists of a strip of treated animal membrane.

The instrument will record relative humidity between 20 and 90 percent with a maximum error of ±5 percent. The calibration should be adjusted initially by means of a psychrometer. If the hair element is exposed to humidities over 90 percent or to free moisture, the calibration should be checked.

Hair elements can be used at temperatures as high as 160°F or 70°C

Fig. 5-15. Hair element.

without damage. However, in this event, they should be calibrated under these temperature conditions.

The hair element is extremely responsive to changes in humidity, and will follow gradual or small changes with good speed. However, following a sudden large change, at least 30 minutes should be allowed for the measuring element to reach a state of equilibrium.

The psychrometer consists of two temperature measurements. One, called the dry bulb, measures the ambient temperature, while the second, the wet bulb, is provided with a wick or sleeve saturated with water that is evaporating. The wet bulb temperature will normally be lower than that of the dry bulb. The lowering of this wet bulb temperature (wet bulb depression) is an indication of the moisture content of the surrounding air, or of its humidity. See Appendix for convenient conversions of wet and dry bulb readings into relative humidity.

Example: dry bulb reads 90°F; wet bulb 80°F; difference is 10°F. Following the coordinates to their intersection, we find the relative humidity to be 65 percent.

A portable version of this instrument, called a sling psychrometer, provides a convenient way to check the calibration of relative humidity instruments. Their air velocity past the bulb should be 15 feet (4.6 metres) per second or higher for accurate results.

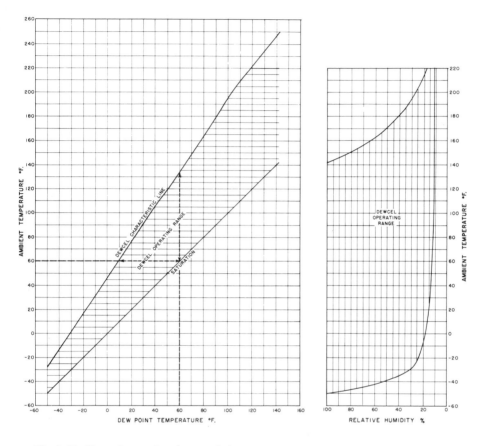

Fig. 5-16. Dewcel operating characteristics.

The Foxboro dew point measuring system consists of an element, the DEWCEL, a source of power to energize it, and a temperature instrument to sense the dew point temperature. The dew point element consists of a cylinder of woven glass fabric with two windings of carefully spaced gold wires. The two windings do not contact each other. The fabric is impregnated with lithium chloride. The windings carry low-voltage power applied through an incandescent lamp ballast resistor.

Moisture determination by this lithium chloride element is based on the fact that, for every water vapor pressure in contact with a saturated salt solution, there is an equilibrium temperature at which this solution neither absorbs nor yields moisture to the surrounding atmosphere. This equilibrium temperature is shown in Figure 5-16 as the "DEW-

CEL® Characteristic Line.'' Below this equilibrium temperature, the salt solution absorbs moisture. Above this equilibrium temperature, the saturated salt solution dries out until only dry crystals are left.

The temperature thus determined may be used as a measure of dew point temperature, absolute humidity, or vapor pressure.

The relative humidity of the atmosphere can be determined from a knowledge of the dew point and reference to a vapor pressure curve or chart (see Appendix, Table A-29).

Absolute measurements are less common in process applications than relative humidity or dew point determinations, but blast furnace control is one common application of absolute humidity measurement.

Questions

5-1. To obtain the most accurate temperature record for the range 100° to 150°F, the filled thermal system selected would be:
  <u>a</u>. Class I   c. Class III
  b. Class II   d. Class V

5-2. A resistance thermometer would be chosen because of:
  a. Ability to measure high temperatures
  b. Economy
  c. Higher accuracy
  d̄. Simplicity

5-3. Thermocouples are often chosen because of:
  a. High accuracy
  b̠. Ability to measure high temperatures
  c. Economy
  d. Ability to measure an extremely narrow span of temperature

5-4. The Class II (vapor-pressure) thermometer employs a nonlinear scale because:
  a. It is easier to read
  b. It may be ratioed to flow measurements
  c̠. The vapor pressure curve is nonlinear
  d̄. The mechanism employed causes nonlinearity

5-5. Which filled thermal system, or systems, would you select for measuring the room temperature in your house?
  <u>a</u>. Class I   c. Class III
  b. Class II   d. Class V

5-6. A Class IIIB (gas-filled) system employs a large bulb to:
  <u>a</u>. Minimize the ambient effects on the connecting capillary
  b. Obtain the optimum dynamic performance

    c. Make a maximum bulb area available for measurement
    d. Reduce the power produced in the element

5-7. A temperature range between 300°F and 310°F or 149°C to 154°C must be measured with the greatest possible accuracy. The best choice of system would be:
    a. A copper-constantan thermocouple
    b. A copper RTD
    c. A nickel RTD
    d. A Class IA filled thermal system

5-8. A hair element is used because it:
    a. Measures absolute humidity
    b. Is the most accurate type of humidity measurement
    c. Is simple and inexpensive
    d. Measures dew point

5-9. A lithium chloride element is usually calibrated to read:
    a. Relative humidity   **c.** Absolute humidity
    b. Wet bulb temperature   **d.** Dew point

5-10. Lithium chloride is used because:
    a. It is nonpoisonous
    b. It is inexpensive
    c. It is extremely hygroscopic
    d. It does not corrode the equipment

5-11. When a wet and dry bulb psychrometer is read to determine relative humidity:
    a. The dry bulb will read lower than the wet bulb
    b. The two thermometers may read the same
    c. The wet bulb will read lower than the dry bulb
    d. A formula may be employed to relate the wet bulb reading to relative humidity

5-12. The purpose of using extension lead wires that have the same thermoelectric characteristics as the thermocouple is to:
    a. Prevent corrosion at all junctions
    b. Extend the reference junction back to the instrument
    c. Prevent creating an unwanted reference junction
    d. Make the thermocouple system operate in standard fashion

5-13. Relative humidity is:
    a. The moisture present in a body of air expressed as a percentage of saturation at the existing temperature
    b. The moisture in a body of air, in grams per cubic meter
    c. The temperature at which moisture will condense from a body of air
    d. The ratio of actual moisture in a volume of air to the moisture that would exist at optimum comfort in a similar volume

5-14. A psychrometer is:
  a. A hair element instrument
  b. A "wet and dry bulb" humidity instrument
  c. An instrument that senses psychological disturbances
  d. An instrument that reads directly in dew point

5-15. A hygrometer is:
  a. Convenient for measuring specific gravity
  b. An instrument that measures gas weight
  c. Any instrument that measures moisture content
  d. Another name for psychrometer

5-16. A certain thermocouple has a specified time constant of 2 seconds. If the process temperature changes abruptly from 800° to 900°C, the temperature readout on the indicator attached to the thermocouple after 6 seconds elapse will be approximately:
  a. 860°C   **c.** 900°C
  b. 835°C   **d.** 895°C

5-17. The air velocity past the sensors of a "wet and dry" bulb instrument should be:
  a. 50 feet per second, minimum
  b. 2 metres per second, maximum
  c. Approximately 4.6 metres per second
  d. Any value

5-18. The advantage of using a three-wire cable to connect an RTD to its associated instrument is that:
  a. Reference junction errors are eliminated
  b. The effect of ambient temperature on the cable will be minimized
  c. Potential failures will be minimal
  d. Resistance in the external circuit is reduced

5-19. A thermocouple instrument has an input resistance of 50,000 ohms and is used with an IC Type J couple. The lead wire is #18 gauge, 120 feet long. What is the approximate error contributed by the lead wire?
  a. ±5.0 percent   **c.** ±0.05 percent
  b. ±0.5 percent   **d.** ±0.1 percent

5-20. The difference between an RTD calibrated to the NR 226 curve and one calibrated to the NR 227 curve is:
  a. A different resistance-to-temperature relationship
  b. Nonexistent
  c. Greater accuracy with the NR 226 curve
  d. Greater accuracy with the NR 227 curve

Analytical Measurements

Analytical measurements seek to define the contents of a process stream and thus enable control of its composition. Some analytical measurements are physical; others are electrochemical. The physical types include humidity, density, differential vapor pressure, boiling point rise, chromotography, ultraviolet, infrared, and turbidity, some of which have been discussed previously. Still others, such as spectroscopy, osmometry, and polarography, are laboratory techniques and not usually associated with automatic control. Some are occasionally used, but are beyond the scope of this book.

Many other analytical measurements, such as conductivity, pH (hydrogen ion concentration), ORP (oxidation-reduction potential), and specific ion concentration, are electrochemical. These measurements, along with capacitance, will be covered in this chapter.

Electrical Conductivity

While dissociation into ions and the resulting ion concentration have been adequate concepts in the past, the fact is that not all the ions present are necessarily effective. Some of them may be "complexed,"

151

that is, "tied up" to other ions and unavailable for reaction. The concept of activity covers this situation. In short, a given compound will dissociate to some degree, described by the dissociation constant, into ions, and some portion of these ions will be active, described by the activity coefficient. However, in many common reactions, the activity coefficient is so near unity that concentration and activity may be used interchangeably. In a growing number of processes, the actual ion activity is different enough from the ion concentration to make it necessary to use the proper term. In general, electrochemical measurements measure activity rather than concentration, and it is always desirable to refer to the measurements as ion activity rather than ion concentration.

Electrochemical measurements all rely upon the current-carrying property of solutions containing ions. Some techniques measure all ions present (electrolytic conductivity). Others respond mainly to particular types of ions—hydrogen ions (pH); oxidizing/reducing ions (ORP); selected ions (ion-selective). These will be discussed separately below.

The ability to conduct electricity, or the reciprocal of electrical resistance, is called *conductance*. The unit in which it is measured is the reciprocal ohm, commonly called *mho*. The conductance of any conductor depends on the nature of the material, the shape of the conducting path, and the temperature. In analytical work, only the nature of the material, in this case the type and activity of the ions present, is important. Thus, the term *conductivity*, the conductance of a volume of the material of unit length and area, is generally used. (Conductivity has largely replaced an earlier term, specific conductance.)

In actual measurement, a conductivity cell of known geometry is immersed in the material, and the resistance (or conductance) across the cell is measured. This gives a measurement that can be calibrated directly in conductivity due to the known shape of the cell. Conductance and conductivity are related as follows: the greater the length of a material of given conductivity, the higher its resistance (the lower the conductance); but the greater the area of the material, the lower the resistance (the greater the conductance). That is,

$$\text{conductance} = \text{conductivity} \, \frac{\text{area}}{\text{length}} \qquad \text{(6-1)}$$

Since the unit of conductance is mho, the unit of conductivity must be

$$\text{conductivity} = \text{mho} \times \frac{\text{centimeters}}{\text{square centimeters}} \qquad \text{(6-2)}$$

or,

$$\text{conductivity} = \text{mho} \cdot \text{cm}^{-1}$$

It is unfortunately common practice to omit the dimensional unit, so that conductivity is then referred to as mhos. This practice leads to confusion with conductance, and should be avoided.

The conductivity of most electrolytes at the concentrations and the temperature ranges normally encountered fall well below unity. For this reason, the micromho per centimetre, the millionth part of the mho per centimetre, is normally used.

In the SI system of units, the siemens replaces the mho. One mho equals one siemens. In conductivity units, one micromho per centimetre (μmho/cm) equals one microsiemen per centimetre (μS/cm). However, the actual SI unit is the microsiemens per metre, since the unit of length in the SI system is the metre rather than the centimetre. One μS/cm equals 100 μS/m.

The conductivity of material of unit length and area which has a resistance of 1,000 ohms is 0.001 mho $\cdot$ cm^{-1} or 1,000 micromho $\cdot$ cm^{-1} (sometimes called simply 1,000 micromhos).

Since the measurement depends on the geometry of the cell used, a cell constant (F) has been defined to describe this geometry simply:

$$F = \frac{\text{length (cm)}}{\text{area (cm}^2)} \tag{6-3}$$

When the length is 1 centimetre and the area 1 square centimetre, $F = 1.0$ cm^{-1}. Here, also, the dimensional unit is omitted in common use, so that the cell is said to have a cell constant of 1.0. Note that the critical dimensions are length and area, not volume. Thus while the above example has a volume of 1 cubic centimetre if the length were 0.5 centimetre and the area were 0.5 square centimetre, the cell constant would still be 1.0. But the volume would only be 0.25 cubic centimetres. Thus it is not correct to say that the resistivity is the resistance of unit volume, but only of unit length and cross-sectional area.

A cell with greater area and the same length will have a lower cell constant. But for the same solution (that is, for a given conductivity) the conductance will be greater with this cell. If we solve Equation 6-1 for conductivity,

$$\text{conductivity} = \text{conductance} \; \frac{\text{length}}{\text{area}} \tag{6-4}$$

and substituting Equation 6-2,

conductivity = conductance × cell constant

A given measuring instrument will have a certain conductance (resistance) range. But it may have a variety of conductivity ranges by simply using cells of different cell constants. For example, an instrument with a range of 0 to 100 micromhos per centimeter with a cell having a constant of 0.01 will have a range of 0 to 1,000 micromhos per centimeter if a cell with a constant of 0.1 is substituted.

Types of Calibration

Conductivity instruments can normally be furnished with
1. Calibration in conductivity ($\mu S \times cm^{-1}$ or $\mu mho \times cm^{-1}$)
2. Calibration in terms of concentration of electrolyte
3. Calibration in terms of conductivity difference

Calibration in Conductivity

Instruments of this type are measured in absolute units, ohms, mhos, siemens, or microsiemens. Such instruments can be used to measure the conductivity of any electrolyte at any solution temperature using a measuring cell with known cell factor. The conductivity of most solutions increases as the temperature increases. Therefore, if a conductivity instrument calibrated in $\mu mho \times cm^{-r}$ is used with a solution of given concentration, the instrument reading will change if the temperature of the solution changes.

Compensation for this effect of temperature on the conductivity of the solution is possible only if the solution's conductivity temperature coefficient is known. For example, if an instrument calibrated from 0 to 1,000 $\mu mho \times cm^{-r}$ were provided with temperature compensation for sodium chloride (NaCl), the instrument would not correctly compensate in any other solution. For this reason, instruments calibrated in absolute units are generally not furnished with temperature compensation. Instruments of this kind are generally the simplest to calibrate and lend themselves to more flexibility of application, since they are not limited to any particular electrolyte. If compensation is desired, it can be supplied, but only for one particular solution.

Calibration in Terms of Concentration of Electrolyte

Instruments calibrated in terms of concentration of electrolyte read percent concentration, grams per litre, parts per million, and the like, for the range specified. This type of calibration is made to the conductivity values within the specified range of concentration, and at a specified temperature of a given electrolyte. The instrument can therefore be used only under the conditions for which it was calibrated.

However, special cases exist which sometimes make conductivity useful in multiple-electrolyte solutions. If the material of interest is much more conductive than others in solution, the contaminants may have a negligible effect on readings. For example, contaminants slowly build up in a sulphuric acid (H_2SO_4) bath for treating textiles. The conductance of the contaminants is very low compared with that of the acid. Thus, the instrument may read directly in concentration of sulfuric acid. Laboratory tests determine when concentration of contaminants is too high and a new bath is then made up.

Figures 6-1 and 6-2 show typical conductivity curves of NaCl and H_2SO_4. Note that the H_2SO_4 curve reverses itself in the region of 30 to

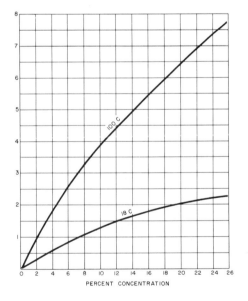

Fig. 6-1. Percent concentration. Conductivity of NaCl at 100°C (212°F) and 18°C (64.4°F).

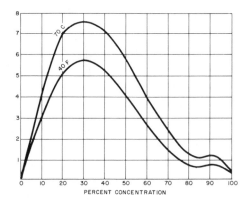

Fig. 6-2. Percent Concentration. Conductivity of H_2SO_4 at 70°F (21.1°C) and 40°F (4.4°C).

32 percent concentration and again in the vicinity of 84 percent and 93 percent. Obviously, it is impossible to calibrate for ranges that include these points of inflection. However, in electrolytes having such maxima they shift with temperature. By controlling sample temperature to a different value, certain ranges may sometimes be measured which would be impossible to measure at process temperature.

Polarization

When an electric current is passed through a solution, electrochemical effects known as polarization occur. These effects, if not minimized, will result in inaccurate measurement. One of the polarization effects is electrolysis. Electrolysis generally produces a gaseous layer at the electrode surface, increasing the apparent resistance of the solution. For this reason, direct current voltage is not practical in conductivity measurements. If the current is reversed, the layer will tend to go back into solution. Thus, if an alternating current is applied, the polarization effect decreases, since the gases and other polarization effects produced on one-half of the cycle are dissipated on the other half cycle.

Cell Construction

The sensitive portion of the cells shown in Figure 6-3 consists of two platinum electrodes mounted in an H-shaped structure of Pyrex glass tubing. The electrodes, located in separate sections of the tubing, are

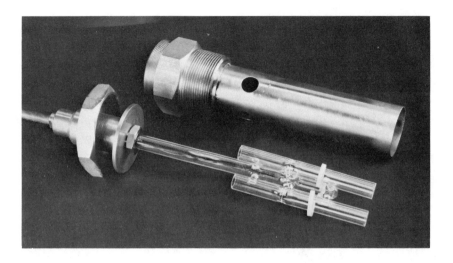

Fig. 6-3. Type H conductivity assembly.

concentrically mounted platinum rings and are flush with the inside surface of the tubing. Fouling or damaging the electrodes is thus minimized, and the cells may easily be cleaned chemically or with a bottle brush. The platinum electrodes in these cells are coated with platinum black to minimize polarization effects.

The cell shown in Figure 6-4 employs graphite, rather than metallic electrodes. The type of graphite used has the same surface properties with respect to polarization as metallic electrodes. These cells require no platinization. They are cleaned chemically or by wiping the surface of the electrode with a cloth or brush.

Conductivity cells (Figure 6-5) are used for detecting impurities in boiler feedwater, for concentration of black liquor (in a pulp digester for kraft paper), for determination of washing effectiveness by measurement of pulp wash, and in many other applications where the presence and concentration of a known salt, base, or acid must be determined.

Electrodeless Conductivity Measurements

In addition to the conventional conductivity measurements just described another approach called an electrodeless technique may be employed. This technique differs from conventional conductivity measurement in that no electrodes contact the process stream. Instead, two

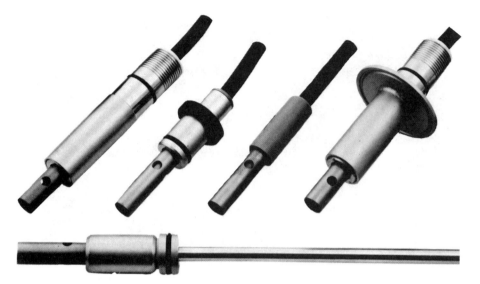

Fig. 6-4. Conductivity cells.

toroidally wound coils encircle an electrically nonconductive tube that carries the liquid sample. The first coil is energized by an ultrasonic oscillator (approximately 20 kH$_z$) which induces an alternating current in the liquid. This current in turn induces a current in the second coil.

Fig. 6-5. Gate valve insertion type conductivity cell.

The induced current in the second coil is directly proportional to the electrical conductivity of the liquid carried by the tube. No direct contact between the coils and the solution is required, thus eliminating potential maintenance problems. Figure 6-6 shows a typical electrodeless conductivity measuring system which transmits a 4 to 20 mA dc signal linearly related to the measured conductivity.

Electrodeless conductivity is especially applicable to the higher conductivity ranges such as 50 to 1,000 millimhos per centimetre. Lower conductivity ranges such as 0.01 to 200,000 micromhos per centimetre are best handled by the conventional techniques previously described.

Hydrogen Ion Activity (pH)

The term pH means a measure of the *activity* of hydrogen ions (H^+) in a solution. It is, therefore, a measure of the degree of acidity or alkalinity of an aqueous solution. The effective amount of a given ion actually present at any time is called the activity of that ion. Activity values can vary from 0 (0 percent) to 1 (100 percent). The measurement of pH, as discussed in this chapter, is a potentiometric measurement that obeys the Nernst equation. This chapter contains an explanation of the measurement technique employed to determine pH along with the Nernst equation.

To aid further in a thorough understanding of pH measurement, some fundamentals of the properties of aqueous solutions must be understood.

Ionization or Dissociation

Stable chemical compounds are electrically neutral. When they are mixed with water to form an aqueous solution, they dissociate into positively and negatively charged par-

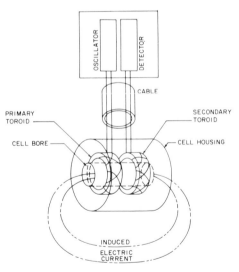

Fig. 6-6. Electrodeless conductivity measuring system.

ticles. These charged particles are called ions. Ions travel from one electrode to the other if a voltage is impressed across electrodes immersed in the solution.

Positive ions, such as H^+, Na^+, and so on migrate toward the cathode, or negative terminal, when a voltage is impressed across the electrodes. Similarly, negative ions, such as OH^-, Cl^-, SO_4^{-2}, and so on, migrate toward the anode, or positive terminal.

The freedom of ions to migrate through a solution is measured as the electrical conductivity of the solution. Chemical compounds that produce conducting solutions are called electrolytes. Not all electrolytes completely dissociate into ions. Those that do (strong acids, strong bases, and salts) are strong electrolytes. Others dissociate, but produce fewer than one ion for every element or radical in the molecule. These are poor electrical conductors, and, hence, weak electrolytes. All weak acids and weak bases fall into this class.

At a specified temperature, a fixed relationship exists between the activity of the ions and undissociated molecules. This relationship is called the dissociation constant (or the ionization constant).

For hydrochloric acid (HCl), the dissociation constant is virtually infinite, which means that for all practical purposes, the HCl is completely composed of positively charged hydrogen ions and negatively charged chloride ions. Because of the essentially complete dissociation into ions, hydrochloric acid is a strong acid:

$$HCl \rightarrow H^+ + Cl^-$$

Conversely, acetic acid has a low-dissociation constant. It breaks up in the following way:

$$CH_3COOH \leftrightarrows H^+ + CH_3COO^-$$

Few hydrogen ions show up in the solution—less than one for every 100 undissociated molecules—therefore, acetic acid is a weak acid.

It follows that the strength of an acid solution depends on the number of hydrogen ions available (the hydrogen ion activity). This depends not only on the concentration of the compound in water, but also on the dissociation constant of the particular compound.

When free OH^- ions predominate, the solution is basic, or alkaline. The dissociation of such a compound is illustrated below.

$$NaOH \rightarrow Na^+ + OH^-$$

Sodium hydroxide is, for all practical purposes, completely dissociated and is a strong base.

Conversely, ammonium hydroxide, NH_4OH, dissociates very little into NH_4^+ ions and OH^- ions, and is a weak base.

The OH^- ion activity, or strength, of a base depends on the number of dissociated OH^- ions in the solution. The number available depends, again, not only on the concentration of the compound in water, but also on the dissociation constant of the particular compound.

Pure water dissociates into H^+ and OH^- ions, but is very weak in the sense used above. That is, very little of the HOH breaks up into H^+ ions and OH^- ions. The number of water molecules dissociated is so small in comparison to those undissociated that the activity of (HOH) can be considered 1 (100 percent).

$$HOH \leftrightarrows H^+ + OH^-$$

At 77°F or 25°C, the dissociation constant of water has been determined to have a value of 10^{-14}. The product of the activities $(a_H+)(a_{OH}-)$ is then 10^{-14}.

If the activities of hydrogen ions and hydroxyl ions are the same, the solution is neutral; the H^+ and OH^- activities must both be 10^{-7} mols per litre.

It must be remembered that, no matter what compounds are present in an aqueous solution, the *product* of the activities of the H^+ ions and the OH^- ions is always 10^{-14} at 77°F or 25°C.

If a strong acid, such as HCl, is added to water, many hydrogen ions are added. This must reduce the number of hydroxyl ions. For example, if HCl at 25°C is added until the H^+ activity becomes 10^{-2}, the OH^- activity must become 10^{-12}.

The pH Scale

It is inconvenient to use terms as 10^{-7}, 10^{-12} and 10^{-2}. Therefore, it has become common to use a special term to represent degree of acidity, or activity of hydrogen ions. This term is pH, originally derived from the phrase "the power of hydrogen."

The pH is defined as the negative of the logarithm of the hydrogen ion activity, or as the log of the reciprocal of the hydrogen ion activity.

$$pH = -\log a_H+ = \log \left(\frac{1}{a_H^+}\right) \tag{6-5}$$

Table 6-1

Temperature °C	$-Log\ (a_H{}^+)(a_{OH}{}^-)$	Neutral pH
0	14.94	7.47
25	14.00	7.00
50	13.26	6.63
75	12.69	6.35
100	12.26	6.13

If the hydrogen ion activity is 10^{-x}, the pH is said to be x. For example, in pure water (at 25°C), the activity of hydrogen ion is 10^{-7}. Therefore, the pH of pure water is 7 at 25°C.

A point frequently overlooked is that the pH for neutrality varies as the solution temperature changes due to a change in the dissociation constant for pure water, as shown in Table 6-1.

An acid solution contains more hydrogen ions than hydroxyl ions.

Table 6-2

	pH	Hydrogen Ion Activity Mols/Litre	Hydroxyl Ion Activity Mols/Litre '
	0	1	0.00000000000001
	1	0.1	0.0000000000001
	2	0.01	0.000000000001
	3	0.001	0.00000000001
Acidic	4	0.0001	0.0000000001
	5	0.00001	0.000000001
	6	0.000001	0.00000001
Neutral	7	0.0000001	0.0000001
	8	0.00000001	0.000001
	9	0.000000001	0.00001
	10	0.0000000001	0.0001
Basic	11	0.00000000001	0.001
	12	0.000000000001	0.01
	13	0.0000000000001	0.1
	14	0.00000000000001	1

Therefore, the activity of hydrogen ions will be greater than 10^{-7}, that is, 10^{-6}, 10^{-5}, 10^{-4}. The pH of an acid solution then, by definition, must be lower than 7, that is 6, 5, 4.

If the number of OH^- exceeds H^+ ions, the hydrogen ion activity must be less than 10^{-7}, that is, 10^{-8}, 10^{-9}, 10^{-10}. Therefore, the pH will be higher than 7, that is, 8, 9, 10.

Table 6-2 demonstrates that a change of just one pH unit means a tenfold change in strength of the acid or base. The reason for this is that there is an exponential relationship between pH numbers and hydrogen ion activity. With so large a change in acidity or alkalinity taking place with a change of just one pH unit, the need for sensitive pH measuring and control equipment cannot be overemphasized.

Table 6-3 shows the nominal pH values for a number of common solutions.

Table 6-3

pH

	14 ← Caustic soda 4% (1.0N)
	13
	← Calcium hydroxide (sat'd sol.) (lime)
	12 ← Caustic soda 0.04% (0.01N)
	11 ← Ammonia 1.7% (1.0N)
	← Ammonia 0.017% (0.01N)
Milk of Magnesia ⟶	10
Borax ⟶	9 ← Potassium acetate 0.98% (0.1N)
	8 ← Sodium bicarbonate 0.84% (0.1N)
Egg white ⟶	
Pure water ⟶	7
Milk ⟶	6
Cheese ⟶	5 ← Hydrocyanic acid 0.27% (0.1N)
Beer ⟶	4
Orange juice ⟶	3 ← Acetic acid 0.6%
Lemon juice ⟶	2
	1 ← Hydrochloric acid 0.37% (0.1N)
	← Sulfuric acid 4.9% (1.0N)
	0 ← Hydrochloric acid 3.7% (1.0N)
	−1 ← Hydrochloric acid 37% (10N)

Fig. 6-7. Schematic diagram of glass-tipped pH measurement electrode.

Measurement of pH—The Glass Electrode

A number of different methods for measuring pH are available for laboratory use, but only one has proved sufficiently accurate and universal for industrial use: the pH glass electrode measuring system.

A special kind of glass, sensitive to hydrogen ions, has been found to be the most useful medium for measuring pH. Figure 6-7 shows a

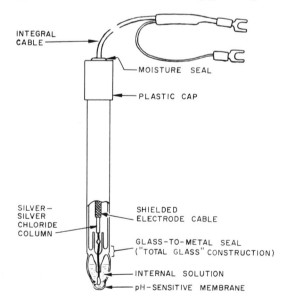

Fig. 6-8. Actual measuring electrode.

schematic diagram of the measuring electrode in which this special glass is used. A thin layer of the special glass covers the tip of the electrode. Figure 6-8 shows an actual measuring electrode.

This glass contains a chamber filled with a solution of constant pH. An internal electrode conductor element immersed in the internal solution is connected to the electrode lead. The conductor element is discussed in the section on temperature compensation.

If the hydrogen ion activity is greater in the process solution than inside the electrode, a positive potential difference will exist across the glass tip. That is, the external side of the glass will have a higher positive potential, and the internal a lower positive potential. If the process is lesser in hydrogen ion activity, a negative potential difference will exist.

The relationship between the potential difference and the hydrogen ion activity follows the Nernst equation.

$$E = E° + \frac{2.3RT}{nF} \log \frac{{}^aH^+outside}{{}^aH^+ inside} \qquad (6-6)$$

where:

E = potential difference measured

$E°$ = a constant for a given electrode system at a specified temperature (25°C)

R = the gas law constant

T = Absolute temperature

n = charge on ion (+1)

F = Faraday's number, a constant

${}^aH^+$ = hydrogen ion activity

In the most common type of pH electrode, that with an internal buffer solution of 7 pH, the voltage across the membrane will be zero at 7 pH.

Reference Electrode

The potential inside the glass is the output of the measuring electrode. It must be compared to the potential in the solution outside the glass to determine potential difference and, hence, pH. The sensing of the potential in the solution must be independent of changes in solution composition. Platinum or carbon would act as ORP, or redox-measuring electrodes. They would be responsive to oxidants or reductants in the solution. They would not yield a true solution potential with solution composition changes.

A reference electrode is the answer. Figure 6-9 is a schematic diagram of a typical reference electrode.

The connecting wire is in contact with silver that is coated with silver chloride. This, in turn, is in contact with a solution of potassium chloride saturated with silver chloride (AgCl). The saturated KCl, called the salt bridge, in turn, contacts the process solution.

Because the concentration of all of the components from the connecting wire to the KCl solution is fixed, the potential from wire to KCl is fixed. The potential between the KCl and the process solution (called the liquid junction potential) is normally small, and will vary only insignificantly with process changes. The *overall* potential of the reference electrode is thus essentially constant, virtually independent of process solution changes, to meet the requirement mentioned earlier.

As the need for high accuracy or repeatability becomes greater, protection of the reference electrode's internal environment becomes more important. Also, prevention of coatings over the electrode tip becomes more critical. Leaking in of process solution must be prevented. Noxious chemicals, or even distilled water, would change the concentration, contaminating the electrolyte. The result would be an unpredictable change in the reference potential. Conductive coatings

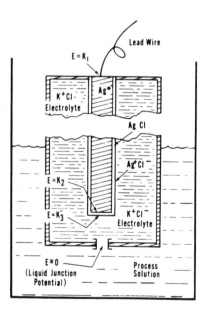

Fig. 6-9. Schematic diagram of reference electrode.

on the electrode tip may cause spurious potentials, and nonconductive coatings may literally open the measurement circuit.

Two types of reference electrodes are available—flowing and non-flowing. The flowing version (Figure 6-10) is usually selected when the highest possible accuracy or repeatability is needed. To prevent electrolyte contamination and to minimize coatings or deposits on the tip, there is provision for a flow of KCl electrolyte out through the porous electrode tip. Instrument air is applied inside the electrode. It maintains a pressure on the electrolyte slightly higher than that of the process liquid. A 3 psi (20 kPa) differential pressure forces a trickle of 1 or 2 ml per day of electrolyte out into the process solution.

The nonflowing version (Figure 6-11) of a reference electrode is often considered standard. No external pressure is applied to this type of electrode. The electrolyte is a paste of KCl and water that actually does flow, or diffuse, into the process solution. The flow rate may be as low as 0.01 ml per day; this is a function of diffusion properties. After

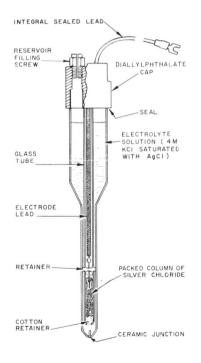

Fig. 6-10. Flowing reference electrode.

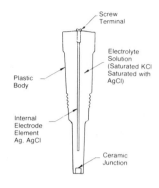

Fig. 6-11. Construction details of nonflowing reference electrode.

depletion of the electrolyte, the nonflowing electrode is usually either recharged or replaced.

Temperature Compensation

The major potential difference exists in the reference electrode, between the metallic silver and the silver ions in the AgCl solution. It follows from the Nernst equation:

$$E = \frac{2.3RT}{nF} \log \frac{{}^{a}Ag^{\circ}}{{}^{a}Ag^{+}} \tag{6-7}$$

Because R, n, and F are constants, and ${}^{a}Ag^{o}$ and ${}^{A}Ag^{+}$ are fixed, this potential (E) will vary only with the *absolute temperature* of the electrode. Since there is a definite chemical relationship between the ionic silver and the activity of chloride in the KCl electrolyte, the above expression may also be written as:

$$E = E^{\circ} - kT \log {}^{a}Cl^{-} \tag{6-8}$$

To compensate for this temperature sensitivity, another silver-silver chloride electrode is inserted into the top of the glass measuring electrode, the internal conductor element mentioned previously. As the temperature of the electrode changes, the potentials of the reference electrode and the conductor element will vary, but will effectively cancel each other, assuming similar values for ${}^{a}Ag^{+}$ (or ${}^{a}Cl^{-}$) in each electrode as is usually the case.

A remaining temperature effect influences the potential across the membrane of the glass measuring electrode. This will vary with the absolute temperature and is greatest at high- and low-pH values. At values around 7, the variation with temperature is zero. This point is called the isopotential point. Figure 6-12 shows the magnitude of the temperature error, away from the isopotential point, with glass electrodes.

In general, some type of temperature compensation is essential for accurate, repeatable pH measurement. If the process temperature is constant, or if the measured pH is close to a value of 7, manual temperature compensation may be used. Otherwise, for good results, automatic temperature compensation will be required.

Reading the Output of the pH Electrodes

The high resistance of the glass measuring electrode results in the need for a millivolt meter with very high internal resistance or sensitivity to measure the cell output. The pH electrodes may be compared to a pair of flashlight cells in series, but unlike the flashlight cell, the electrodes

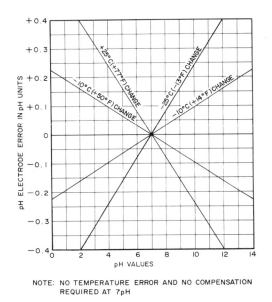

NOTE: NO TEMPERATURE ERROR AND NO COMPENSATION
REQUIRED AT 7pH

Fig. 6-12. Graph of pH measurement errors versus pH values at various solution temperatures due to temperature differences across the measurement electrode tip.

are characterized by extremely high internal resistance (as high as a billion ohms). Unless the millivolt meter employed to measure the voltage created by the electrodes has almost infinite resistance, the available voltage drop will occur across the internal resistance and result in no reading, plus possible damage to the electrodes.

In addition to high impedance or sensitivity, the instrument must have some provision for temperature compensation. It must also have available adjustments that provide for calibration and standardization. The Foxboro pH transmitter shown in Figures 6-13 and 6-14 is typical of an instrument designed to make this measurement. Its output is typically 4 to 20 mA dc which can be fed to a control system. When pH is measured in the process plant, it is generally for the purpose of control.

pH Control

Experience has shown that it is virtually impossible to control pH without considering the process composition as well as more obvious parameters of flow rate, pressure, temperature, and so on. This consideration is frequently even more critical where plant waste neutralization is being controlled; in such applications, there are usually wide, relatively uncontrolled variations in composition as opposed to the normally better defined, and controlled, process stream.

A pH control system can be a relatively simple on/off control loop in some batch processes. At the other extreme, it can be a complex, feedforward/feedback system with multiple sequenced valves for continuous neutralization over a wide, dynamic pH range.

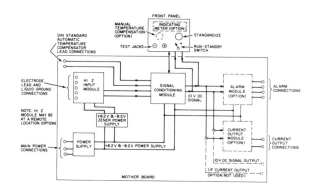

Fig. 6-13. Block diagram of typical transmitter.

Fig. 6-14. Actual transmitter.

The rate of change of inlet pH and flow is a major consideration in determining whether feedforward control is necessary. Since changes in pH reflect logarithmic changes in composition, they can affect reagent demands tremendously and rapidly. Since reagent addition varies directly and linearly with flow, doubling the flow requires twice as much reagent at a given pH. But a change of one pH unit at a given flow requires a tenfold change in reagent. Flow rate compensation is often required in complex pH control loops, but usually only when flow changes are greater than 3 : 1.

The amount of reagent addition for buffered and unbuffered solutions, as seen on conventional acid-base titration curves (Figure 6-15), is also an extremely important consideration in the design of a pH control system. Reaction time of the various reagents must also be

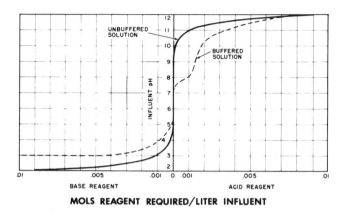

Fig. 6-15. Typical neutralization curves for unbuffered solutions (strong acid or strong base) and buffered solutions. Examples of buffered solutions are (1) weak acid and its sale with the addition of a strong base, and (2) strong acid or base, concentrated, 0.01 M.

given serious consideration in designing both the process and the control system.

A pH measurement and control system, in conjunction with a well-designed process, can be a most successful, reliable system when proper consideration has been given to the various design parameters.

Summary

Continuous pH measurement is a valuable tool for industry and gives information that cannot be obtained economically in any other fashion. It is generally more complex than, say, temperature or pressure measurements, but with constant use, it becomes a routine analytical tool.

Oxidation-Reduction Potential

Oxidation-reduction potential or redox measurements determine the oxidizing or reducing properties of a chemical reaction. In this application, the term oxidation is used in its electrochemical sense and applies to any material which loses electrons in a chemical reaction. By definition, a reduction is the opposite of oxidation, or the gaining of external electrons. There can be no oxidation without an attending reduction.

For example, a ferrous ion may lose an electron and become a ferric ion (gaining increased positive charge) if a reduction, say, of stannic to stannous ions (which is the reverse of this operation) occurs at the same time.

This measurement uses electrodes similar to those in pH measurement (except that metal is used instead of glass), but the two types of measurement should not be confused. The measurement depends on the oxidizing and reducing chemical properties of reactants (not necessarily oxygen). The inert metal electrode versus the reference electrode will produce a voltage that is related to the ratio of oxidized to reduced ions in solution.

This measurement is similar to that of pH in the requirements that it places on the voltmeter used with it. It is useful for determinations in waste treatment, bleach production, pulp and paper bleaching, and others.

The application of ORP measurement or control depends on a knowledge of what goes on within the particular process reaction. The ORP measurement can be applied to a reaction only under the following conditions:
1. There are present in the solution two reacting substances: one that is being oxidized and one that is being reduced.
2. The speed of reaction, following an addition of one of the substances above, is sufficiently fast for good measurement or control.
3. Contaminating substances are held to a minimum, especially those capable of causing side reactions of oxidation or reduction.
4. The pH of the solution is controlled in those areas of the applicable curve where variations in pH can affect the ORP measurement.

Ion-Selective Measurement

Certain applications require that the activity of a particular ion in solution be measured. This can be accomplisehd with an electrode designed to be sensitive to the particular ion whose concentration is being measured. These electrodes are similar in appearance to those employed to measure pH, but are constructed as glass-membrane electrodes, solid-membrane electrodes, liquid-ion-exchange-membrane electrodes, and silicone-rubber-impregnated electrodes.

A common example of ion-selective measurement is found in many municipal water treatment plants. In this application, a measurement of fluoride ion activity can be related to concentration. The measuring electrode is made of plastic and holds a crystal of lanthanum fluoride through which fluoride ions are conducted. The reference electrode is the same as that used for pH. The electrode output is read on a high impedance voltmeter very similar to that used with pH electrodes.

Many applications are possible using the ion-selective technique. At present, measurements in water treatment include hardness, chlorine content, waste, and pollution. Many other industrial applications have been suggested and many have proven effective.

Chromatography

The original application of chromatography consisted of studying the migration of liquid chemicals through porous material, usually paper. The word chromatography means color writing, which occurs when certain extracts or dyes applied to paper produced colored bands. Modern gas chromatography dates from 1952, when James and Martin first used the principle to separate a mixture of volatile fatty acids having nothing to do with color.

The modern chromatograph is generally used for gas analysis. It consists of a column, a tube or pipe packed with materials that will absorb the gases being analyzed at different rates. The gas to be analyzed is carried through the columns by an inert carrier gas, usually helium. As the gas mixture passes through the column, different components are delayed for varying increments of time. Thus, as the gas stream leaves the column and is passed through a gas detector, a chromatogram, or "fingerprint," is formed which may be used to determine the components in the gas as well as the quantity present. The packing material for the column must be properly selected to provide the separation desired for the sample under analysis. The operating parameters, such as temperature, carrier gas flow and pressure, sample valve timing and detector sensitivity, all influence output. Therefore, in a working chromatograph, these factors must be carefully controlled.

Chromatographs are widely used in process work; in fact, they are among the most popular analytical tools. They are used for both liquids and vapors, and adaptations are available to provide an output in conventional analog form such as a 3 to 15 psi (20 to 100 kPa) pneumatic signal. The details of this instrument's operations are beyond the scope of this book. Several references suitable for further study are listed at the end of this chapter.

Table 6-4. Comparison of Capacitance, Conductivity, pH, and ORP Techniques

	Capacitance	Conductivity	pH or ion-selective	ORP
Specificity	Poor	Poor	Excellent	Poor
Sensitivity	Fair	Good	Excellent	Excellent
Conducting fluids	Not applicable	Good	Good	Good
Nonconducting fluids	Good	Not applicable	Not applicable	Not applicable
Maintenance	Low	Low	High	Medium
Installation problems	Low	Low	High	Low
Cost	Low	Low	Medium	Low

Capacitance

Capacitance is not, as applied, an electrochemical measurement. However, a measurement of some characteristic in a nonconducting liquid is frequently required, and in these applications, capacitance may provide the answer.

Electrical capacitance exists between any two conductors separated by an insulator (dielectric). The amount of capacitance depends on the physical dimensions of the conductors and the dielectric constant of the insulating material. The dielectric constant (K) for a vacuum is 1; all other dielectric materials have a K greater than 1. For example, for air, $K = 1.00588$; for dry paper, $K = 2$ to 3; for pure water, $K = 80$. The Table of Dielectric Constants of Pure Liquids (NBS Circular 514), available from the U.S. Government Printing Office, lists the dielectric constants of nearly all common liquids.

Mixtures of materials have a composite value of K that can be directly related to composition. This approach is readily applied to the determination of water content in materials such as paper and crude oil. Capacitance is also applicable to level, interface, octane, and other measurement problems.

In general, the capacitance technique has provided a solution to many measurement problems that cannot be solved easily by more conventional means. Table 6-4 compares the application of pH, ORP, ion-selective, conductivity, and capacitance measurements to process situations.

References

Conductivity cells. *Technical Information Sheet* 43-10a. Foxboro, MA., The Foxboro Company.

Fluoride measuring systems—potable water. *Technical Information Sheet* 43-21a. Foxboro, MA., The Foxboro Company.

Fundamentals of pH measurement. *Technical Information Sheet* 1-90a. Foxboro, MA., The Foxboro Company.

pH electrodes and holders. *Technical Information Sheet* 43-11a. Foxboro, MA., The Foxboro Company.

Theory and application of oxidation-reduction potentials. *Technical Information Sheet* 1-61a. Foxboro, MA., The Foxboro Company.

Pneumatic composition transmitter. *Technical Information Sheet* 37-130a. Foxboro, MA., The Foxboro Company.

Lipták, B. G. *Instrument Engineer's Handbook,* Sec. 81, Volume I. Radnor, PA: Chilton.

Shinskey, F. G. *pH and pION Control in Process and Waste Streams.* New York: John E. Wiley and Sons, 1973.

Questions

6-1. Conductivity and pH measurements are:
  a. Two different techniques
  b. Similar in operation
  c. Identical but given different names
  d. Two techniques that use the same equipment

6-2. The three factors that control the conductivity of an electrolyte are:
  a. Specific gravity, density, and volume
  b. Concentration, material in solution, and temperature
  c. Color index, turbidity, and temperature
  d. Hydrogen ion concentration, temperature, and pressure

6-3. pH is a measure of:
  a. Effective acidity or alkalinity of a liquid
  b. The oxidation or reduction properties of a solution
  c. Specific conductance of an electrolyte or total ionic activity
  d. Purity in an aqueous solution

6-4. A buffer solution is used with pH-measuring instruments to:
  a. Standardize the equipment
  b. Protect the equipment
  c. Clean the electrodes
  d. Platinize the reference electrode

6-5. Oxidation-reduction potential (ORP) is a measurement of:

a. The oxidizing or reducing chemical properties of a solution
b. The oxygen present in any quantity of a given gas mixture
c. The hydrogen ion concentration in a given solution
d. The degree of ionization for a particular solution

6-6. Capacitance measurements are usually applied to:
a. Conducting liquids   c. Gas measurements
b. Nonconducting liquids   d. Ionized gases

6-7. Which of the following are electrochemical measurements:
a. Humidity and density
b. Turbidity and differential vapor pressure
c. pH and ORP
d. Dew point and boiling point rise

6-8. A salt (NaCl) solution must be controlled at a concentration of 12 percent. The best choice of measurement for the control system would be:
a. Conductivity   c. ORP
b. pH   d. Capacitance

6-9. An industrial effluent stream is to be neutralized by adding a sodium hydroxide solution. The best choice of analytical measurement for the control system would be:
a. Conductivity   c. ORP
b. pH   d. Capacitance

6-10. Ion-selective measurements:
a. Are similar to conductivity in operation but use a different cell
b. Are similar to capacitance measurements but use a different instrument
c. Are similar to pH measurements but use different electrodes
d. Are similar to density but use more exact techniques

6-11. A pH control system is used to neutralize a chemical waste stream being dumped into a municipal sewage system. The desired pH is 7 or complete neutrality. An unfortunate accident short circuits the cable connecting the electrodes to the measuring transmitter. The result will be:

a. The control valve admitting the neutralizing agent will fully open
b. The control valve will close
c. The control valve will remain approximately at its half open position
d. The system will cycle

6-12. The concentration of salt in a liquid used to carry a slurry must be monitored. The best choice of measurement will be:
a. Electrodeless conductivity   c. ORP
b. pH   d. An ion-selective system

6-13. A pH system is to be selected. It is required that the system function with little or no maintenance. The reference electrode selected would be:

    a. A flowing type   **c.** A nonflowing type

    b. A pressurized flowing type   **d.** A pressurized nonflowing type

6-14. The pH of a stream is to be monitored accurately. It is discovered that the temperature of this stream varies from 40 to 60°F:

    a. A measuring system with automatic temperature compensation is indicated

    b. Temperature compensation is not necessary

    c. A manual temperature compensation will be adequate

    d. The pH range will determine the need for temperature compensation

6-15. The most popular carrier gas used in gas chromatographs is:

    a. Helium   **c.** Hydrogen

    b. Air   **d.** Oxygen

The Feedback Control Loop

The feedback control loop, introduced in Chapter 1, requires further discussion. This chapter is devoted to a more detailed description of loop operation and its implicit mathematical relationships.

In the feedback loop, the variable to be controlled is measured and compared with the desired value—the set point. The difference between the desired and actual values is called error, and the automatic controller is designed to make the correction required to reduce or eliminate this error.

Within the controller, an algebraic "summing point" accepts two inputs—one from measurement, the other from set. The resulting output represents the difference between these two pieces of information. The algebraic summing point is shown symbolically in Figure 7-1, where c represents measurement; r represents set point; and e represents error. Both r and c must be given their proper signs ($+$ or $-$) if the value of e is to be correct.

A typical closed-loop feedback control system is shown in Figure 7-2. Assume that the algebraic summing point is contained within the feedback controller (*FBC*). In order to select the proper type of feedback controller for a specific process application, two factors—time and gain—must be considered. Time consists of dead time, capacity,

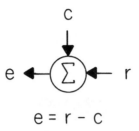

$$e = r - c$$

Fig. 7-1. Algebraic summing point.

and resistance. These factors cause phase changes that will be described in this chapter.

Gain appears in two forms—static and dynamic. Gain is a number that equals the change in a unit's output divided by the change in input which caused it.

$$\text{Gain} = \frac{\Delta \text{ output}}{\Delta \text{ input}}$$

The static gain of an amplifier is easily computed if, for a given step change in input, the resulting change in output can be monitored. Case A in Figure 7-3 shows an amplifier with a static gain of one. In Case B, the output is magnified when compared with the input by a factor greater than one. In Case C, the output change is less than the input. Here, the input has been multiplied by a number less than one.

The dynamic gain of an amplifier may be computed by inducing a sine wave on the input and observing the resulting output. Figure 7-4 illustrates this procedure with an amplifier that has no time lag between its input and output. If the amplitude of the output is only half as high as

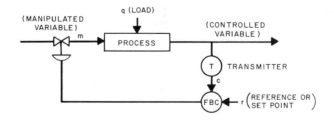

Fig. 7-2. Closed-loop control system.

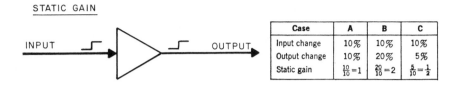

Case	A	B	C
Input change	10%	10%	10%
Output change	10%	20%	5%
Static gain	$\frac{10}{10}=1$	$\frac{20}{10}=2$	$\frac{5}{10}=\frac{1}{2}$

Fig. 7-3. Static gain.

the amplitude of the input sine wave, the amplifier is said to have a dynamic gain of 0.5 for the particular frequency of the input wave, for example, 1 Hz.

By monitoring the output amplitude for many different input frequencies, a series of dynamic gain numbers may be found. A plot of the amplitude ratio (gain) as a function of frequency of the input sine wave is the gain portion of a Bode diagram, or a frequency-response curve (Figure 7-4). Note that the higher the frequency, the lower the gain.

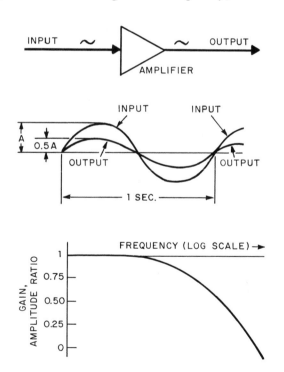

Fig. 7-4. Dynamic gain.

This is true for nearly all processes and instruments. The frequency scale is normally a logarithmic scale. Gain or amplitude ratio is also normally expressed in decibels, where 1 decibel equals 20 times the logarithm to the base 10 of the amplitude ratio.

As the frequency becomes lower and lower, and finally approaches zero, a measure of the amplifier's static gain can be obtained.

Just as the static or dynamic gain of an individual amplifier can be computed, so can the static or dynamic gain of a process control loop. Figure 7-5 shows a temperature control loop. The static loop gain has been computed by multiplying the static gains of each of the individual components of the loop. Likewise, the dynamic loop gain could be calculated by multiplying the dynamic gains of each element at a particular frequency. Pure dead time has a gain of 1.

Each element in the control loop contributes gain to the total loop. Increasing the size of a control valve or narrowing the span of a transmitter has exactly the same effect as increasing the gain of a controller.

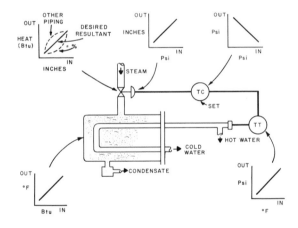

Fig. 7-5. Response characteristics of each element in a feedback control loop-heat exchanger process.

TT Transmitter gain = psi/°F

TC Controller gain = psi/psi

Diaphragm actuator gain = inches/psi

Valve gain = Btu/inches (lift)

Heat exchanger gain = °F/Btu

Static loop gain = $\left(\dfrac{psi}{°F}\right)\left(\dfrac{psi}{psi}\right)\left(\dfrac{inches}{psi}\right)\left(\dfrac{Btu}{inches}\right)\left(\dfrac{°F}{Btu}\right)$

Proper selection of valve size and transmitter span is as important as selecting the gain range in the controller.

In practice, the gain of the components in the loop may not be uniform throughout their operating range. For example, control valves having several characteristics will exhibit linear, equal-percentage (logarithmic), or other types of gain curves. Unless special precautions are taken, the gain of many control loops will change if the operating level shifts. The gain must then be readjusted at the controller to provide satisfactory control.

Let us investigate how a feedback control system functions under closed-loop conditions.

The Closed-Loop Control System

The process consists of multiple capacitances (to store energy) and resistances (to energy flow), as shown in Figure 7-6. Similarly, the measuring system generally contains multiple capacitances and resistances. The controller also consists of capacitance and resistance networks. The final operator, or valve motor, likewise contains capacitance and resistance, and the control valve is essentially a "variable resistance." Thus, the analysis of the system is simply an analysis of how signals change as they pass through resistance-capacitance and dead time networks of various types.

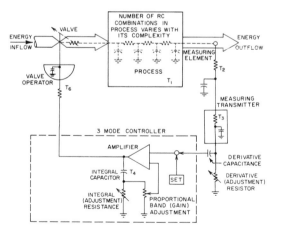

Fig. 7-6. Closed-loop process control system illustrates the *RC* combinations that could exist around a feedback loop. Each affects process stability.

Phase Shift Through RC Networks

Figure 7-7 illustrates the phase shifts that occur when a sine wave is applied to a network of resistances and capacitances. Each succeeding *RC* combination contributes its own phase shift and attenuation. In Figure 7-7, bottom, a phase shift of 180 degrees (a full half cycle) takes place. That is, the output (e_{R_3}) is 180 degrees out of phase with the input (*e* applied). The signal has also been attenuated because each resistor causes a loss in the energy level in the system. This phase change can lead to instability in a closed-loop system.

Oscillation

In a physical system employing feedback, instability will occur if energy is fed back in such direction (phase) as to sustain the instability or oscillation.

Oscillators are often divided into two major categories, those that utilize a resonant device and those that do not. The resonant oscillator uses a device that requires a minimum of added energy each cycle to maintain oscillation. On the other hand, the nonresonant oscillator is simply an amplifier with some type of phase-shifting network between output and input. A public address system with the gain so high that it howls is a nonresonating type.

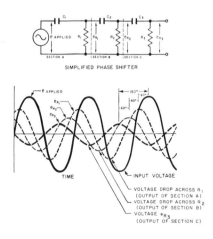

Fig. 7-7. Phase of signal across R_1, R_2, and R_3 is different. At one frequency, the output signal can be exactly 180 degrees out of phase with the input signal, as shown in the graph.

Academically, both the resonant and nonresonant oscillators follow the same mathematical law, but the amplifier gain requirements are quite different for the two categories. With a nonresonant circuit, oscillation will occur (1) if the feedback is positive (inphase) and (2) if the gain is unity or greater. These two conditions are necessary for sustained oscillation.

Figure 7-8 shows a simple mechanical oscillator. In this system, energy stored in the spring is transferred to the weight, back to the spring, to the weight again, and so on. The oscillations soon die out because some of the energy is dissipated in each cycle. If the oscillation is to be continuous, energy must be added to make up the losses.

In the bottom part of the figure, this energy is added by depressing and releasing the weight. If the energy is to enforce the oscillation, it must be in the proper phase with the motion. That is, enforcing energy must be added at a time when both the energy and the oscillation are moving in the same direction.

Figure 7-9 shows a free-running oscillator in which the required energy is introduced through feedback. This circuit is called a phase shift oscillator. It consists of a transistor amplifier and a feedback path comprised of three resistance-capacitance combinations. This circuit is important because it is a simple illustration of the manner in which an ordinary amplifier can be made to oscillate simply by use of feedback.

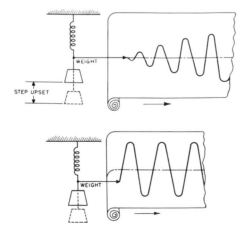

Fig. 7-8. At top, the mass on the spring received a single downward displacement and the oscillation is delayed. Below, after the initial displacement, the mass receives periodic small displacements and the oscillation is continuous.

The transistor amplifier not only makes up for circuit losses, but also contributes a phase shift of 180 degrees; that is, the collector is 180 degrees out of phase with the base signal. Each *RC* element contributes additional phase shift, depending on the values of the components and the frequency involved. If each *RC* combination contributes a phase shift of 60 degrees, the three together will result in a 180-degree phase shift. This, added to the 180-degree phase shift through the amplifier, results in 360 degrees of phase shift. The feedback signal is now in phase with the input signal. This inphase feedback produces continuous oscillation.

Since only one frequency will be shifted in phase exactly 360 degrees by the *RC* networks, the oscillator output is of this one frequency.

The closed-loop control system (Figure 7-6) also has phase-shifting networks, made up of all process and controller components, together with an amplifier capable of contributing sufficient gain to overcome the system's losses. This closed-loop system contains the same basic ingredients as the phase-shift oscillator. Each resistance-capacitance combination will shift the phase of the energy flowing around the control loop in the same manner as in Figure 7-9.

Oscillation (instability) will occur whenever (1) the phase relationships through the various resistance-capacitance combinations provide feedback in proper phase, and (2) the system gain is unity, or greater at the frequency at which the phase shift is 360 degrees.

Figure 7-10 illustrates a plot of system gain (output/input signal-Bode diagram) and output/input phase shift versus frequency for a three-mode controller. The phase shift (expressed in degrees of lead or

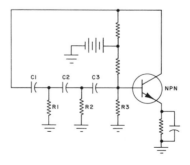

Fig. 7-9. Shift of phase in three *RC* sections maintains oscillation in phase-shift oscillator.

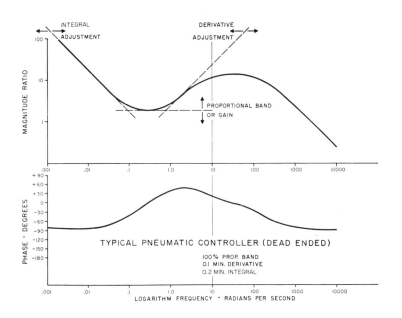

Fig. 7-10. Bode diagram shows gain of system (*top*) and phase change (*bottom*) at various frequencies. The gain (output signal/input signal) and phase (output/input) are measured with the system in open-loop operation.

lag) is plotted linearly against the logarithm of frequency. Note that the integral is phase lagging and the derivative is phase leading.

The proportional adjustment in the controller adjusts its overall gain and hence the loop gain. The integral adjustment governs the low-frequency response, and the derivative adjustment governs the high-frequency gain of the control mechanism.

In addition to controlling gain, these adjustments also affect the phase shifts. Thus, an improper setting of any one of the three control modes can cause the control system to satisfy the two criteria for oscillation and become unstable.

Stability in the Closed-Loop System

Under normal operating conditions, the control system should produce stable operation. That is, the controller should return the system to set point, in the event of an upset, with minimum overshoot and oscillation.

Too much overall gain (too narrow a proportional band) can cause the system to oscillate if the feedback reinforces the oscillation. Too little gain can cause the system to deviate too far from the desired set point.

The damped oscillation shown in Figure 7-11 is characteristic of the curve a closed-loop control system will produce when it is subjected to a step change. If the phase and gain relationships are proper, energy is not fed back to sustain the oscillation, and the cycles die out. The pattern of decay generally can be controlled by the adjustments available in the control mechanism. Engineering judgment and skill are required to select the recovery curve best suited to the process being controlled.

A high-gain setting is desired because this gives the fastest and most accurate control action. However, too much gain produces oscillation. The best compromise is to use enough gain to produce a damped oscillation, as shown in the middle of Figure 7-11 (0.25 damping ratio).

Nonlinearities

Thus far it has been assumed that the capacitances and resistances found in the process control loop have a fixed value that does not change with process conditions. This is not always true in practice. Frequently, process conditions vary the value of the resistances and capacitances involved and, as a result, the phase and gain relationships are in constant transition from one value to another. At other times, these values change, limiting or restricting the natural behavior of the

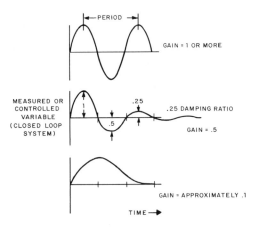

Fig. 7-11. Closed-loop response to an upset depends on gain of loop.

system. Such changes are often considered as a group and are called *nonlinearities*.

A thorough understanding of the operation of the linear system is a prerequisite to an understanding of the nonlinear type. It is common practice to assume a system is linear for the purpose of basic control, and then deal with the nonlinear characteristics as problems arise.

Controllers and Control Modes

A wide variety of response characteristics involving gain and time create the different modes of process control. The controller mode selected for a given process will depend on:

Economics

Precision of control required

Time response of the process

Process gain characteristics

Safety

Now let us investigate the most common control modes.

Two-Position Control

Within this classification there exist several specific methods—on/off, differential gap, and time-cycle control.

On/off control is the most common. As soon as the measured variable differs from the desired control point, the final operator is driven to one extreme or the other.

In the usual sequence (Figure 7-12), as soon as the measured variable exceeds the control point, the final operator is closed. It will remain closed until the measured variable drops below the control point, at which time the operator will open fully. The measured variable will oscillate about the control point with an amplitude and frequency that depend on the capacity and time response of the process. As the process lag approaches zero, the curve will tend to become a straight line. Then the frequency of the final operator open/close cycle will become high. The response curve will remain constant (amplitude and frequency) as long as the load on the system does not change.

On/off control is found in many household applications, such as the refrigerator, heating system, and air-conditioning system. A simple

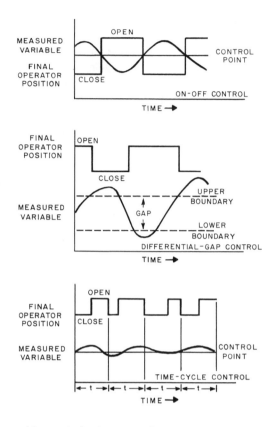

Fig. 7-12. Two-position control acts on operator.

thermostatically controlled electric heater is another familiar example. The room in which the heater is placed determines the capacity of the process, and, hence, the response curve.

The following requirements are necessary for on/off control to produce satisfactory results:

1. Precise control must not be needed.
2. Process must have sufficient capacity to allow the final operator to keep up with the measurement cycle.
3. Energy inflow is small relative to the energy already existing in the process.

Differential-gap control is similar to on/off except that a band, or gap, is created around the control point. In Figure 7-12, note that, when the measured variable exceeds the upper boundary of the gap, the final operator is closed and remains closed until the measured variable drops below the lower boundary. It now opens and remains open until the measured variable again exceeds the upper boundary. In a process plant, differential-gap control might be used for controlling noncritical levels or temperatures.

A time-cycle controller is normally set up so that when the measured variable equals the desired control point, the final operator will be open for half the time cycle, and closed the other half. As the measured variable drops below the control point, the final operator will remain open longer than it is closed.

Two-position control is nearly always the simplest and least expensive form of automatic control. Any of the forms discussed can be implemented with commercially available mechanical, pneumatic, or electronic instrumentation. On the other hand, two-position control may not meet the requirements often demanded by today's sophisticated processes. Now let us investigate a typical on/off control loop. Assume we have a liquid-level process as shown in Figure 7-13.

Throttling Control

Proportional or throttling control was developed to meet the demand for a more precise regulation of the controlled variable. Throttling control indicates any type of control system in which the final operator is purposely positioned to achieve a balance between supply to and demand from the process.

The basic type of throttling control is proportional-only control. This term applied to a control action wherein the position of the final operator is determined by the relationship between the measured variable and a reference or set point. The basic equation for a proportional only controller is:

$$m = \text{Gain (Error)} + \text{Bias} \tag{7-1}$$

where:

m = controller output or valve position
Error = difference between set (r) and measurement (c)

or

$$e = r - c \tag{7-2}$$

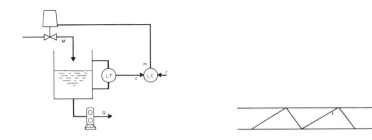

Fig. 7-13. PROBLEM: DIFFERENTIAL GAP CONTROL, CALCULATE PERIOD.

Question 1: What is the period of oscillation of the control loop?

Given:

Controller: on/off with 7 percent differential gap.

Instrumentation response: assume instantaneous.

Load = Q = 60 gpm.

Manipulated Variable = M = 0 gpm or 90 gpm.

Tank: 6 feet diameter, 12 feet high.

Level transmitter: 0 to 8 feet.

Set Point = 50 percent

Question 2: What is the period of oscillation if the load is 50 gpm?

Question 3: At which load will the period be shortest?

Solution:

1. Volume of transmitted signal = area × height

$$V = \frac{\pi D^2}{4} \times H = \frac{\pi (6 \text{ ft.})^2}{4} \times 8 \text{ ft.} = 226.2 \text{ ft}^3 \times \frac{7.48 \text{ gal.}}{\text{ft}^3} = 1692 \text{ gal.}$$

2. Volume within 7 percent differential

$(0.07)(1692) = 118.4$ gal.

3. Level cycle:

3a. Rate of rise = $(90 - 60)$ gpm = 30 gpm

Time to rise = $\dfrac{118.4 \text{ gal.}}{30 \text{ gpm}}$ = 3.95 min.

3b. Rate of fall = $(0 - 60)$ gpm = -60 gpm

Time to fall = $\dfrac{-118.4 \text{ gal.}}{-60 \text{ gpm}}$ = 1.973 min.

3c. Total time = $(3.95 + 1.973)$ min. = 5.92 min.

Solution to no. 2:

3. Time to rise = $\dfrac{118.4 \text{ gal.}}{(90 - 40) \text{ gpm}}$ = 2.96 min.

Time to fall = $\dfrac{-118.4 \text{ gal.}}{(0 - 50) \text{ gpm}}$ = 2.37 min.

Period of oscillation = 5.33 min.

Solution to no. 3:

When $Q = \dfrac{M_{\text{on}} + M_{\text{off}}}{2} = \dfrac{90 \text{ gpm} + 0 \text{ gpm}}{2} = 45$ gpm

Bias = usually adjusted to place the valve in its 50 percent open position with zero error.

The term proportional band is simply another way of expressing gain.

$$\% \text{ Proportional Band} = \frac{100}{\text{Controller Gain}} \tag{7-3}$$

or

$$\text{Controller Gain} = \frac{100}{\% \text{ Proportional Band}}$$

Substituting the error formula ($e = r - c$), Equation 7-2, and the proportional band formula $\left(\% \text{ } PB = \dfrac{100}{\text{Gain}} \right)$, Equation 7-3, into Equation 7-1, it may be expressed as:

$$m = \frac{100}{\%PB} (r - c) + 50\% \tag{7-4}$$

Another approach to visualizing the effect of varying the proportional band is shown in Figure 7-14. Each position in the proportional band dictates a controller output. The wider the band, the greater the input signal (set point - measurement) must change in order to cause the output to swing from 0 to 100 percent. Manual adjustment of the bias shifts the proportional band so that a given input signal will cause a different output level.

Application of Proportional Control

Proportional control attempts to return a measurement to the set point after a load upset has occurred. However, it is impossible for a proportional controller to return the measurement exactly to the set point, since, by definition (Equation 7-4), the output must equal the bias setting (normally, 50 percent) when measurement equals the set point. If the loading conditions require a different output, a difference between measurement and set point must exist for this output level. Proportional control may reduce the effect of a load change, but it cannot eliminate it.

The resulting difference between measurement and set point, after a new equilibrium level has been reached, is called offset. Equation 7-5

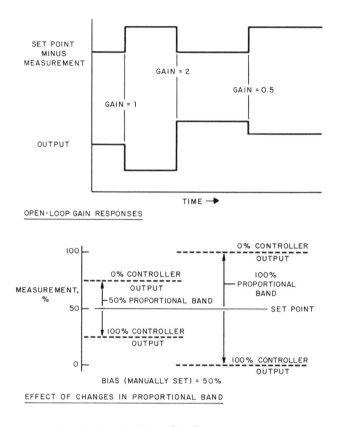

OPEN-LOOP GAIN RESPONSES

EFFECT OF CHANGES IN PROPORTIONAL BAND

Fig. 7-14. Output response reacts to change in gain.

indicates proportional response to a load upset, and the resulting offset. The amount of offset may be calculated from:

$$\Delta e = \frac{\%\ \text{Proportional Band}}{100}\Delta M \tag{7-5}$$

where Δe is change in offset, and ΔM is change in measurement required by load upset.

From Equation 7-5, it is obvious that, as the proportional band approaches zero (gain approaches infinity), offset will approach zero. This seems logical, because a controller with infinite gain is by definition an on-off controller that cannot permit a sustained offset. Conversely, as the proportional band increases (gain decreases), proportionately more and more offset will exist. Offset may be eliminated by manually adjusting the bias until the measured variable equals the set point.

Narrow-band proportional-only controllers are often used in noncritical, simple temperature loops, such as in maintaining a temperature in a tank to prevent boiling or freezing. A low, dynamic gain allows a narrow band to be used. Controllers of this nature are typically field-mounted and pneumatically operated. Many noncritical, level-control applications having long time constants also use proportional-only control. Now let us apply proportional-only control to the level problem shown in Figure 7-15.

Proportional-Plus-Integral Control

If offset cannot be tolerated, another control mode must be added. Integral action will integrate any difference between measurement and set point, and cause the controller's output to change until the difference between measurement and set point is zero.

The response of a pure integral controller to a step change in either the measurement or the set point is shown in Figure 7-16. The controller output changes until it reaches 0 or 100 percent of scale, or the measurement is returned to the set point. Figure 7-16 assumes an open-loop condition where the controller's output is dead-ended in a measuring device, and is not connected to the process.

Figure 7-16 also shows an open-loop proportional-plus-integral response to a step change. Integral time is the amount of time required to repeat the amount of change caused by the error or proportional action. In Figure 7-16, integral time equals t_r, or the amount of time required to repeat the amount of output change (x_1). (Some instrument manufacturers define integral as the inverse of the above, or the number of times per minute the amount of change caused by proportional action is repeated.)

Another way of visualizing integral action is shown in Figure 7-17. In this example, a 50 percent proportional band is centered about the set point. If all elements in the control loop have been properly chosen, the controller's output should be approximately 50 percent. If a load upset is introduced into the system, the measurement will deviate from the set point. Proportional response will be immediately seen in the output, followed by integral action.

Integral action may be thought of as forcing the proportional band to shift, and, hence, causing a new controller output for a given rela-

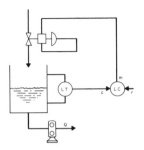

Fig. 7-15. *Sample problem:* PROPORTIONAL CONTROL, CALCULATE MEASUREMENT.

Question: What is the level under steady-state control?

Given:

Range of level transmitter: 0 to 70 inches

Level controller: Proportional only, PB = 75 percent, bias = 50 percent, Set Point = 40 inches

Load, q: Fixed at 3.5 gpm

Valve: Pressure drop, ΔP, is constant, Linear characteristic, delivers 6 gpm at 100 percent stroke.

Solution:

1. Under steady state, inflow must equal outflow.

2. Inflow $= \dfrac{3.5 \text{ gpm}}{6.0 \text{ gpm}} = 58.3\%$

3. Inflow % = Controller output
   m = 58.3%

4. Set point $= r = \dfrac{40 \text{ inches}}{70 \text{ inches}} = 57.1\%$

5. Equation (7-4) for proportional control:

$$m = \frac{100}{\%\text{PB}} (r - c) + 50\%$$

$$58.3 = \frac{100}{75} (57.1 - c) + 50$$

Solving for c: $c = 50.9\%$
Converting to inches: $0.509 \times 70 = 35.6$ inches

tionship between measurement and set point. Integral action will continue to shift the proportional band as long as a difference exists between measurement and set point. Integral action has the same function as an operator adjusting the bias in the proportional-only controller. The width of the proportional band remains constant, and is shifted in a

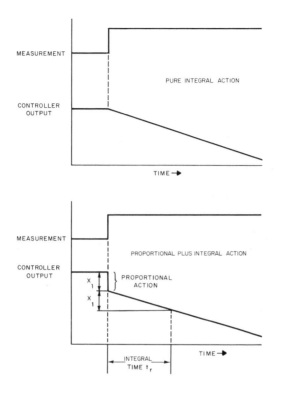

Fig. 7-16. Output responds to step change in input.

direction opposite to the measurement change. Thus, an increasing measurement signal results in a decreasing output, and vice versa.

If the controller is unable to return the measurement to the set point, integral action will drive the proportional band until its lower edge coincides with the set point (see Figure 7-17); hence, a 100 percent output. If the measurement should then start to come back within control range, it must cross the set point to enter the proportioning range, and cause the output to begin throttling. If the system has any substantial capacity, the measurement will overshoot the set point, as shown on the right side of Figure 7-17.

This basic limitation is called integral windup. It must be seriously considered on discontinuous or batch processes where it is common for the controller output to become saturated and overshoot. A pneumatic mechanism called a batch switch designed to prevent this overshoot is described in Chapter 8.

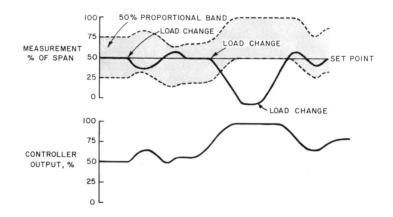

Fig. 7-17. Integral action shifts proportional band in reference to difference between input signal and set point.

Integral setting is a function of the dead time associated with the process and other elements in the control loop. Integral time should not be set faster than the process dead time. If it is set too fast, the controller's output will be capable of changing faster than the rate at which the process can respond. Overshoot and cycling will result. An alternate definition of integral action is the fastest rate at which the process can respond in a stable manner.

Proportional-plus-integral controllers are by far the most common type used in industrial process control. Pneumatic and electronic analog, and digital hardware are available with proportional-plus-integral action.

The classical equation or expression for a proportional-plus-integral controller is:

$$m = \frac{100}{\%PB} \left[e + \frac{1}{R} \int e \ dt \right] \tag{7-6}$$

m = output or valve position

$\dfrac{100}{\%PB}$ = gain

e = error (deviation)

R = integral time

If you compare this equation with Equation 7-4, the proportional-only expression, you will find the bias term replaced by the integral term. In operation, then, the output of this controller will continuously change in the presence of an error signal. Once the error has been reduced to zero, the output will stop changing and stay fixed until an error redevelops. At this time, the output again will change in such direction as to eliminate the error once again.

Adding Derivative (Rate)

In controlling multiple-capacity processes, a third mode of controls is often desirable.

By definition, derivative is the time interval by which derivative action will advance the effect of proportional action on the final operator. It is the time difference required to get to a specified level of output with proportional-only action as compared to proportional-plus-derivative action. Figure 7-18 illustrates derivative response to a ramp change, where t_d equals the controller's derivative time.

Derivative action occurs whenever the measurement signal changes. On a measurement change, derivative action differentiates the change and maintains a level as long as the measurement continues to change at the given rate. Under steady-state conditions, the derivative acts as a 1 to 1 repeater. It has no influence on a controller's output. By reacting to a rate of input change, derivative action allows the controller to inject more corrective action than is initially necessary in order to overcome system inertia. Temperature presents the most common application for derivative. Derivative action should not be used on processes that are characterized by predominant dead times, or processes that have a high noise content, that is, high-frequency extraneous signals such as are present in the typical flow application.

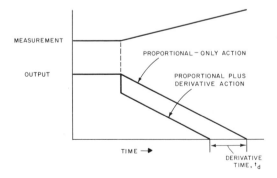

Fig. 7-18. Derivative action increases rate of correction.

The equation or expression for the proportional-plus-integral-plus derivative controller is:

$$m = \frac{100}{\%PB} \left[e + \frac{1}{R} \int e \, dt + D \frac{de}{dt} \right] \tag{7-7}$$

where:

e = error
R = integral time
D = derivative time

Question: How long will it take for the output to change 5 percent if the measurement remains constant?

Given:
Proportional band = 50 percent
Integral time = 3 minutes
Derivative time = 3 minutes
Set point = 45 percent
Measurement = 35 percent
Control action: increase-decrease

Solution:
1. Error stays constant, therefore no output change due to derivative.
2. No output change due to proportional action. However, the proportional action affects the integral response.
3. Error = r − c = 45% − 35% = 10%
4. R = 3 minutes = time to change an amount equal to the error
5. Because the proportional band = 50%, the output in Step 4 is multiplied by $\frac{100}{PB} = 2$.

Therefore 2(10%) = 20% change in 3 minutes

or

5% change in ¾ minute.

Selecting the Controller

Now that the variety of available control modes has been described, the next logical question is: which should be selected to control a particular process? Table 7-1 relates process characteristics to the common control modes. Let us apply the chart to the heat exchanger process.

The heat exchanger acts as a small-capacity process; that is, a small change in steam can cause a large change in temperature. Accu-

Table 7-1. Process Characteristics

Control Mode	Transfer Lag	Dead Time	Capacitance	Reaction Rate	Load Changes	Self-Regulation
On/Off	Min	Min	High	Slow	Any	—
Floating	Small	Min	Low	High	Slow	Must have
Prop.	Small	Small	Moderate	Slow	Slow Small	—
Prop. + Integral	Moderate	Moderate	Moderate	Any	Any	—
Prop. + Integral + Deriv.	Any	Any	Any	Any	Any	—

rate regulation of processes such as this calls for proportional rather than on/off control.

Variations in water rate cause load changes that produce offset, as described previously. Thus, the integral mode should also be used.

Whether to include the derivative mode requires additional investigation of the process characteristic. Referring to the reaction curve (Figure 7-19), notice that the straight line tangent to the curve at the point of inflection is continued back to the 150°F (starting) level. The time interval between the start of the upset and the intersection of the tangential line is marked T_A (in Figure 7-12, this was dead time); the time interval from this point to the point of inflection is T_B. If time T_B exceeds time T_A, some derivative action will prove advantageous. If T_B is less than T_A, derivative action may lead to instability because of the lags involved.

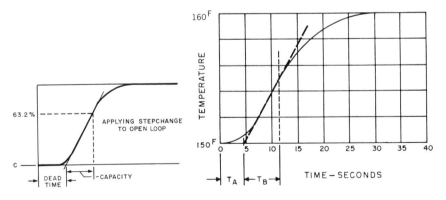

Fig. 7-19. The process reaction curve is obtained by imposing a step change at input.

Table 7-2.

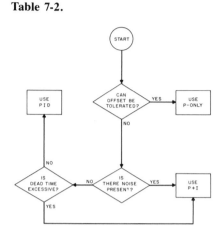

The reaction curve of Figure 7-19 clearly indicates that some derivative action will improve control action. Thus a three-mode controller with proportional, integral, and derivative modes satisfies the needs of the heat exchanger process. Table 7-2, as applied to this problem, will yield the same result.

Questions

7-1. With adequate gain and inphase feedback, any system will:
a. Drift
b. Oscillate
c. Increase amplitude
d. Degenerate

7-2. The natural frequency at which a closed-loop system will cycle depends upon:
a. The amplifier gain
b. The attenuation provided by the process
c. The phase shift provided by the resistance-capacitance and dead time networks that exist in the system
d. Resonance

7-3. The Bode diagram describes:
a. Gain and phase shift through the usable frequency range
b. The system's linearity
c. The reaction to a step change
d. The recovery curve that will result from a load change

7-4. All systems may be assumed to be:
a. Linear
b. Nonlinear

   c. Linear for the purpose of initial consideration but with full knowledge that this may not be the case

   d. Nonlinear for purposes of analysis with the exception that the system may prove to be linear

7-5. A closed-loop control system that employs a three-mode controller:
   a. Can oscillate or cycle at only one frequency
   b. Can oscillate or cycle at several frequencies, depending on controller adjustment
   c. Will not oscillate because of the stability provided by derivative
   d. Will produce only damped oscillations with a 0.25 damping ratio

7-6. We have a closed-loop control system which is cycling. We should:
   a. Increase the proportional band   **c.** Check and adjust both
   b. Increase integral action   **d.** Immediately shut it down

7-7. A proportional controller is being used to control a process, and the offset between set point and control point must be held to a minimum. This would dictate that the proportional band:
   a. Be as narrow as possible
   b. Be as wide as possible
   c. Be of moderate value
   d. Does not relate to the problem

7-8. The system gain of the closed-loop control system:
   a. Refers to the process gain
   b. Refers to the gain of the measurement and control devices
   c. Refers to total gain of all components, including measurement, controller, valve operator, valve, and process
   d. Relates only to the gain of the controller

7-9. If a closed-loop control system employs a straight proportional controller and is under good control, offset:
   a. Will vary in magnitude
   b. Will not exceed one-half of the proportional band width
   c. Will exceed the deviation
   d. Is repeated with each reset

7-10. Any closed-loop system with inphase feedback and a gain of one or more will:
   a. Degenerate   **c.** Exhibit a 0.25 damping ratio
   b. Cycle or oscillate   **d.** Produce square waves

7-11. How long will it take for the output to change 5 percent if the measurement remains constant?
   Given:
   Proportional band = 50 percent
   Integral time = 3 minutes
   Derivative time = 3 minutes

Set point = 45 percent
Measurement = 35 percent
Control action = increase/decrease

7-12. If a closed-loop control system is adjusted to produce a 0.25 damping ratio when subjected to a step change, the system gain is:
a. 0.1   c. 0.5
b. 0.25   d. 1.0

7-13. A straight proportional controller is employed to control a process. Narrowing the proportional band will:
a. Not change the offset   c. Always cause cycling
b. Decrease the offset   d. Never cause cycling

7-14. The type of process that most often can benefit from derivative is:
a. Flow   c. Temperature
b. Level   d. Pressure

7-15. Pure dead time in a process contributes a gain of:
a. Zero   c. Depends upon dynamics
d. Infinite   d. One

7-16. Referring to Figure 7-5, the transmitter span is 200°F; the controller proportional band is adjusted to 150 percent; the equal percentage valve delivers saturated steam containing 40,000 Btu per minute at full open; and water enters the exchanger at 50°F and is heated to 200°F at a 40 gpm maximum flow rate. The static gain of this control loop is approximately:
a. 0.5   c. 1.5
b. 1.0   d. 2.0

7-17. If the flow rate of heated water in Problem 7-16 is reduced to 5 gpm, you would expect the gain to:
a. Increase   c. Remain the same
b. Decrease

7-18. The integral control mode is:
a. Phase-leading   c. Inphase
b. Phase-lagging   d. Phase-reversing

7-19. The derivative control mode is:
a. Phase-leading   c. Inphase
b. Phase-lagging   d. Phase-reversing

7-20. The most common combination of control modes found in the typical process plant is:
a. Proportional-only
b. Proportional, integral, and derivative
c. Proportional-plus-integral
d. On/off

Pneumatic Control Mechanisms

The basic pneumatic control mechanism is the flapper-nozzle unit. This unit, with amplifying relay and feedback bellows, is a simple, rapid-acting control mechanism.

The basic pneumatic mechanism converts a small motion (position) or force into an equivalent (proportional) pneumatic signal. Since pneumatic systems may use a signal of 3 to 15 psi (20 to 100 kPa) (3 psi or 20 kPa at 0 percent and 15 psi or 100 kPa at 100 percent scale), the instrument must have the ability to convert a position or small force into a proportional pneumatic span of 12 psi (3 to 15 psi or 20 to 100 kPa).

The Flapper-Nozzle Unit

Figure 8-1 shows the principle of the flapper-nozzle device. Input air (regulated at 20 psi or 138 kPa) is fed to the nozzle through a reducing tube. The opening of the nozzle is larger than the tube constriction. Hence, when the flapper is moved away from the nozzle, the pressure at the nozzle falls to a low value (typically 2 or 3 psi, or 10 or 20 kPa); when the flapper is moved close to the nozzle, the pressure at the

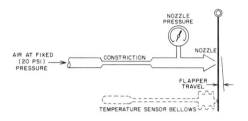

Fig. 8-1. Flapper-nozzle is the basic pneumatic control element. Flapper can be positioned by temperature (as shown), pressure element, or any sensor.

nozzle rises to the supply pressure (20 psi or 138 kPa). Flapper movement of only a few thousandths of an inch (Figure 8-2) produces a proportional pneumatic signal that may vary from near zero to the supply pressure. Some pneumatic control mechanisms use the air at the nozzle (nozzle pressure) to operate a control valve.

The simple flapper-nozzle unit shown in Figure 8-1 has several basic limitations. The output air must all come through the input constriction if it is used directly to operate a control valve. Hence, the output pressure can change only slowly, causing sluggish action. Just as the rate of increase in pressure is limited by the input constriction, the rate of decrease in pressure is similarly limited by the slow rate of air passage through the nozzle to the atmosphere.

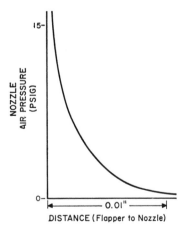

Fig. 8-2. Flapper need move only a few thousandths of an inch for full range of output (nozzle) pressure.

Also, the measuring element that positions the flapper must be strong enough to overcome the blast of air leaving the nozzle. Since many measuring sensors are relatively low-force elements, they could not be used to position the flapper accurately and positively if the nozzle were made large enough to pass sufficient air to overcome this limitation.

A second limitation is that the flapper moves only a few thousandths of an inch for complete action from minimum to maximum nozzle pressure. This small travel for a change in output from zero to full value makes the entire element susceptible to vibration and instability.

All these limitations are overcome by employing (1) a relay, or other amplifier, which amplifies the nozzle pressure; and (2) a feedback system which repositions the flapper or nozzle.

Valve Relay or Pneumatic Amplifier

Figure 8-3 shows a relay connected to a nozzle. The nozzle pressure is applied to the pneumatic relay, which contains a diaphragm operating a small ball valve. Because the diaphragm has a large area, small pressure changes on its surface result in a significant force to move the ball valve. The ball valve, when open, permits the full air supply to reach the output; when closed, it permits the nozzle pressure to bleed to atmosphere.

If the flapper position is changed with respect to the nozzle, the air pressure acting on the relay diaphragm changes and either opens or closes the relay ball valve, thus either increasing or decreasing the flow of supply air, which now can flow directly from the supply to the output, overcoming the first deficiency (slow action) of the flapper nozzle.

The relay is often called a pneumatic amplifier because a small

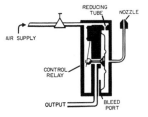

Fig. 8-3. Relay is an amplifier. That is, small change in nozzle pressure (the input to the relay) causes a large change in output pressure (to the control valve).

change in nozzle pressure causes a large change in output pressure and flow. The pneumatic relay shown in Figure 8-3 (used in the Foxboro Type 12A temperature transmitter and the Model 130 controller) amplifies the nozzle pressure by a factor of 16; that is, a change in flapper position of 0.0006 inch, which produces a change in nozzle pressure of ¾ psi, and results in a change of relay output of 12 psi (from 3 to 15 psi).

The relay shown in Figure 8-3 increases output pressure as nozzle pressure increases. Relays can also be constructed to decrease the output pressure as nozzle pressure increases.

The Linear Aspirating Relay or Pneumatic Amplifier

Flapper-nozzle detectors employed in the set-point transmitter and derivative sections of the Model 130 controller use a different type of pneumatic amplifier. If the flow or volume output requirements are small, an aspirating relay may be employed. The pneumatic amplifier, shown in Figure 8-4, makes use of the Venturi tube principle and resembles a small Venturi tube. With a 20 psi supply, the throat pressure of the Venturi can vary 3 to 15 psi or 20 to 100 kPa for a change in flapper position with respect to the nozzle of less than 0.001 inch. The aspirating relay accomplishes this with almost perfect linearity and does not require parts that are subject to wear. This recent develop-

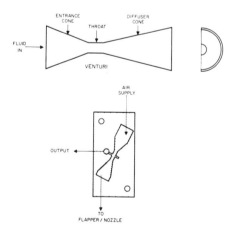

Fig. 8-4. Aspirating cone, showing Venturi tube principle.

ment offers advantages over conventional pneumatic amplifiers, provided that a high volume output is not required.

Proportional Action

If a controller had only the units described thus far, flapper-nozzle and relay, it would only have on/off action. On/off control is satisfactory for many applications, such as large-capacity processes. However, if on/off action does not meet the control requirements, as in low-capacity systems, the flapper-nozzle unit can easily be converted into a stable, wide-band proportional device by a feedback positioning system that repositions the moving flapper.

An example of a proportional action mechanism is the Foxboro 12A pneumatic temperature transmitter (Figure 8-5). This device, called a "force-balance pneumatic transmitter," develops a 3 to 15 psi (20 to 100 kPa) output signal proportional to the measured temperature. Thus, it is functionally a proportional controller.

Fig. 8-5. Pneumatic temperature transmitter, with tubing and sensor.

Figure 8-6 shows the principle of the Model 12A transmitter. The forces created by the bellows are automatically balanced as follows. When the temperature sensor is subjected to an increase in temperature:

1. The increased temperature expands the gas contained in the sensor and increases the force exerted by its bellows. This increases the moment of force, tending to rotate the force bar clockwise.

2. The flapper now approaches the nozzle and the nozzle pressure increases. This pressure is applied to the pneumatic relay (Figure 8-3), increasing its output and the pressure applied to the feedback bellows, thus increasing the counterclockwise moment of force sufficiently to restore the force bar to equilibrium.

The force bar (flapper) is now repositioned slightly closer (less than 0.001 inch) to the nozzle, and the output pressure has reached a new level linearly related to the measured temperature.

In the actual unit, two additional forces act on the force bar. One is that applied by the zero elevation spring. Adjustment of this spring determines the constant force it adds to that supplied by the feedback bellows. This allows a given span of temperature measurement to be raised or lowered. The other force is applied by the ambient-temperature- and barometric-pressure-compensating bellows. This force compensates for ambient temperature or atmospheric pressure changes acting on the gas-filled thermal system, thus minimizing errors caused by these changes.

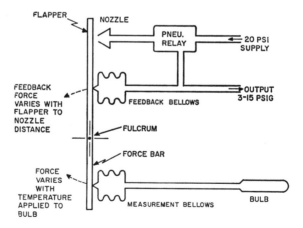

Fig. 8-6. Principle of pneumatic temperature transmitter. Forces created in bellows are balanced.

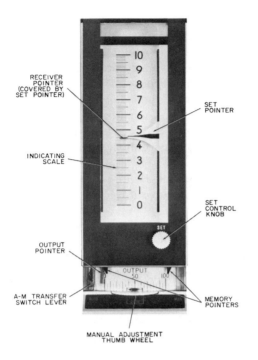

Fig. 8-7. General purpose controller.

The Model 12A temperature transmitter produces an output proportional to measurement and is a proportional control mechanism. However, a useful general purpose controller should incorporate additional modes, and also have the ability to adjust the parameters of the mechanism along with other convenient features. For these reasons the unit described, although a proportional controller, is used primarily as a transmitter to send an input signal to a general purpose controller (such as the 130 Series), as shown in Figure 8-7.

Control Mechanism Requirements

A control mechanism should be able to control the process and ease the job of the operator. To achieve these goals, it must meet the following requirements:

1. The required control modes must be available and easily adjustable through the required range.

2. The controller should have an easily adjusted set point.
3. The mechanism should clearly display what is going on for the benefit of the operator.
4. It should be convenient for the operator to switch over to manual control either for the purpose of operating the process manually during startup or in other unusual situations, or to perform maintenance or adjustments on the automatic control unit.
5. After manual operation, it should be a simple matter to switch back to automatic control smoothly without shocking or bumping the process.
6. If for any reason the unit requires maintenance or adjustment, it should be simple to remove it for this purpose without interfering with the process.
7. In many situations, it may also be convenient and practical to have a continuous record of the controller variable and an alarm system to signal the operator when some predetermined limit is exceeded. Figure 8-8 shows the complete Model 130 controller in block diagram form.

The total control mechanism consists of the following functional parts:
1. A set-point unit that produces a 3 to 15 psi or 20 to 100 kPa pneumatic signal to be fed into the controller mechanism. This unit is generally located within the controller, but it may be remote.
2. The derivative unit that acts directly on the measurement signal.
3. The automatic controller itself—headquarters for the control action.
4. The manual control unit, which performs two functions—switching from automatic to manual and from manual to automatic, and manual adjustment of the control valve (regulated variable).
5. The automatic balancing unit, consisting of a simple floating controller through which the operator can switch from automatic to manual and back without any balancing adjustments and yet without causing a bump or sudden valve change.
6. A measurement indicator that displays with pointers and scale the measurement being fed into the controller, and the set point.
7. An output indicator—a display of controller output that may be interpreted in terms of valve position.

The Automatic Controller

Figure 8-9 develops the mechanism of a proportional controller in schematic form. This controller, like the Model 12A shown in Figure 8-7, has measurement and feedback bellows and a flapper-nozzle relay

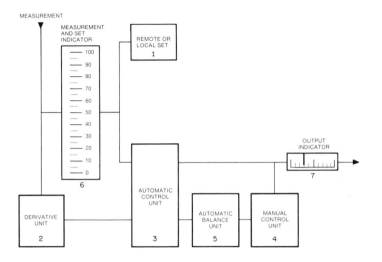

Fig. 8-8. Block diagram of Foxboro pneumatic Consotrol 130M controller.

unit. The basic added features are a set bellows, which permits adjustment of the set point; and an integral (R) bellows.

In the actual Model 130 controller (Figure 8-10), a floating disk acts as the flapper of the flapper-nozzle system. The resultant moments of force due to the four bellows determine the position of the disk with

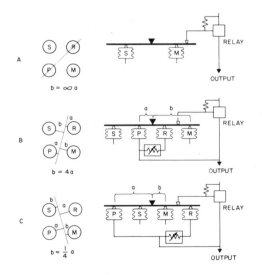

Fig. 8-9. Schematic development of proportional controller.

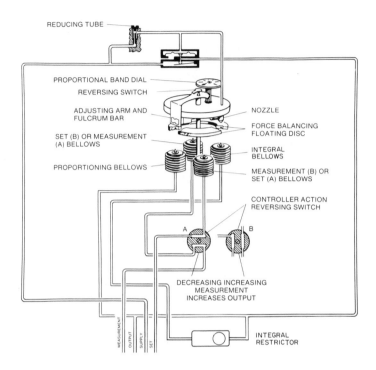

Fig. 8-10. Control unit with floating disk flapper nozzle.

respect to the nozzle. Therefore, the relay output pressure varies with changes in pressure in any of the bellows. The mechanism is aligned to produce a midrange output of 9 psi (60 kPa) when the error signal is 0. This is called bias.

The width of the proportional band is adjusted by the position of the proportional-band-adjusting lever. (The term gain is also used in place of proportional band. Gain is the reciprocal of proportional band.) This lever positions the fulcrum about which the moments of force created by the pressure in the four bellows act. These moments of force tilt the floating disk upward or downward, thereby causing a change in nozzle pressure. The result is an increased or decreased output pressure from the controller. This output pressure, which is proportional to changes in measurement or set pressure, acts to reposition the control valve, thus influencing the process and bringing about a change in the measurement bellows to establish a new equilibrium in the system.

Derivative and Integral

The general purpose controller needs derivative and integral action, as described in Chapter 1. These are added to the pneumatic controller. The derivative function is added to the measurement signal before it reaches the controller (block 4, Figure 8-8). This eliminates derivative response to set-point change and practically eliminates any interaction between derivative and the other control modes.

During steady-state conditions, the unit acts as a 1 : 1 repeater, but upon a change in the measurement the unit adds a derivative influence to the measurement change.

In the derivative unit (Figure 8-11), the force moment (bellows area times distance from fulcrum) of bellows A is 16 times that of bellows B, and the force moment of bellows B plus bellows C equals that of bellows A. As the measurement signal increases, the immediate change in feedback pressure in B is 16 times the change in pressure in bellows A. Simultaneously, air starts to flow through the restrictor to bellows C, gradually reducing the pressure needed in bellows B to restore equilibrium. Thus, the output of the derivative unit, which is the signal to the automatic control unit measurement bellows, reflects the change in measurement plus a derivative response added to that change. The graph at the bottom of Figure 8-11 shows the signal that the measure-

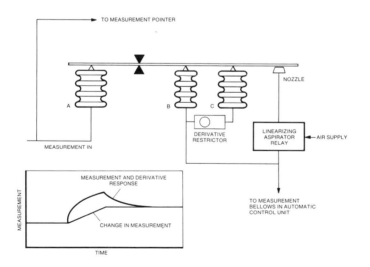

Fig. 8-11. Derivative unit, schematic diagram and response curve.

ment bellows in the automatic control unit receives from the derivative unit at the measurement change shown.

The integral function takes place within the automatic control unit itself. The integral bellows opposes the feedback bellows; thus, if the feedback bellows introduces negative feedback, the force created by the feedback bellows will act like positive feedback; that is, it tends to move the flapper in the same direction as error. This would lead to instability if the integral restriction were not used. Only very slow signals (minutes in duration) can affect the integral bellows because of the very small integral restriction. The integral action occurs only after the proportional and rate actions have affected the process. If an error remains (such as offset due to load change), the integral action takes place slowly. Because integral action will continue in the presence of a continuous error, such as we may find in a batch process, it can cause the output to go to an extreme. This is called integral windup.

Integral time is adjusted by setting an adjustable restrictor, or needle valve. The dial on this pneumatic valve is calibrated in minutes of integral time.

Figure 8-12 illustrates the Foxboro Model 130 controller. The control modes—proportional, integral, and derivative—are in the controller shown. The signal level with respect to time may be changed by appropriate adjustments of proportional, integral, and derivative modes. The proportional band adjustment is calibrated from 5 to 500 percent, while the integral and derivative dials are calibrated from 0.01 to 50 minutes. These adjustments are calibrated in "normal" times. The effective times are somewhat different.

In any controller, some interaction between control actions may exist because change in one action (proportional, integral, or derivative) can create change in the others. This unavoidable interaction, even if small, should be kept in mind when adjusting or tuning a control ler.

Manual Control Unit

Manual operation is achieved by using the manual control unit shown schematically in Figure 8-13 and depicted in Figure 8-14. With the transfer switch in the manual position, the thumbwheel is engaged, and by its action the flapper-nozzle relationship governs the pneumatic output. The operation of the manual control unit is similar to that of any pneumatic transmitter in that it employs a feedback bellows. Full-scale manipulation of the thumbwheel corresponds to a 3 to 15 psi (20 to 100 kPa) pneumatic output.

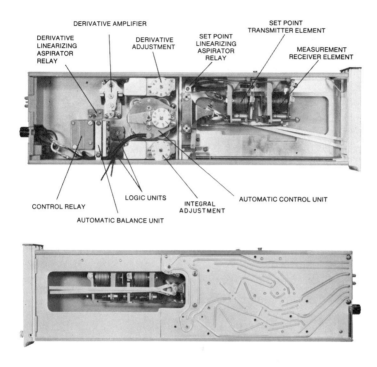

DERIVATIVE AMPLIFIER

DERIVATIVE
LINEARIZING
ASPIRATOR
RELAY

DERIVATIVE
ADJUSTMENT

SET POINT
LINEARIZING
ASPIRATOR
RELAY

SET POINT
TRANSMITTER ELEMENT

MEASUREMENT
RECEIVER ELEMENT

CONTROL RELAY

LOGIC UNITS

INTEGRAL
ADJUSTMENT

AUTOMATIC CONTROL UNIT

AUTOMATIC BALANCE UNIT

Fig. 8-12. Pneumatic Consotrol 130M controller. (*Top*) right side cover removed and manual controls. (*Bottom*) the left side cover is removed to show pneumatic printed circuit board.

Transfer

Automatic to Manual

With the transfer switch in the automatic position, the thumbwheel in the manual control unit is disengaged and the automatic controller's pneumatic output is fed into the manual control unit bellows. The manual control unit remains balanced at this output. At the time of transfer to manual, the manual unit instantly starts transmitting this same output. Thus, the response record of the transfer from automatic to manual (Figure 8-15) is smooth and bumpless.

Manual to Automatic

The transfer from manual to automatic requires the use of an automatic-balancing unit, which consists of a single-pivoted diaphragm with four air pressure compartments. This is shown in schematic form

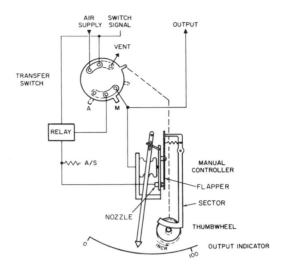

Fig. 8-13. Schematic of manual control unit.

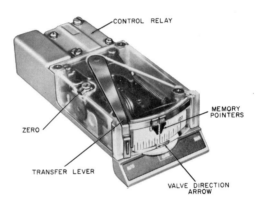

Fig. 8-14. Manual control unit.

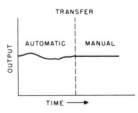

Fig. 8-15. Response record of transfer from automatic to manual.

218

in Figure 8-16. At the left, the unit is in the automatic position; at the right, it is in manual. It will act as a simple proportional controller with a fixed proportional band of approximately 30 percent. The basic problem in transferring from manual to automatic lies in the fact that the output of the automatic controller must equal that of the manual unit at the moment of transfer, and then a change at a predetermined integral rate is necessary to bring the measurement to the set point. The output of the manual station is the control unit's output. Full air supply is sent to the three pneumatic switches, closing two and opening one, as shown in Figure 8-16.

The proportional bellows of the automatic control unit is disconnected from the output of the controller. The integral restrictor is bypassed, and the integral bellows is disconnected from the proportional bellows. The input signals to the automatic balancing unit shown represent the automatic control unit proportional bellows pressure in bellows D, and the manual control output in bellows A. Bellows B is the balancing pressure and bellows C is the output to the integral bellows.

If either the measurement or the set point to the automatic control unit changes, the pressure in the proportional bellows must also change, because it is operating as a proportional-only control unit. When the change in pressure in the proportional bellows is sensed by the balancing unit, the unit's output will change the pressure in the controller's integral bellows. This, in turn, will cause the proportional bellows pressure to change in the opposite direction until it once again equals the output of the manual control relay, or until the supply pressure limits are reached.

Any difference between the set and measurement bellows is thus balanced by the difference between the integral and proportional bellows.

If the output of the manual control unit changes, a similar action occurs, forcing the proportional bellows pressure to equal the manual control unit output.

When transferring from manual to automatic (Figure 8-17), the output will remain at the level determined by the operator when the controller is in manual. If the measurement input is equal to the set point, the output remains constant until corrective action is required. If the measurement does not equal the set point at the moment of transfer, the output will ramp from the level of manual operation to the level necessary to make the measurement equal to set point a function of the controller's integral rate.

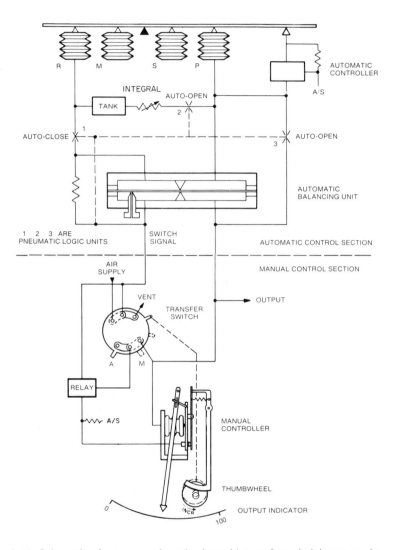

Fig. 8-16. Schematic of auto-manual mechanism with transfer switch in automatic position (*above*) and manual position (*opposite page*).

Set Point

The set-point knob is attached to a pneumatic transmitter, which relates the transmitter's output position of the set-point pointer (Figures 8-18 and 8-19). The output of the transmitter is applied to the set bellows of the controller. If the automatic controller is operating, normal

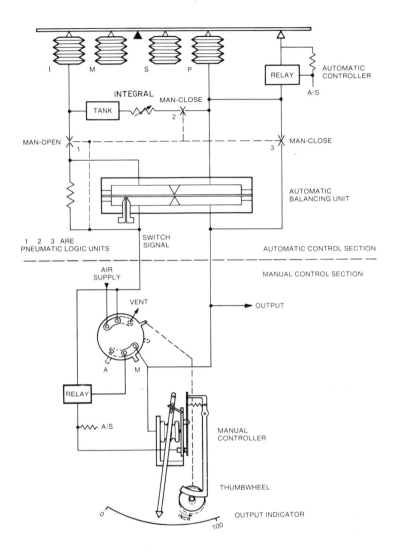

changes can be made by turning the set-point knob. Proportional-plus-integral action will occur. Since the derivative amplifier exists only in the measurement circuit, derivative action will not occur. If it is desired to bring the process slowly to the new set point with integral action only, the controller is simply switched to manual, the set-point change is made, and the controller is switched back to automatic. Now the measurement will approach the new set point, with integral action only, and no overshoot will occur.

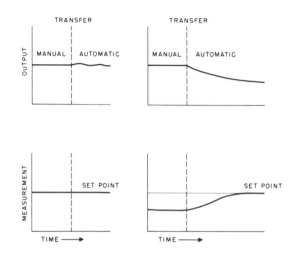

Fig. 8-17. Transfer from manual to automatic (*left*) when measurement equals set point and (*right*) when measurement does not equal set point.

Fig. 8-18. Foxboro pneumatic Consotrol 130M controller pulled out to show access-to-mode adjustments.

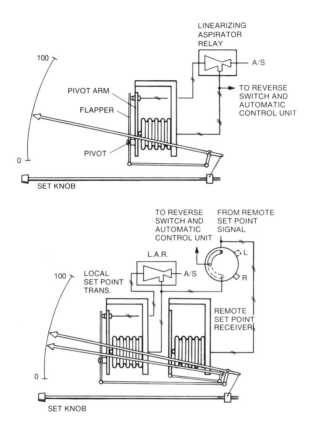

Fig. 8-19. Schematic of local and remote set-point mechanisms.

The Closed-Loop Pneumatic Controlled System

Now let us apply the pneumatic controller to a process control loop. The objective of a control system is to maintain a balance between supply and demand over time. (See Chapter 1 for a discussion of supply and demand.) The closed-loop control system achieves this balance by measuring the demand and regulating the supply to maintain the desired balance.

Figure 8-20 illustrates a typical and familiar controlled system in which the temperature of heated water (controlled variable) is regulated by control of input steam. The control system comprises a

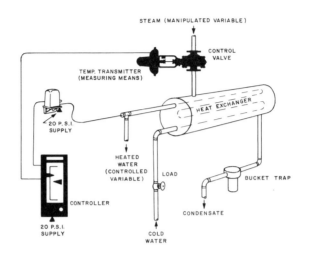

Fig. 8-20. The process to be controlled occurs in a heat exchanger. All elements of the pneumatic control system are shown—transmitter, controller, valve, input water, output water, and steam.

pneumatic temperature transmitter (Foxboro Model 12A), a force-balance pneumatic controller (Foxboro 130 Series), a pneumatic (diaphragm) valve actuator, and a control valve (wide-range V-port).

Figure 8-21 shows an accepted manner of representing a system in a block diagram. All the elements of the actual system are included.

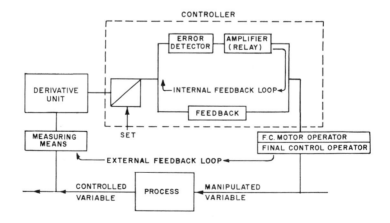

Fig. 8-21. Block diagram showing all the elements of Figure 8-20.

The Process

The process to be controlled is a shell-and-tube heat exchanger. Low-pressure steam applied to the shell heats water flowing through the tubes. The exchanger has 33.5 square feet of heat-transfer surface, and the time required for the heat exchange to take place across this surface causes the exchanger to have a delayed response. The response characteristic is that of a multiple-capacitance, multiple-resistance circuit.

The response or reaction curve of the heat exchanger was obtained by (1) allowing the heat exchanger to stabilize with a constant flow of both steam and water, (2) holding the water flow constant, (3) increasing the steam flow suddenly by opening the control valve, and (4) plotting the increase in water temperature. The resulting curve is the reaction curve illustrated in Figure 8-22.

Since the water temperature was measured with the pneumatic temperature transmitter, the transmitter output depends not only on the response of the heat exchanger, but also on the response of the transmitter, which is not instantaneous.

The Controller Set Point

The set point of the controller is established by setting an air pressure in a "set" bellows by means of an air regulator. The setting dial (Figure 8-23) is spread over the span of the measured (controlled) variable.

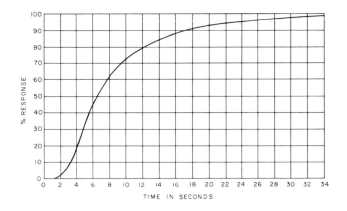

Fig. 8-22. Response of outlet temperature to steps in steam valve position test results.

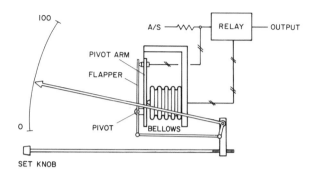

Fig. 8-23. Schematic of set-point mechanism.

Valve and Actuator

The controller output positions a control valve which, in turn, governs the energy or material inflow to the process. The speed of response of the valve depends on the type of valve actuator, that is, the device that transduces the controller output signal into a valve position.

The actuator can be pneumatic or electric. The system under study uses a familiar pneumatic actuator, which consists of a large diaphragm (110 square inches) to which the pneumatic signal is applied. When 3 psi (20 kPa) is applied to an area of 110 square inches, a force of 330 pounds is developed; when 15 psi (100 kPa) is applied, the thrust will be 1,650 pounds. This downward diaphragm force is opposed by a large coil spring that pushes upward with a force of 330 to 1,650 pounds. Increasing or decreasing the pneumatic signal to the diaphragm will cause the diaphragm to move until it reaches a position where diaphragm force and spring force are in equilibrium. Since the relationship between spring thrust and excursion is a linear one (Hooke's law), valve travel is related linearly to controller output.

Valve actuators may be arranged either to open the valve or close it on increasing air pressure. The choice of action is usually dictated by the process being controlled. For a heat exchanger process, the air-to-open valve usually is selected because the spring will close the valve and cause the temperature to drop in event of air failure. This is called fail-safe action. If the process were one that became hazardous when the valve closed, the action would be reversed to make the valve open on air failure.

The valve actuator has a response time because it has the capacity to hold air and the connecting tubing offers resistance to air flow. The time constant for a pneumatic operator is typically several seconds with normal lengths of connecting tubing. Adding tubing lengthens the time constant. Thus, the valve actuator contributes time lag or phase shift to the control loop, depending on the length of connecting tubing used.

Final Control Element

The function of the pneumatic valve actuator is to position the control valve. The valve can be one of many types, depending upon the process to be controlled. For the heat exchanger process being discussed, a single-seat wide-range equal percentage valve is used (Figure 8-24).

The *single-seat equal percentage* valve has a contoured inner valve that provides an exponential relationship between valve stroke and valve capacity.

Fig. 8-24. Cutaway view of single-seat, wide-range, equal percent valve.

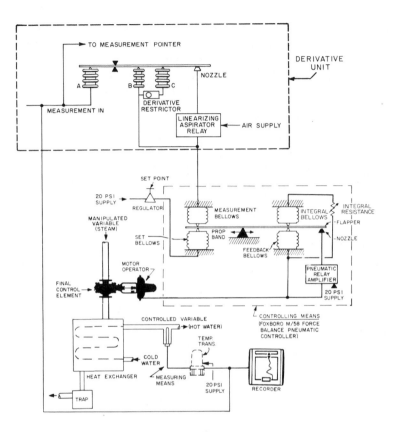

Fig. 8-25. Closed-loop control system, showing all basic components.

Dynamic Behavior of Closed-Loop Control Systems

When all components are connected to form the closed-loop control system (Figure 8-25), each component contributes to control system operation. For example, the total gain (output signal/input signal) is affected by the gain of every one of the loop components. However, the only loop component that has an adjustable gain is the controller. Adjustment of the proportional band results in adjustment of the total loop or system gain.

Let us assume that the control system has stabilized with the outflow temperature at 150°F, 66°C and that the system is subjected to an upset by raising the set point suddenly to 160°F (71°C).

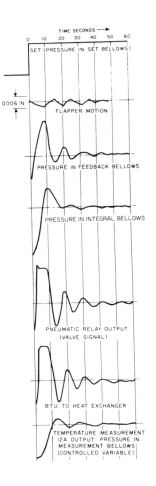

Fig. 8-26. Dynamic behavior of closed loop.

The action around the control loop (Figure 8-26) will be as follows:

1. Increasing the set point increases air pressure in the set bellows, thereby causing the flapper to approach the nozzle.
2. Increased nozzle pressure is amplified and becomes output to the control valve. Because the valve actuator is a resistance-capacitance element, the actual change in valve position will be delayed behind the *change* in output signal.

3. The added heat (Btu) will be nearly in step or phase with the valve position.
4. The added heat causes the temperature of the heat exchanger to rise, and as it rises, the pressure in the measurement bellows will increase.
5. Any unbalance of the moments of force contributed by the measurement and set bellows will cause the integral bellows to continue to change through the integral restricter, and the output pressure to increase until the temperature reaches the set point. The moments of force are now balanced, and the flapper is within its throttling range (a band of travel less than 0.001 inch wide).
6. After approximately 2 minutes have elapsed, the forces exerted by the various bellows will be in equilibrium; the flapper is within its throttling range; and the process has stabilized at 160°F, 71°C. These steps are illustrated in Figure 8-26.

(Note: If the upset were caused by a load change, derivative action would have been added to the functions described.)

Adjusting the Controller

To accomplish effective control and cause the process to react in an optimum fashion, the controller behavior, with respect to time, must be matched to that of the process. This is done by adjusting the proportional band, integral time, and derivative time.

Tuning or adjusting the controller for optimum performance can be achieved either by trial-and-error, or by more exacting methods. The usual trial-and-error procedure in adjusting the controller settings to the process conditions is to set the integral restrictor at maximum and the derivative restrictor at minimum, and then to adjust the proportional band to produce the minimum process stabilization time. Then the derivative restrictor is increased gradually, and the proportional band narrowed, until a combination of proportional and derivative action is obtained which produces a shorter stabilization time than normal proportional, and with less upset to the process.

To eliminate offset, the integral restrictor is then set to a value that will bring the control point to the set point in a minimal time without upsetting the stability of the system.

Other methods of adjustment by mathematical and analog analysis are discussed in Chapter 12.

Suppose a set point of 20 percent scale change is made in the Foxboro 130 Series controller with the controller on automatic. Note

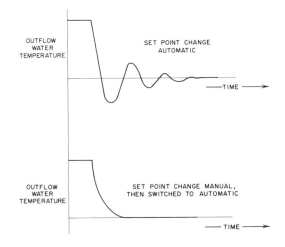

Fig. 8-27. Controlled set-point change made in automatic mode (*top*) and in manual mode, then switched to automatic (*bottom*).

the damped oscillations that occur as the system restabilizes (Figure 8-27, top). The instability lasts for several minutes. If the controller is first switched to manual for the set-point change, a reaction will occur when switched back to automatic. This is shown in the lower part of Figure 8-27 and discussed previously. The controlled variable returns to the new set point at the integral rate. Any adjustments tend to cause a temporary upset; therefore, the technique of switching to manual, making any required adjustments and switching back to automatic, can be used to advantage with the Foxboro 130 Series controller.

Batch Controller

In Chapter 7, reference was made to integral saturation or integral windup. This occurs when the integral mode is used on a batch or discontinuous process. During the time the process is shut down, the integral circuit of a conventional two- or three-mode controller will saturate at the supply pressure.

After the process is restarted, measurement overshoot of the set point can occur unless precautions are taken. The batch switch is a special pneumatic mechanism designed to prevent this overshoot.

During the time a process is off control, the measurement will be

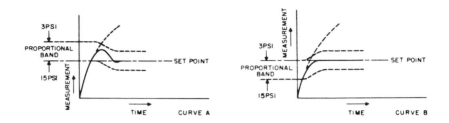

Fig. 8-28. (*Left*) Recovery from sustained deviation; proportional plus integral controller without batch feature. (*Right*) With batch feature.

below the set point and the controller's output will be at its maximum value. If the time of this deviation is long enough, the integral circuit of a conventional controller will also reach this value. When process conditions return to normal, no change in controller output can occur until the measurement reaches the set point. Figure 8-28, curve A illustrates this control action.

The batch switch eliminates this integral circuit saturation and conditions the integral bellows to permit output to start to change before the measurement reaches the set point. If the controller output starts to change before the measurement reaches the set point, overshoot can be prevented and the measurement can return to the set point smoothly. Figure 8-28, curve B illustrates this control action.

Principle of Operation

A schematic diagram of a proportional-plus-integral batch controller is shown in Figure 8-29. The only addition to a conventional two-mode controller is a specially designed pressure switch between the relay output and the integral circuit.

The switch is actuated by the output pressure of the controller relay. Its trip point is adjusted by a spring, or an external pneumatic signal, to trigger at 15 psi (100 kPa). When the controller's output is below this pressure, the ball valve closes the vent port and permits passage of the relay output to the integral circuit. This allows normal integral response as long as the measurement remains within the proportional band (below 15 psi or 100 kPa). Should the measurement reach the 15 psi or 100 kPa limit of the proportional band throttling range, the force on the batch switch diaphragm will seat the ball valve in the opposite direction. This cuts off the relay output pressure to the integral circuit and simultaneously vents the circuit to the atmosphere.

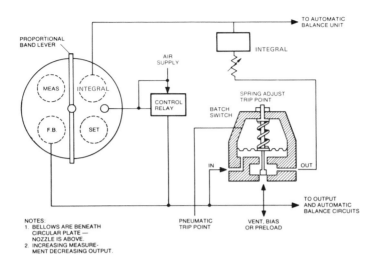

Fig. 8-29. Schematic of proportional-plus-integral batch controller.

This causes the pressure in the circuit to drop. As long as the relay output pressure is above the trip point, the integral circuit pressure will drop until there is no longer any pressure in the integral bellows. At this point, the proportional band throttling range will have shifted completely *below* the set point as indicated in Figure 8-28B, curve B. During this operation, the controller's output pressure is at 15 psi or 100 kPa (or more) and the valve is wide open. When the batch process is restarted, the control valve will begin to throttle as soon as the measurement reaches the 15 psi or 100 kPa limit of the proportional band throttling range. Note that as soon as the controller output drops below 15 psi or 100 kPa, normal integral response is restored. This overcomes the tendency to overshoot the set point.

Load Bias (Preload)

Due to the time characteristics of certain processes, shifting the proportional band completely below the set point will cause an intolerable delay in bringing the batch to the set point. In such cases, an adjustable back pressure may be applied to the vent port to "bias" the batch switch. Thus, the pressure in the integral circuit may be prevented from dropping below a preselected amount and the proportional band throttling range will shift only partially below the set point. Although this will allow faster recovery, a slight overshoot will occur if the "bias" is increased too much. Increasing the bias to 15 psi or 100 kPa would obviously completely eliminate batch action.

One limitation that applies to all pneumatic control systems is the distance that may be accommodated between components. The maximum distance depends on two major factors: the pneumatic signal travels at the speed of sound, and the loop components along with the tubing all have capacity and resistance. Thus, an *RC* time constant must exist. Volume boosters often are applied to lengthen this working distance. Unfortunately, a volume booster will do nothing to speed up signal velocity. A booster will help if a large volume, such as that found in a pneumatic valve actuator, is involved.

The distance limit for pneumatic control systems is approximately 200 feet (60 meters). If this distance factor is ignored, the dynamics of the control system will suffer and result in poor control operation. If it becomes necessary to lengthen these distances substantially, the only practical solution is to utilize electrical signals that may work over an almost unlimited distance.

Questions

8-1. The integral dial in a pneumatic controller is calibrated in:
- **a.** Minutes or repeats
- **b.** Integral units
- **c.** Gain
- **d.** Percentage
- **e.** Offset

8-2. True or false: In a proportional-only controller, if the measurement equals the set point the output will equal the bias.

8-3. True or false: In an integral controller, the rate of change of the output is proportional to the error.

8-4. True or false: The larger the number on the integral dial the greater the effect of the integral action.

8-5. True or false: In a batch operation, if a controller has wound up, it is quite possible that the valve may stay in an extreme position until the measurement actually goes beyond the set point before the valve begins to change its position.

8-6. True or false: What some manufacturers call rate others call derivative.

8-7. Indicate all the correct statements:
- **a.** Gain is the reciprocal of the proportional band.
- **b.** The proportional band is the reciprocal of gain.

c. The proportional band times the gain equals 1.

d. The gain divided by the proportional band equals 1.

e. The narrower the proportional band, the higher the gain.

f. The wider the proportional band, the lower the gain.

8-8. In a process controlled by a proportional-plus-integral controller, the measurement was at the set point and the output was 9 psi (60 kPa). The measurement then quickly decreased to a certain value below the set point and leveled out there. The output responded by changing to 8 psi (55 kPa). As time progresses, what would you expect of the output pressure?

a. Increase to 9 psi (60 kPa) to bring the measurement back to the set point.

b. Remain at 8 psi (55 kPa) as long as the measurement stays where it is.

c. Decrease and continue to decrease to 3 psi (20 kPa).

d. Continue to decrease until the measurement reaches the set point, or if it does not return to the set point, decrease to 0 psi.

8-9. The range of the temperature measuring system used in conjunction with a Model 130 proportional-only controller is 0 to 150°F (66°C). The output is 9 psi (60 kPa) when the set point and indicator are both at 75°F (24°C). If the proportional band is 200 percent, what is the output when the measurement is 150°F (66°C)?

a. 9 psi (60 kPa)   c. 12 psi (83 kPa)

b. 3 psi (20 kPa)   d. 15 psi (100 kPa)

8-10. If the span of a measuring transmitter in a control system is made one-half of its value, the proportional-band adjustment in the controller must be _____ to maintain the same quality of control.

a. Cut in half   d. Narrowed

b. Doubled   e. None of the above.

c. Squared

8-11. With a proportional-plus-integral controller, a sustained error will result in:

a. Windup

b. A fixed offset

c. A temporary narrowing of the proportional band

d. A delay in the process

e. None of the above

8-12. By locating the derivative function in the input measurement circuit, which of the following advantages can be realized?

a. Smooth bumpless transfer

b. No derivative bump with a set-point change

c. No proportional response to a set-point change

d. Integral adjustment is isolated from response

8-13. If proportional-plus-integral control is good, the addition of derivative:
  a. Will anticipate changes and speed up corrections
  b. Will always improve control
  c. Will make the controller adjustments easier to accomplish
  d. May create stability problems in some systems

8-14. The advantage of adding derivative to a controller is always:
  a. Increased stability
  b. The ability to overcome a big pure dead time lag
  c. The ability to react more quickly to measurement change
  d. A decrease in the pure dead time of the process

8-15. For fail-safe action the control valve should, upon energy (air) failure:
  a. Open
  b. Close
  c. Move in such direction as to make the process nonhazardous
  d. Stay in its previous position

8-16. If the closed-loop control system has too much gain, it will cycle. The only loop component that has conveniently adjustable gain is the:
  a. Measuring transmitter   **c.** Process
  b. Valve operator   **d.** Controller

8-17. Adjusting the controller for optimum performance:
  a. Is not required, because it adjusts itself
  b. Requires a special tool
  c. Is usually done by trial and error
  d. Always requires a very involved mathematical analysis of the process

8-18. A process is to be controlled using an all pneumatic system. The maximum distance between loop components will be:
  a. 1,000 feet   **c.** 200 feet
  b. 500 feet   **d.** 20 feet

8-19. If the distance between loop components in the all pneumatic control system must be increased it will require:
  a. A pneumatic volume booster
  b. Larger sized tubing
  c. Smaller sized tubing
  d. Conversion to an electrical signal

8-20. Pneumatic signals travel through the signal tubing at:
  a. 100 feet per second
  b. Approximately the speed of sound
  c. A rate that depends on tubing size
  d. Approximately the speed of light

Electronic (Analog) Control
Systems

Electronic control systems, already widely accepted, are gaining rapidly in popularity for a number of reasons:
1. Electrical signals operate over great distances without contributing time lags.
2. Electrical signals can easily be made compatible with a digital computer.
3. Electronic units can easily handle multiple-signal inputs.
4. Electronic devices can be designed to be essentially maintenance free.
5. Intrinsic safety techniques have virtually eliminated electrical hazards.
6. Generally, electrical systems are less expensive to install, take up less space, and can handle almost all process measurements.
7. Electronic devices are more energy efficient than comparable pneumatic equipment.
8. Special purpose devices such as nonlinear controllers and analog-computing units are simplified.

This chapter will describe the electronic instrument components that make up a typical multiloop system. The process is one found within the boiler room of most process plants—a feedwater control system for a boiler drum.

Feedwater Control Systems

A sample electronic control system is shown in Figure 9-1. The water level in the drum is to be controlled. This is accomplished by measuring not only the water level, but the steam flow out of the boiler and the water flow into the drum. The three measured variables are fed to an electronic control system, which controls the water to the drum.

Transmitters

Two differential pressure transmitters (Fig. 9-2) are used, one to measure water flow; the other to measure drum water level.

Referring to Fig. 9-3, in operation the difference in pressure between the high and low side of the transmitter body is sensed by a twin diaphragm capsule (1) which transforms the differential pressure into a force equal to the differential pressure times the effective area of the diaphragm. The resultant force is transferred through the C-flexure (2) to the lower end of the force bar (3). Attached to the force bar is a cobalt-nickel alloy diaphragm which serves as a fulcrum point for the force bar and also as a seal to the process in the low-pressure cavity side of the transmitter body. As a result of the force generated, the

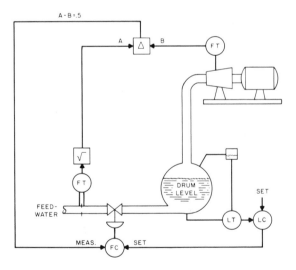

Fig. 9-1. Feedwater control system. (1) Drum water level (2) Steam flow from boiler, via turbine steam pressure, and (3) Water flow into drum.

Fig. 9-2. Differential pressure transmitter.

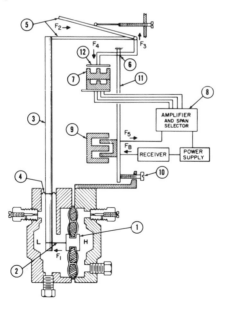

Fig. 9-3. Operation of electronic force balance differential pressure transmitter.

239

force bar pivots about the CoNi alloy seal, transferring a force to the vector mechanism (5).

The force transmitted by the vector mechanism to the lever system (11) is dependent on the adjustable angle. Changing the angle adjusts the span of the instrument. At point (6), the lever system pivots and moves a ferrite disk, part of a differential transformer (7) which serves as a detector. Any position change of the ferrite disk changes the output of the differential transformer determining the amplitude output of an oscillator (8). The oscillator output is rectified to a d-c signal and amplified, resulting in a 4-20 mA d-c transmitter output signal. A feedback motor (9) in series with the output signal, exerts a force proportional to the error signal generated by the differential transformer. This force rebalances the lever system. Accordingly, the output signal of the transmitter is directly proportional to the applied differential pressure at the capsule.

Any given applied differential pressure within the calibrated measurement range will result in the positioning of the detector's ferrite disk, which, in turn, will maintain an output signal from the amplifier proportional to measurement, thus keeping the force balance in equilibrium. A simplified schematic of the electronic circuitry is shown in Fig. 9-4.

Steam flow out of the drum is measured in terms of steam pressure, which is measured with a pressure transmitter. The operation of the pressure transmitter is similar to the operation of the differential

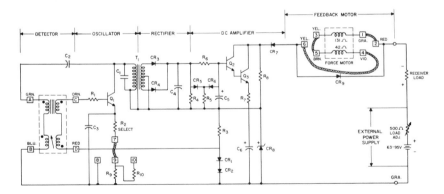

Fig. 9-4. Working schematic of electronic transmitter.

pressure transmitter. The same feedback technique used in the differential pressure transmitter is employed in the pressure transmitter. The major difference lies in the size and construction of the sensor.

The pressure transmitter measures the pressure in the first stage of the power turbine. Since there is a linear relationship between first stage turbine pressure and steam flow, the output signal from the pressure transmitter is linear with steam flow out of the boiler.

A variety of electronic differential pressure transmitters are available from a number of manufacturers. Some of these make use of strain gauge detectors, capacitive detectors, resonant wire detectors, and inductive detectors. Many are motion or open loop devices which are still capable of accuracies within a fraction of a percent. In the application of any electronic transmitter, the user must guard against subjecting the electronics to temperatures which might result in damage. This problem generally can be avoided by observing the precautions recommended by the manufacturer.

The Controllers

The controllers described here are the Foxboro SPEC 200 type. The SPEC 200 is generally a split-architecture system. In this system two areas may be used, a display area and a nest area. Field equipment, such as measuring transmitters, electrical valve actuators, and the like, generally operates on 4 to 20 mA dc. Within the nest, SPEC 200 operates on 0 to 10 volts.

The *display* area contains control stations, manual stations, recorders, and indicators to provide the necessary operator displays and controls. These units are shelf-mounted and contain only the electronic circuitry required to communicate the display and adjustments necessary for an operator to control and monitor a process.

The *nest* area contains the analog control, computing, alarm, signal conditioning, and input and output signal converter units. These units are in the form of "modules" and "circuit cards." The nest itself is basically an enclosure provided for the mounting of "modules."

System power is supplied to the nest. This power supply must deliver $+15$ and -15 V dc for operation of the display and nest-mounted instruments. Recorder chart drives and alarm lights require 24 V ac.

When a single location is required, the nest and display areas may

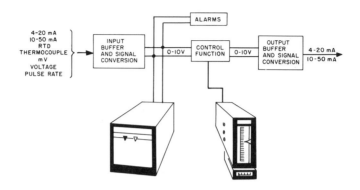

Fig. 9-5. Basic arrangement of SPEC 200 loop.

be combined into one unit. In fact, a single-unit controller with display, controller card, and power supply may be combined into a single case. This is called a SPEC 200/SS (single-station) controller. The basic arrangement of the conventional SPEC 200 system is shown in Figures 9-5 and 9-6. In the single-station unit (Figure 9-7), all components shown in Figure 9-5 are contained in the single unit. For the boiler drum level control, either the single-unit construction or the split system could be used. Typically, the choice would be to split the system into nest and display areas. This will provide added flexibility. The typical boiler room would have many control loops in addition to the boiler drum level. This could all be combined into one nest and display area.

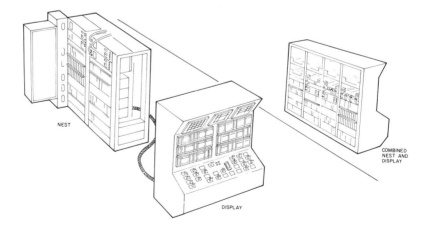

Fig. 9-6. Display and nest areas.

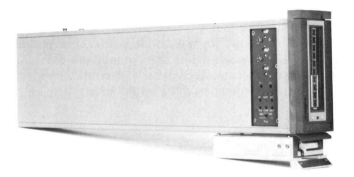

Fig. 9-7. Single station control unit. Electronic display: The measured variable and set point are continuously displayed as vertical bar graphs on a dual gas-discharge unit. The right-hand bar represents the measured variable; the left-hand bar represents the set point. Transparent scales are easily changed without recalibration.

Whether the controller is nest-mounted or case-mounted, it is essentially the same controller. Now let us take a closer look at its operation.

Principle of Operation

A SPEC 200 card-tuned control component consists of a printed wiring assembly (circuit card) in a module. The component operates in conjunction with a control station in the system display area.

The basic circuits of the proportional-plus-integral-plus-derivative control action model is shown in Figure 9-8. This diagram, with the

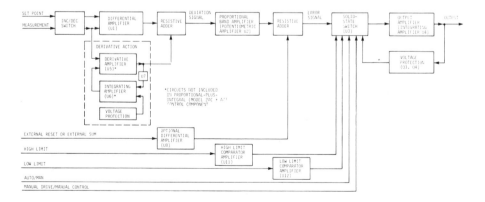

Fig. 9-8. Foxboro PID control component circuit diagram.

derivative action circuits removed, also applies to the proportional-plus-integral control action.

The output signal in automatic mode is a function primarily of the measurement and set-point inputs. The output and control action are also affected by the increase/decrease switch on the control component and by the manual/automatic switch on the display station.

Increase/Decrease Switch

An INC/DEC switch on the front panel provides polarity reversing of both the set point and measurement signal inputs. Reversing the polarity of these input signals effectively reverses the action of the control component. With this switch in the INC position, control action is a function of measurement minus set point (an increase in output is caused by an increase in measurement). In DEC, control action is a function of set point minus measurement (a decrease in output is caused by an increase in measurement).

Deviation Signal Generation

The 0 to 10 V dc measurement and set-point signals are applied to the input of differential amplifier U1. The amplifier has a gain of 1 and high common mode rejection. An output signal is produced by the amplifier any time the measurement differs from the set point.

The output signal from amplifier U1 passes through a resistive adder to the proportional band amplifier as the deviation (error) signal. The resistive adder allows combination of the amplifier output with the derivative signal (in P + I + D control components). In proportional-plus-integral control components, this adder is part of the input resistance of the proportional band amplifier.

Derivative Action

When derivative action is included in the control component, the measurement signal from the INC/DEC switch is applied in parallel to the deviation signal generator amplifier and derivative amplifier U5. Derivative action occurs in proportion to the rate-of-change of the measurement signal.

Derivative amplifier U5 has a fixed gain of 9. Also included in these circuits are integrating amplifer U6 (in the feedback loop of U5) and solid-state (J-FET) switch U7. The input network to the integrator consists of fixed resistors and a potentiometer. The effective resistance, and, hence, the derivative action time, can be controlled by the potentiometer.

Derivative action can be switched off by rotating the derivative potentiometer fully counterclockwise. Both derivative and integral time constants are adjustable from 0.01 to 60 minutes in three ranges (determined by card jumper).

Proportional Band Action

The proportional band stage consists of potentiometric amplifier U2 arranged so that the output is proportional to the input. The gain of the amplifier is determined by the P control on the front panel.

Final Summing and Switching

Signal summing and switching is provided at the input of the output amplifier by a resistive adder and by solid-state switch U3.

The external integral/summing and deviation signals, and also the output signal, are connected to the summing junction of the output amplifier through separate capacitors and field effect transistors (FET). These three signals are connected together through separate resistors at a point called the resistor junction. The resistor junction is separated from the summing junction by a field effect transistor.

The field effect transistor is turned off when the control component is in manual mode. This disconnects the deviation signal from the summing junction, and connects the dc voltage to the input of the output amplifier. Releasing the manual drive thumbwheel from the slow change position causes a spring return to the center position. As a result, the FET is turned off to open the integrating input, and the component is placed in the "hold" condition.

The output stage of the control component is integrating amplifier U4. The amplifier drives the composite signal produced by the final summing and switching element.

High and Low Limits

High and low limits are established for the 0 to 10 V output signal by appropriate circuits. Each limit circuit consists of comparator amplifiers U11 and U12 with associated potentiometers, R77 (HI) and R87 (LO) on the front panel. The output signal from each comparator amplifier is applied to solid-state switch U3. The signal serves as a clamp on the manual or automatic signal input to the output amplifier.

When summed with either the automatic or manual signal, the comparator amplifier outputs prevent the control component output from exceeding the limit reference voltages.

Test jacks on the front panel allow monitoring of the limit adjustments for setting.

External Integral (−R) Option

Differential amplifier U8 is added to the standard control component card for an external signal to reset the output. Amplifier U8 accepts external signals and sums them with the control component deviation signal. This provides an integral input to solid-state switch U3. Anytime the external integral input and the control component are unequal, an error signal is generated. The error signal biases the output in proportion to the external input value, and windup is thus prevented. As long as the control component output tracks the external integral signal, normal integral action takes place. The control component output is biased by the external integral signal in all other conditions, and only proportional action of the control component deviation signal is obtained. This option is used in this application.

External Summing (−S) Option

External summing allows biasing the output directly with an external signal. Differential amplifier U8 is inserted into the "proportional path" solid-state switch U3. The proportional signal is summed with the external bias signal in amplifier U8. The control component output responds directly to changes in the external summing signal in addition to the ordinary control functions.

Power Supply Fault Protection Circuit

This circuit preserves the control component output during short periods of power failure. After such a failure, the output will be essentially at the last value prior to the power failure.

The maximum duration of the fault condition for which this preservation will hold is 1 second. In the manual mode, the drift is less than 1 percent for this 1 second maximum duration. In the automatic mode, additional considerations will dictate the control component output if the fault is corrected within 1 second. Principally, there can be a new measurement signal value when power is restored. This, coupled with the tuning adjustments, will determine the output.

The power supply fault protection circuit can be bypassed by jumper selection.

Automatic/Manual Switch

When the automatic/manual switch in the display area is moved to the MAN position, dc voltages are applied to the solid-state switch (high impedance module U3). This affects the circuitry in two ways. First, the output amplifier is disconnected from all "automatic" signals, and the output remains at the last value. Second, the output amplifier is connected to dc voltages that cause a ramp up or ramp down output. Releasing the manual drive (thumbwheel) will cause a spring return to center position, thereby opening a J-FET switch on the integrating input. This places the circuit in a "hold" condition.

When the automatic/manual switch is transferred to the AUTO position, the solid-state switch connects the deviation signal to the output amplifier. Normal control action then begins. This switch from the manual to the automatic mode occurs without a "bump" (drastic change) in output signal due to the action of the balance circuits.

Controller Adjustments

This controller has a proportional band adjustable from 2 to 500 percent, integral adjustable from 0.01 to 60 minutes (equal to 100 to 0.017 repeats per minute integral rate) and available, but not required for this application, derivative adjustable from 0.01 to 60 minutes.

General Description of the Feedwater Control System

The purpose of this electronic control system is to maintain drum level at a manually set value with minimum fluctuation. The system continuously matches feedwater flow (supply) to steam flow (demand) to maintain the proper relationship of these variables. This relationship is trimmed by the drum level control unit to maintain drum level at any desired, manually set value over the load range of this unit.

This control system is a three-element electronic cascade system in which the primary or master control unit functions to control drum level and the secondary, or slave control unit, functions to maintain the balance between feedwater and steam flow. The primary control unit positions the set point of the secondary control unit, which functions to control the feedwater valve.

The primary control unit compares a measurement of drum level with its manually adjusted set point to develop an output signal that feeds to the external set point input of the secondary control unit. The output of the primary control unit trims the demand for feedwater to maintain the drum level at the set point.

The secondary control unit compares the feedwater/steam flow relationship with its set point to develop an output signal to regulate the feedwater valve.

To maintain drum level, feedwater flow must essentially equal steam flow. For this condition, the measurement input to the secondary control unit is at midscale.

Whenever the drum level is not at the control set point, the primary control unit will alter this secondary control unit set point to trim the flow of feedwater to the drum. Trimming action will continue until the drum level returns to the control set point.

Any change in steam or feedwater flow is immediately sensed as a discrepancy in the actual feedwater/steam flow relationship by the secondary control unit. This control unit, tuned to the response of the feedwater loop, will correct the feedwater flow to restore the feedwater/steam flow relationship.

The measurement input to the secondary control unit is fed to the integral circuit of the primary control unit to prevent integral windup of the primary control unit when the control station is in manual control, or when the secondary measurement cannot follow its set point.

Control Station

The Foxboro Model 230SW cascade set control station is used with a module (Figure 9-9) containing two electronic control units to provide a 0 to 10 volt valve control signal in a single station three-element cascade feedwater control.

The control station is a shelf-mounted display instrument located in the display panel of the SPEC 200 system. The control station contains two nitinol drive units for indicating the primary and secondary measurements of a cascade system. A SET knob on the front plate positions the set-point index on the scale and provides the set-point signal from the drive unit to the control unit.

A front panel cascade/secondary switch allows the operator to select cascade or secondary operation. The set-point and valve control output signals are adjustable.

In cascade (transfer switch in CAS position) operation, this station

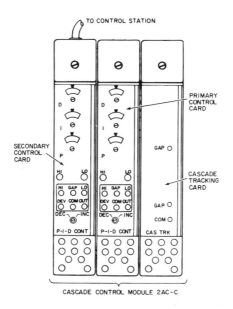

Fig. 9-9. Cascade control module. This is a dual unit module that provides the circuitry to configure a dual unit cascade feedwater control subsystem. This module is wired internally to provide tracking by external reset.

indicates drum level set point (B/W pointer), valve control output signal (black pointer), drum level measurement (red pointer), and feedwater steam flow relationship (green pointer).

In secondary operation (transfer switch in secondary position) the function of the set knob is transferred from the drum level (primary) to the feedwater (secondary) control unit. The secondary switch position allows tuning of the secondary loop with only the feedwater control unit active. The switch should normally be in the CAS position.

Either of two operating modes, automatic or manual, may be operator selected. In either mode, the control station indicates the selected set point (PRI or CAS), valve control signal, and both measurement values.

M/2AX + A4-R Drum Level Control Unit

This is the primary or drum level control unit with circuitry to provide for the secondary measurement (feedwater/steam flow balance) to reset the drum level control unit output. This prevents the drum level control unit from winding up when the secondary control unit is in secondary or manual control. The secondary measurement drives the drum level control unit output to follow. As long as the secondary measurement tracks the secondary set point, normal integral action takes place. In all other conditions the drum level control unit output is driven (biased) by the external secondary measurement signal.

M/2AX + A4 Feedwater Control Unit

This is the secondary, or feedwater control unit. Measurement input to this unit is from the feedwater/steam flow computing unit. When feedwater flow matches steam flow, the measurement input signal is midscale. In cascade operation, the midscale set point is received from the drum level control unit. In secondary operation, the set point is received from the control station set-point circuit.

M/2AP + SUM Feedwater/Steam Flow Computing Unit

Steam flow is compared with feedwater flow in the feedwater/steam flow computing unit. When feedwater flow and steam flow are equal, the output of this computing unit is midscale. The summing unit is a function card that slides into a module located in the nest. Its output may be the algebraic sum of two, three, or four signals scaled to achieve the proper relationship. In this application, only two inputs—steam flow and feedwater flow—are used. In this application, the summing unit is adjusted to accept two 0 to 10 volt signals, one representing feedwater flow (A) and the other steam flow (B). The summer also has an adjustable bias, which is set to produce 50 percent scale or 5 volts output when A = B. This output becomes the measurement input to the feedwater flow controller and is indicated on the control station by the green measurement pointer (Figure 9-10A).

It should be emphasized that all analog computing devices deliver scaled values. This is a major advantage of working with the 0 to 10 volt signal level within the nest. For example, if the summer were to add two 10 volt signals, the output would not be 20 volts, since 10 volts is the maximum output. Depending on the calibration, the output signal will be between 0 and 10 volts, but will represent the addition of the two applied signals.

Square Root Extractor

The square root extractor is a nest-mounted card that accepts a signal input from the differential pressure transmitter and produces a 0 to 10

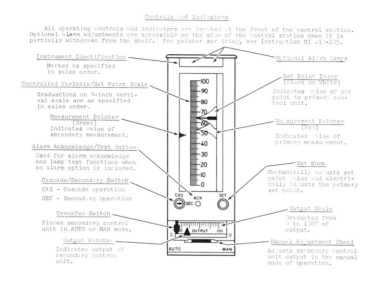

Fig. 9-10A. Foxboro SPEC 200 control station. All operating controls and indicators are located at the front.

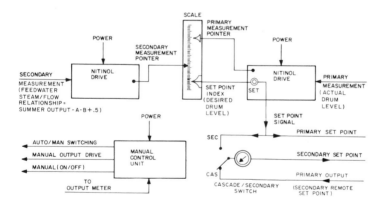

Fig. 9-10B. The diagram is a simplified block diagram of the control station using the nitinol drive units, the CAS/SEC switch and the manual control station.

volt output signal proportional to the square root of the input. It is used primarily in conjunction with differential pressure transmitters to produce an output directly proportional to flow. Another version of the square root extractor can be part of the differential pressure transmitter's electronics. In either case, the square root extractor is designed to create an input/output characteristic such that the output represents the square root of the input.

In Chapter 4, it was established that any restriction flowmeter has questionable accuracy at very low flow rates. Making the signal linear does not solve this problem. Therefore, this device features a low-signal cutoff circuit that automatically forces the output to zero scale level when the input signal falls below 0.75 percent of input span.

The square root extractor is employed in this application to translate water flow into linear units that can be compared with linear units of steam flow. If the summer received one linear signal and one square root signal, the difference would be meaningless.

Feedwater Control

Figure 9-1 illustrates how the three signals—steam flow, feedwater flow, and water level—are combined. The first two (steam flow and feedwater flow) are introduced directly into the sum-computing unit.

The drum water level comes from a level controller. The measurement signal to the flow controller is the difference between steam flow out and water flow in. The resultant error signal is fed through the feedwater controller to provide the feedwater flow signal to the final operator on the feedwater control valve.

The output from the feedwater flow controller is 4 to 20 mA dc. This is converted into a pneumatic signal capable of operating the control valve by means of a current-to-air valve transducer operating on the signal received from the controller. The operation of this transducer is covered in Chapter 10. The transducer provides 3 to 15 psi (20 to 100 kPa) pneumatic output linearly related to a 4 to 20 mA dc input. This pneumatic output is used to position the control valve, thus governing the inflow of feedwater to the boiler. Valves are discussed in Chapter 11.

Closed-Loop Operation

Theoretically, the automatic control system need only balance steam from the boiler (demand) against feedwater into the boiler (supply). If the measurements and control were perfect, such a system would meet all requirements. However, in practice, the drum water level must be used as an overriding control signal, not only for safety, but also because there is always some loss of water through blowdown. (Blowdown is a continuous removal of contaminated water from the drum.) Variation in blowdown rate requires a change in the drum-level controller output to maintain correct level. The blowdown compensation readjusts the drum level. Thus, a measurement of boiler drum level is made. This information, used to enforce a more perfect balance between supply and demand, is called feedback trim.

Assume that the drum level can vary ±15 inches from the desired level. The differential pressure transmitter is calibrated −15 to +15 inches of water head (span = 30 inches) and creates an output of 4 mA dc at −15 inches, 12 mA dc at 0, and 20 mA dc at +15 inches. Assume also that the level control set dial is graduated −15 to 0 to +15 (Figure 9-10A). If the dial is set at 0, optimum level would be desired in the drum.

The usual procedure for adjustment of a multiple-control loop, such as the one described, is to isolate the loops and adjust them one at a time. The feedwater flow controller will normally have a wide proportional band (typically 250 percent) and a fast integral (1/10 minute). Having adjusted the flow controller, the level controller proportional band and integral are adjusted to give maximum response with full stability. No derivative is employed on the feedwater flow controller

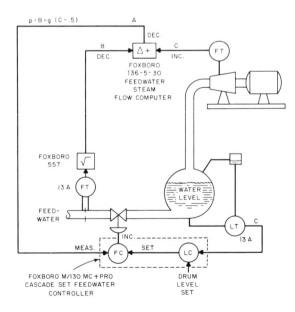

Fig. 9-11. Drum level control—pneumatic.

because derivative creates high-frequency response, and flow ''noise'' would upset the controller.

A large number of power plants have been instrumented in the manner shown and are currently working satisfactorily.

Comparison of Electronic and Pneumatic Systems

Figure 9-11 shows a pneumatic drum level control system. You will note the pneumatic system is essentially identical to the electronic system. Both systems use panel control stations that look and operate alike—occasionally the systems are combined in the same panel. In the drum level control process, the electronic system proves advantageous over a pneumatic system for several reasons. First, it is possible to mix the multiple signals easily. Second, the installation cost is greater for pneumatic because of piping. The pneumatic system also has less flexibility. The two systems are compared in detail in Table 9-1.

Table 9-1. Comparison of Pneumatic and Electronic Control Systems

Feature	Pneumatic	Electronic
Transmission distance	Limited to a few hundred feet	Practically unlimited
Standard transmission signal	3–15 psi practically universal	4–20 mAdc practically universal
Compatibility between instruments supplied by different manufacturers	No difficulty	Occasionally nonstandard signals may require special consideration and may not be compatible
Control valve compatibility	Controller output operates control valve operator direct	Pneumatic operators with electropneumatic converters or electro-hydraulic or electric motor operator required
Compatibility with digital computer or data logger	Pneumatic-to-electric converters required for all inputs	Easily arranged with minimum added equipment
Reliability	Superior if energized with clean dry air	Excellent under usual environmental conditions
Reaction to very low (freezing) temperatures	Inferior unless air supply is completely dry	Superior
Reaction to electrical interference (pickup)	No reaction possible	No reaction with dc system if properly installed

Table 9-1. Comparison of Pneumatic and Electronic Control
Systems *(continued)*

Feature	Pneumatic	Electronic
Operation in hazardous locations (explosive atmosphere)	Completely safe	Intrinsically safe equipment available—equipment must be removed for most maintenance
Reaction to sudden failure of energy supply	Superior—capacity of system provides safety margin— backup inexpensive	Inferior—electrical failure may disrupt plant; backup expensive, battery backup available
Ease and cost of installation	Inferior	Superior
System compatibility	Fair—requires considerable auxiliary equipment	Good—conditioning and auxiliary equipment more compatible to systems approach
Instrument costs	Lower if installation costs are not considered	Higher—becomes competitive when total including installation is considered
Ease and cost of maintenance	Fair—procedures more readily mastered by people with minimum of training	Good—depends upon training and capability of personnel
Dynamic response	Slower but adequate for most situations	Excellent—frequently valve becomes limiting factor
Operation in corrosive atmospheres	Superior—air supply becomes a purge for most instruments	Inferior, unless special consideration is given and suitable steps taken
Measurement of all process variables	A few measurements are difficult but can be made available	Excellent
Performance of overall control systems	Excellent, if transmission distances are reasonable	Excellent—no restrictions on transmission distance
Politics (the unmentioned factor that frequently pops up)	Generally regarded as acceptable but not the latest thing	Often regarded as the latest and most modern approach

Questions

9-1. The SPEC 200 control unit (2AC + A4) has a deviation and is in the manual position; when the transfer switch is moved to automatic the controller reacts to the deviation:
   a. With an initial large change and then a gradual recovery
   b. Without a bump
   c. At the rate at which the deviation occurred
   d. At a rate determined by the proportional band

9-2. Performance parameters of most transmitters are rated as a function of:
   a. Full-scale range   **c.** Span
   b. Upper range limit   **d.** Reading

9-3. The greatest advantage of an electrical over a pneumatic control system is:
   a. Price   **c.** No transmission lag
   b. Safety   **d.** Accuracy

9-4. The electronic type of controller may be considered to be:
   a. An analog of a pneumatic type
   b. An entirely different concept of control
   c. A more economical way to obtain automatic control
   d. A more accurate method of providing control

9-5. First stage turbine pressure:
   a. Is linearly related to steam flow
   b. Is a measure of efficiency
   c. Fluctuates only with temperature
   d. Does not relate to load

9-6. Both the pneumatic and electrical systems use a live zero because:
   a. It makes it possible to tell the difference between a dead instrument and one reading zero
   b. Zero is the most important point on the scale
   c. A live zero facilitates calibration
   d. It is important to have the line energized at all times

9-7. When we adjust derivative time in a controller:
   a. We determine an RC time constant in the controller's controlled variable input
   b. We adjust the time it will take for integral to equal derivative
   c. We set the process time constant so that it will always equal 1
   d. What happens specifically depends on the type of controller, pneumatic or electronic

9-8. If the control valve in an electronic control system moves in the wrong direction, it may be easily reversed by:

a. Reversing the ac line connections at the controller
b. Reversing the signal leads
c. Changing the reversing switch position
d. Adjusting the valve

9-9. A plant is being designed for an area that has tremendous temperature extremes. The process is spread out and requires that most instrument lines run both indoors and outdoors to and from the control room. The best choice of instrumentation type would be:
a. Pneumatic
b. Electronic
c. A combination of electronic outside and pneumatic inside
d. A combination of pneumatic outside and electronic inside

9-10. In the electronic boiler drum level system described, measurement input to the secondary control unit is fed to the integral circuit of the primary control unit.
a. To lock the two control actions together
b. Only when in manual control to prevent integral windup
c. To provide bumpless transfer
d. Statement is incorrect

9-11. The 4-20 mA dc output from the electronic differential pressure transmitter is:
a. Linear with differential pressure
b. Linear with flow
c. Unrelated to the frequency of oscillation
d. None of the above

9-12. Assume the d/p Cell signal of feedwater flowrate measurement was acutally 5 percent low. The boiler drum:
a. Would overfill
b. Would go dry
c. Feedback trim would correct for the error
d. Level would be correct if the steam flow was adjusted to compensate for the error

9-13. The pneumatic boiler drum level control system shown in Figure 9-11 uses a square root extractor on the feedwater flow control loop. If the signal from the differential pressure transmitter to the square root extractor is 9 psi (60 kPa), the output from the square root extractor should be:
a. 3 psi   c. 6 psi
b. 11.5 psi   d. 15 psi

9-14. In the SPEC 200 boiler drum level control system there is a summing unit. If two inputs A = 6 volts and B = 3 volts are applied, the output will be:
a. 3 volts
b. 9 volts

    c. 18 volts

    d. A voltage between 0 and 10, representing the scaled value of the inputs

9-15. A boiler delivers 50,000 pounds of steam per hour. The steam pressure is 750 psi (5,171 kPa) and the temperature is 510°F (266°C). The feedwater pump delivers water at 900 psi (6,206 kPa). The size of the linear globe valve controlling feedwater should be: (Hint: A pound of water makes a pound of steam.)

    a. 4 inches   **c.** 1 inch

    b. 2 inches   **d.** ¾ inch

Actuators

Operation of the closed-loop control system depends on the performance of each loop component, including the final control element, whether it be damper, variable speed pump, motor relay, saturable reactor, or valve. Each of these elements requires an actuator that will make the necessary conversion from controller output signal to element input. This controller output may be pneumatic or electric and in some cases hydraulic or mechanical. The first need, then, is a device, an actuator, that will convert this control signal into a force that will position the final control element. From economic and performance standpoints, the most popular final operator is the pneumatic diaphragm actuator. A typical actuator is shown in Figure 10-1.

Valve Actuator

The pneumatic signal is applied to a large flexible diaphragm backed by a rigid diaphragm plate. The force created is opposed by a coil spring with a fixed spring rate. Thus, the stem position is an equilibrium of forces that depends on diaphragm area, pneumatic pressure, and spring characteristic. The spring tension is adjusted to compensate for line

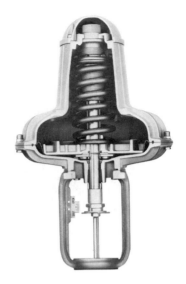

Fig. 10-1. Cutaway of valve actuator showing diaphragm.

pressure on the valve and to produce a full valve stroke with signal changes from bottom to top scale value. The mechanical designs employed vary from one manufacturer to another. Figure 10-1 illustrates the Foxboro Series P actuator.

Pneumatic spring-diaphragm actuators have many applications, the most common of which is the operation of a control valve. They have been adapted to globe, Saunders-patent, butterfly, and ball valves. Spring-diaphragm actuators convert air signal pressure to force and motion and can be adapted to a large number of industrial requirements when precise loading or positioning is required.

On loss of air signal, the spring will cause the actuator to return to the zero pressure position. This feature provides fail-safe action. First, in order to provide maximum safety, the function of the valve is determined. Second, the action—air-to-open or air-to-close—that will allow the spring to put the valve in that position if the energy supply (air pressure) should fail is selected. The actuator shown can be reversed for either air-to-open or air-to-close action by simply removing the cap, turning over the actuator, and replacing the cap.

The motion of the valve stem positioned by a diaphragm actuator is not exactly linear for uniform changes in air pressure (pneumatic signal). The nonlinearity is caused by the diaphragm material, variations in spring rate, the moving pieces, and packing box friction. (Packing box friction varies with fluid pressure and the type and compression of packing.)

One of the effects of friction and nonlinear diaphragm characteristics is hysteresis. This is the difference between position on increasing versus position on decreasing pneumatic signal pressure. With compression on the packing and fluid pressure on the valve, the hysteresis could possibly be as high as 10 percent of the total travel. In control applications with low-gain or high-proportional band, hysteresis can produce an insensitive area or dead zone in the control loop. This would make precise control difficult. The solution to this problem is (1) good design of the actuator; (2) careful choice of low-friction packing, such as Teflon rings; and (3) use of a valve positioner.

Valve Positioner

If the diaphragm actuator does not supply sufficient force to position the valve accurately and overcome any opposition that flowing conditions create, a positioner may be required. If the change in controller air pressure is small, the change in force available to reposition the valve stem might be too small to reposition it accurately. In this situation, a valve positioner (Figure 10-2) will prove helpful. Positioners are used to overcome the factors previously listed, along with other things, such as the effects of highly viscous fluids, gumming, or sedimentation.

The positioner principle is shown in Figure 10-3. It is essentially a fixed-band proportional flapper-nozzle controller. If a difference (error) exists between the actual valve position and the position that the pneumatic signal should produce, the air pressure applied to the diaphragm motor is changed in the right direction (up to full supply pressure or down to zero) to supply added force that overcomes any opposition and precisely positions the valve.

In Figure 10-3, a controller air-signal pressure of between 3 and 15 psi (20 to 100 kPa) is applied to the bellows (A). Since this air signal is applied only to the bellows, rather than to a large-volume diaphragm motor, the response of the positioner is substantially faster than that of the valve actuator alone. The bellows is opposed by the flexure assembly (B). Any unbalance between these forces will cause the flapper to approach or move away from the nozzle. Any motion of the flapper causes a change in pressure on the diaphragm of the control relay (C), which controls the exhaust port and supply port, so that up to full

Fig. 10-2. Type C. Vernier Valvactor positioner provides fast, precise valve position-ing proportional to 20 to 100 kPa or 3 to 15 psi.

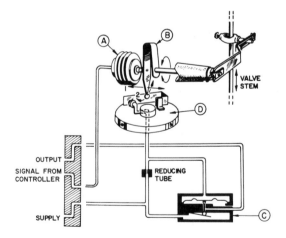

Fig. 10-3. Principle of valve positioner.

supply pressure or down to none can be applied to move the valve to its proper position. The disk (D) may be positioned to cause the valve to stroke on signal pressures other than the conventional 3 to 15 psi (20 to 100 kPa). A Foxboro Valvactor positioner of the type shown in Figure 10-3 can be adjusted to stroke the valve fully on as little as 1-psi (6.9 kPa) change in air pressure. It is also possible to turn the disk and thereby reverse the action of the valve.

Positioners provide precise positioning and also increase the response speed of the valve. There are times, however, when a positioner should not be used: when the process responds faster than the valve (such as in a flow process), and when use of a positioner makes it necessary to set the proportional band of the process controller three-to-five times wider than it would be set if a positioner were not used. In many applications, this is not possible. Two typical applications of valve positioners are temperature control and pH control.

If a diaphragm actuator cannot provide enough force to position the final control element, a piston-and-cylinder actuator may be employed. This type of actuator, when used with high pressure, will deliver force or torque outputs beyond those obtainable with diaphragm actuators. It is suitable for the automatic operation of most dampers, variable speed drives, small sluice gates, blast gates, certain valves, and other similar equipment; and is capable of precise, positive positioning against high-resisting forces. A typical piston-and-cylinder actuator is shown in Figure 10-4. Piston-and-cylinder actuators are generally used in conjunction with a positioner and a supply pressure of approximately 100 psi (690 kPa).

Electrical Signals

When the output of the controller is an electrical signal, additional equipment is required to position the valve or operators. One approach is to convert the electrical signal into a pneumatic signal at the valve location and use a pneumatic actuator. This is a very common solution. The second method is to utilize an all electrical system. Both methods will be discussed below.

Fortunately, the standards for electrical and pneumatic signals are in the same ratio. Both have a live zero and the zero level multiplied by five equals the upper or full-scale value. This simplifies the conversion from one system to the other. The converters may be either rack- or field-mounted. However, when used with valves, the field-mounted

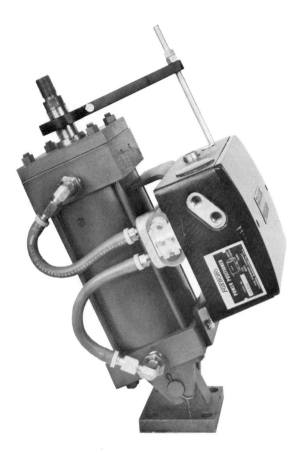

Fig. 10-4. Hannifin cylinder actuator.

type is generally employed. There are two types: one a current-to-pneumatic converter, and the other a positioner. The converter accepts a current signal, usually 4 to 20 mA dc, and converts it into 3 to 15 psi (20 to 100 kPa) or other suitable pneumatic output. The positioner mounted on the valve yoke is a device that converts a current input signal to a proportional stem position. The pneumatic output of the positioner supplies air pressure to a pneumatic actuator. The valve stem is mechanically linked to a shaft on the positioner.

Split input ranges are available when a less than full-scale input current will provide a full-scale pneumatic output or full-scale valve movement; for example, a 4 to 12 mA dc input produces a 3 to 15 psi (20 to 100 kPa) output. Reverse action, in which an increasing current input causes a decreasing output, is also available.

Principle of Operation

These instruments are typical examples of position-balanced systems. The small changes in position generated by a galvanometric motor when a change in the current input signal occurs is balanced by a pneumatically actuated follower system.

The galvanometric motor, shown in Figure 10-5, consists of a wound rectangular coil of fine copper wire surrounding, but not contacting, a cylindrical permanent magnet. The coil is suspended and restricted in movement by flexures that permit it a small amount of rotation about the magnet axis. Input current flowing through the coil interacts with the magnetic field, causing the coil to rotate a maximum of 7 degrees about the axis against the spring rate of the flexure. A shorted turn on the coil provides back-emf damping, which—along with the carefully balanced assembly, the stiff flexures in the feedback mechanism, and the low mass of the moving components—contributes greatly to the vibration and position insensitivity of the instruments. This is an extremely important feature, particularly for field-mounted instruments, when, as the next to the last element in a control loop, they may be subject to vibration from mixers, pumps, or other types of equipment.

As shown in Figures 10-6 and 10-7, a flapper is an integral part of the coil structure. As the coil is rotated by an increasing input signal, the flapper moves to cover a pneumatic nozzle. This increases the nozzle back pressure. The nozzle is connected to a pneumatic relay. As

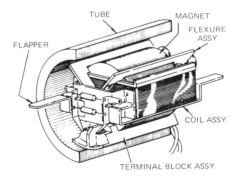

Fig. 10-5. Galvanometric motor.

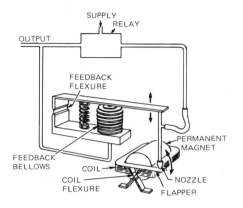

Fig. 10-6. Schematic electric-to-pneumatic converter.

the back pressure from the nozzle increases, the output pressure from the relay is increased. It is at this point that the converters differ from the positioner in the feedback method employed.

As shown in Figure 10-6 for the converters, the output from the relay flows to external devices, such as valves, and also provides pressure to a feedback bellows within the converter. Increasing pressure in this bellows, acting against a spring, moves the nozzle assembly in the same direction as the flapper-coil assembly moves when the input current signal changes. A new equilibrium position is achieved where the *output pressure* is proportional to the input current

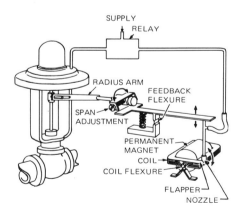

Fig. 10-7. Electric to pneumatic converter with feedback from valve stem.

signal. Thus it can be seen that the converters are current-to-position-to-pneumatic devices with *pneumatic* feedback.

On the other hand, the positioner shown in Figure 10-7 is a current-to-position-to-pneumatic device with *mechanical* feedback. As can be seen, the relay output is connected to the valve actuator. As the valve stem moves to open or close the valve, a radius arm connected to the feedback shaft of the positioner and to the valve stem acts to move the nozzle assembly in the same direction as the flapper-coil assembly. In this device, equilibrium position is achieved where the *valve position* is proportional to the input current signal. Full supply pressure may be applied, if necessary, by the relay to the valve actuator to move the valve to achieve this balance.

Electric Motor Actuators

If compressed air is not available, it may be advantageous to use an electric motor actuator. An example of this mechanism is shown in Figure 10-8. Proportional control of actuator output or stem position is achieved through a feedback slide wire along with an internal servo-amplifier. The 4 to 20 mA dc analog signal from the controller is

Fig. 10-8. Electric motor actuator.

applied directly to the actuator. Depending on the gear ratios used, full stroke may take from 15 to 51 seconds.

Electric motor actuators generally cost more than ten times as much as pneumatic actuators, operate at a much slower speed, and are not, therefore, generally the first choice when one is selecting an actuator.

In addition to electric motor actuators, there are electrical solenoid operators. The solenoid actuator is simple, small, and inexpensive. However, its application is limited to the on/off or two-position action.

Questions

10-1. A valve positioner:
   a. Takes the place of a cascade control system
   b. Provides more precise valve position
   c. Makes a pneumatic controller unnecessary
   d. Provides a remote indication of valve position

10-2. Assume that a control valve regulates steam flow to a process and that high temperature makes the reaction hazardous. The usual pneumatically operated control valve utilizes the following action for fail-safe operation:
   a. Air-to-open
   b. Air-to-close
   c. 3 psi (20 kPa) to fully open
   d. 15 psi (100 kPa) to fully close

10-3. The basic function of the spring in a control valve is to:
   a. Characterize flow
   b. Oppose the diaphragm so as to position the valve according to signal pressure
   c. Close the valve if air failure occurs
   d. Open the valve if air failure occurs

10-4. A diaphragm actuator has a diaphragm area of 50 square inches and is adjusted to stroke a valve when a 3 to 15 psi (20 to 100 kPa) signal is applied. If the signal is 15 psi (100 kPa) the force on the valve stem will be:
   a. 750 pounds
   b. 750 pounds less the opposing spring force
   c. Dependent on hysteresis
   d. None of the above

10-5. A high-pressure flow process requires a valve with tight packing. This would suggest that:
   a. A valve positioner should be employed
   b. The actuator must be sized to provide adequate force

c. Oversized pneumatic signal lines are required

d. The controller supplying the signal to the valve must have a very narrow proportional band.

10-6. An electronic controller creates a 4 to 20 mA dc signal that must actuate a steam valve for temperature control. The best and most economical choice would be to:

a. Use an all electric actuator system

b. Convert to a pneumatic signal at the controller and use a pneumatic actuator.

c. Use pneumatic actuator with an electric-to-pneumatic valve positioner

d. None of the above

10-7. A pressure control process using proportional-plus-integral control has a time constant of 10 seconds. The best choice of actuator would be:

a. An electric motor   c. A piston-and-cylinder

b. A pneumatic diaphragm   d. A solenoid-electrical

10-8. A diaphragm actuator has a diaphragm area of 115 square inches. A valve positioner is attached to the actuator and fed with a 22-psi air supply. If after a 9-psi signal is received from the controller the signal changes to 10 psi and the valve fails to move, what is the force applied to the valve stem?

a. 2,530 pounds   c. 1,035 pounds

b. 1,495 pounds   d. None of the above

10-9. One advantage of an electric-to-pneumatic valve positioner is:

a. It can be used on flow control

b. It produces positive valve position

c. It conserves energy

d. It dampers valve travel

10-10. A single-seated globe valve containing a plug 1½ inches in diameter is used in a line pressurized to 500 psi. What actuator force is required for tight shut-off.

a. 884 pounds

b. 2,000 pounds

c. Depends upon direction of flow through the valve

d. None of the above

11

Control Valves

A control valve regulates the supply of material or energy to a process by adjusting an opening through which the material flows; it is a variable orifice in a line. The formula (Bernoulli's theorem) for flow through an orifice is:

where:

$$Q = CA \sqrt{\Delta P}$$

Q = quantity of flow
C = constant for conditions of flow
A = valve opening area
ΔP = pressure drop across the valve

Flow through the valve is proportional to the area of opening and the square root of the pressure drop across the valve. Both factors vary—the area varies with the percent travel (position) of the valve, and the pressure drop is related to conditions outside the valve and established by the process, such as layout and piping.

Figure 11-1 shows valve percent travel plotted against the resultant flow for the two most popular valve types. This group of curves was plotted at several fixed pressure drops (under actual working condi-

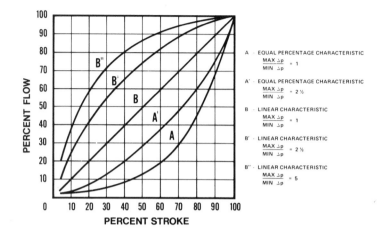

Fig. 11-1. Differential pressure effects.

tions, a constant pressure drop across the valve is seldom encountered). Hence, the valve user or system designer must consider valve *and* process characteristics so that the two may be combined to provide the required overall performance.

The selection of the proper valve characteristic is one of the most important phases of designing a control loop. It is not enough to assume that a stable process with a wide proportional-band-plus-integral controller will cover up any mismatch between the process and the valve characteristic. While it is true that an exact match between valve and process would require a characteristic especially designed for the process, it is possible to choose the better of two standard characteristics (linear or equal percentage) for the process. This is important, since a serious mismatch will cause the system to be unstable and difficult to control effectively.

The linear flow characteristic produces a flow rate that varies in direct proportion to change in valve stem position at a constant pressure drop. That is, for 10 percent of valve stem travel, the valve will pass 10 percent of its rated C_v (see definition in next section); for 20 percent of travel, 20 percent of its rated C_v, and so on.

The equal percentage characteristic is so named because for equal increments of stem travel at constant pressure drop, an equal percentage change in existing flow occurs. This means that the same *percentage* increase in flow will occur when the stem position changes from 40 to 50 percent of travel, as would occur, for example, when stem position changes from 70 to 80 percent of travel.

In actual practice, a valve has two characteristics. One is the design, or inherent characteristic, which the valve exhibits in laboratory conditions where the pressure drop (ΔP) is held constant. The second, and more important, characteristic can be called the resultant, or installed characteristic. This is the relationship between flow and stroke when the valve is subjected to the pressure conditions of the process.

Capacity of a Control Valve

The capacity or flowing rate of a control valve must be matched to the process conditions it is called upon to regulate.

The unit that describes the flow capability of a valve is the C_v rating; all manufacturers of control valves publish C_v ratings of their control valves. The C_v is defined as the number of U.S. gallons of water per minute (at standard conditions of temperature and pressure) that will flow through the wide-open valve when there is a 1-psi (kPa) pressure drop across it.

Valve Sizing

It is essential to size a control valve properly for two reasons:
1. Economics. If the valve is oversized, it does not have enough "resistance," except in a limited part of its stroke, to control the fluid; therefore, it will obviously pass the required flow, but will be more expensive than a properly sized smaller valve. If the valve is too small, it will not pass the necessary flow, even when wide open. The valve will have to be discarded and replaced by a larger, properly sized valve.
2. Control. An undersized valve will never deliver the full flow rate; thus it will sharply narrow the controllable flow range. An oversized valve will be throttling near the closed position, and the full control range of the valve will not be utilized. When the plug throttles very close to the seat, high-fluid velocities occur that can cause erosive damage. The accepted method of valve sizing is the C_v approach. The three basic formulas for C_v calculation are:

1. Liquids: $C_v = Q \sqrt{\dfrac{G}{\Delta P}}$ **(11-1)**

2. Gases: $C_v = \dfrac{Q}{1360} \sqrt{\dfrac{T_f G}{\Delta P (P_2)}}$ (11-2)

3. Steam and Vapors: $C_v = \dfrac{W}{63.3} \sqrt{\dfrac{v}{\Delta P}}$ (11-3)

It should be noted that an important limitation is imposed on the value of ΔP used for vapor and gas sizing. It may never exceed one-half of the absolute inlet pressure (P_1) even though the valve will absorb up to 100 percent of the inlet pressure. If the pressure drop is greater than $\frac{1}{2}P_1$, use $\frac{1}{2}P_1$ for both ΔP and the downstream pressure (P_2). Remember to use this adjusted downstream pressure $(\frac{1}{2}P_1)$ in determining the downstream specific volume (v) under these conditions.

where:
Q or W = flow rate—liquid (gpm), gases (scfh), vapors (lb/h)
G = Specific gravity
T_f = Flowing temperature in degrees rankins (°F + 460)
ΔP = Pressure drop in psi $(P_1 - P_2)$
P_1 = Upstream pressure at valve inlet in psi absolute
P_2 = Downstream pressure at valve discharge in psi absolute
v = Downstream specific volume in cubic feet per pound

(See Appendix)

Determining Pressure Drop Across the Valve

The desired flow rate, specific gravity, temperature, and downstream specific volume are quantities that are easily determined. What is not so readily available is the throttling pressure drop. It is most important to realize that a control valve does not define the pressure drop across it. It will absorb whatever excess pressure is left in the system. This point can be illustrated graphically by the hydraulic gradient method described below.

At maximum flow rate, plot the fluid static pressure versus the physical location system. Then plot the delivered and remaining pressure from left to right, and stop at the control valve. Then plot the required pressure from right to left, stopping at the control valve. The difference between these final points is the pressure drop that the control valve must maintain at maximum flow (Figure 11-2).

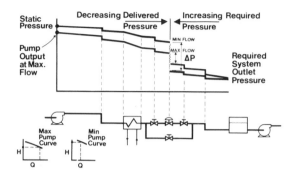

Fig. 11-2. Pressure drop through control valves (maximum and minimum).

A similar analysis can be made for minimum flow. From a pump curve, outlet pressure is higher at low flow. Because of the lower fluid velocity, the pressure loss through the pipe and fittings will be much less than at maximum flow. This greatly increases the delivered pressure at the inlet of the control valve and also decreases the required pressure at the outlet of the control valve. As a result, the pressure drop that must be maintained across the control valve is much greater at the low-flow rate than it is at the high-flow rate (Figure 11-2).

To make sure that valve size is properly calculated, sizing is always done at maximum flow rate and minimum pressure drop.

Cavitation and Flashing

Under normal conditions, fluid passing through a valve will undergo a pressure drop across the valve orifice which, at its lowest pressure, is called the *vena contracta*. Further downstream in the valve, the fluid pressure will partially recover and line pressure is again increased. Figure 11-3 illustrates this pressure drop and recovery when the fluid is a liquid and remains a liquid as it passes through a valve. In this figure P_1 is the fluid pressure at the valve inlet and P_2 is the exit pressure.

When a liquid enters a valve and the static pressure at the vena contracta drops to less than the fluid vapor pressure, and the valve outlet pressure is also less than the fluid vapor pressure, the condition called *flashing* exists. In other words, fluid enters the valve as a liquid and exits as a vapor. This condition is shown in Figure 11-4.

Figure 11-5 illustrates a third condition called *cavitation,* which occurs in a valve when the pressure drop across the orifice first results

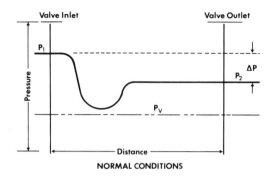

Fig. 11-3. Normal conditions.

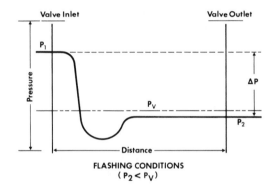

Fig. 11-4. Flashing conditions $(P_2 < P_V)$.

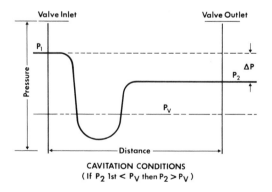

Fig. 11-5. Cavitation conditions (if P_2 1st $< P_V$ then $P_2 > P_V$).

in the pressure being lowered to below the liquid's vapor pressure and then recovering to above the vapor pressure. This pressure recovery causes an implosion, or collapse of the vapor bubbles formed at the vena contracta.

Flashing and cavitation must be considered to ensure proper valve sizing and to allow the selection of a valve that will resist their effects. If the presence of cavitation and flashing is neglected when valves are sized, undersized valves will be selected and rapid valve deterioration can take place. In addition to deterioration, cavitation is a source of loud, unwanted noise.

Valve Rangeability

Rangeability is "the ratio of maximum controllable flow to minimum controllable flow." Thus a valve with a characteristic curve as in Figure 11-6 will have a rangeability of $\frac{100}{2}$, or 50:1.

Valve rangeability must be based on controllable flow. It is impractical, from a manufacturing standpoint, to characterize flow when the valve is barely open. Similarly, at shutoff it is impossible to characterize any leakage flow that may occur. If minimum flow, not minimum controllable flow, were used, a valve with bubble-tight shutoff would have infinite rangeability. Such reasoning is logical, but unrealistic.

Since C_v is an expression of flow capacity, the definition of range-

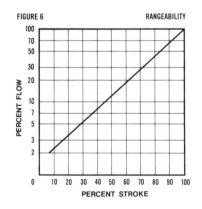

Fig. 11-6. Rangeability of a control valve.

ability can be restated to be "the ratio of rated C_r to minimum controllable C_r."

To determine the required rangeability needed in a control valve for a particular set of flow conditions, we merely take the ratio of maximum required C_r to minimum required C_r. If this ratio is less than the available rangeability in the valve, we will be able to control properly.

A valve has a rangeability as specified by the valve manufacturer. After the valve has been installed in the process, it has an installed rangeability. These two rangeabilities may be quite different. The relationship between these two values may be expressed as follows:

R_I = Installed rangeability

R_S = Shelf rangeability

$$R_I = R_S \sqrt{\frac{\Delta P \ MIN}{\Delta P \ MAX}}$$

$\Delta P \ MIN$ = Minimum pressure drop at full load

$\Delta P \ MAX$ = Maximum pressure drop at minimum load

If the installed rangeability is significantly greater than the manufacturer's rated rangeability, a problem exists. The solution may be to simply select another type of valve that meets the requirement or it may be necessary to use two valves and split range them.

While C_r is an expression for liquid flow at a constant pressure differential, Figure 11-1, flow versus stroke, shows how the design or inherent characteristic is changed by variations in pressure differential. This occurs as the process changes from high flows, where pressure differential is distributed throughout the system, to low flows, where most of the pressure differential is concentrated at the control valve. This variation in pressure differential concentration is one of the most important factors in choosing the proper characteristics of a given process.

Among the normal applications encountered in industrial instrumentation are flow pressure, temperature level, and pH control. The suggested valve characteristics to be used for these applications follow.

For stability over the entire control range, most control loops require that flow be manipulated in uniform proportion to controller output. Because frictional losses through pipes and fittings increase and pump output pressures decrease with increasing flow, the available pressure differential at the control valve usually diminishes as flow increases. To maintain the desired uniform proportionality, an equal percentage characteristic is used in these varying pressure differential applications. When a nearly constant pressure differential exists at the control valve, a linear characteristic is preferable on all but temperature control.

A misapplied equal percentage valve characteristic results in increasing valve sensitivity at high-flow rates and can cause instability unless the controller proportional band is adjusted at the high-flow rate. The control loop will then tend to be overdamped at the low-flow rates, with corresponding sluggish response.

A misapplied linear valve exhibits the opposite effect. A controller properly adjusted at the high-flow rate would have too narrow a proportional band setting for stability at low flows. The proportional band would have to be adjusted at the low-flow rates, and the control loop would then be overdamped and sluggish at higher flow rates.

The recommendations given are only as good as the hydraulic analysis of the pressure conditions in the system. Maximum differential pressure is usually fairly exact, since it is close to the shutoff pressure. Minimum valve pressure drop at full flow is more often a guess than the result of a hydraulic analysis. This guess is usually much too low. Using this imaginary number as a guide to the selection of a valve characteristic will normally result in the choice of an equal percentage valve. The end result is the selection of an oversized valve that even at maximum flow will be operating in the low portions of lift. Consequently, rangeability and response at low-flow conditions are lost.

Selection Factors

In selecting a control valve, many aspects must be considered:
1. Environmental factors, such as corrosion, abrasion, temperature, and pressure. The valve must be able to cope with these factors.
2. Percentage travel versus flow characteristic plus loop-and-process characteristics. This is essential for good control (Table 11-1).
3. Size. A valve of the incorrect size costs more initially and in the long run because the tendency is to oversize the valve. In general, the smallest valve that will pass the necessary maximum flow provides optimum control and maximum economy. The C_v ratings for a given valve size vary. A butterfly would rate highest, while

Table 11-1. Characteristic Selection by Application

Applications (Factor Controlled)	% of System Drop at Max. Flow	Characteristics
Flow—linear w/dp	<20%	equal percentage
Flow—linear w/flow	<40%	equal percentage
Flow—linear w/dp	>20%	linear
Flow—linear w/flow	>40%	linear
Pressure	≈100%	linear
Pressure	<50%	equal percentage
Liquid level	<40%	equal percentage
Liquid level	>40%	linear
pH	<50%	equal percentage
pH	>50%	linear
Temperature	>50%	equal percentage

FEEDFORWARD CONTROL SYSTEMS
Valve characteristics will be determined by system characteristics and resultant function used.
CASCADED CONTROL LOOPS
Take same characteristic required for a normal manual-set feedback loop.

the ball valve would rank next, and the Saunders and globe valves would have the minimum C_v rating for a given size.

4. Rangeability. Both maximum and minimum valves of C_v are calculated. If the ratio of maximum C_v to minimum C_v falls within the rangeability for the valve selected, the valve should function properly. If not, valve sequencing may be necessary.

5. Still another factor to be considered is the valve's ability to cope with high pressure. Assume a pump delivers 500 psi and at times most of this pressure appears across the valve. In this type of application, a globe valve would be the best choice, while a ball valve would be adequate. However, the butterfly, pinch, and Saunders valves would be poor choices.

Sequencing-Control Valves

When the rangeability requirements rule out a valve that otherwise would be suitable for the application, it may be feasible to use two control valves arranged for sequential action. The valves are installed in parallel so that their individual flow rates are additive. Careful selection of each valve size is required to achieve the necessary rangeability. Total valve rangeability will be the ratio of the minimum controllable C_v of the smaller valve to the combined maximum C_v's of both valves. The maximum C_v of the smaller valve must be at least equal to the minimum controllable C_v of the larger valve.

The large valve should not be so much larger than the small valve that its leakage rate affects the total flow rate more than the small valve does. Thus, with a balanced or double-seated globe valve having a leakage capacity of 0.50 percent of maximum C_v, the best possible rangeability using sequenced parallel valves would be 200:1, independent of the minimum controllable C_v of the smaller valve. A single-seated globe valve or a tight shutoff ball, butterfly, or Saunders diaphragm would remove this consideration.

However, it is usually advisable to keep the two valves as close in size as possible while still obtaining the required rangeability. For example, two 3-inch balanced equal percentage globe valves with diaphragm actuators and split-range positioners would be able to pass as much flow as one 4-inch and one 1½-inch balanced equal percentage globe valves. Total rangeability of the two 3-inch valves would be 100:1 as opposed to 200:1 for the 4-inch and 1½-inch valves, limited by the 4-inch leakage capacity. A slightly lower cost would result with the two 3-inch valves, along with a much simpler and less expensive manifold arrangement.

Saunders diaphragm control valves often require sequenced parallel valves to satisfy rangeability requirements. Ordinarily, it is not possible to use two valves of equal size, but this does not cause a control problem, since they are tight shutoff valves. As with any type of valve, of course, the greatest rangeability theoretically possible with two valves is the square of the rangeability of one valve. Thus, the absolute maximum rangeability of two Saunders diaphragm control valves is 15^2 or 225:1. In practice, the maximum rangeability would be in the order of 125:1 due to the problem of matching the valve sizes.

Example: Given a required maximum C_v of 180 and a required minimum C_v of 3 in an application where Saunders diaphragm control valves are desired, determine the proper size selections required. Glass-lined Saunders diaphragm valves are to be used.

For proper control, the combined maximum C_v values of the two valves should be approximately 250. A 2½-inch and 2-inch size combined will deliver a C_v of 253. The minimum C_v of the 2-inch is approximately 6, which would not satisfy the application, and rules out the 2½-inch and 2-inch combination. A 3-inch and 1-inch size combined will deliver a C_v of 257. The minimum C_v of the 1-inch is approximately 1.5, which is satisfactory. The maximum C_v of the 1-inch is 22, and exceeds the minimum C_v of the 3-inch which is 16. This combination in parallel, operating sequentially from a 3-9/9-15 psi control signal will cover a C_v range of 1.5 to 257, which satisfies the application.

The actual sequencing is accomplished by the actuator. If a pneumatic actuator is used, sequencing may be accomplished by spring adjustment. But this is much easier to accomplish with a valve positioner (Figure 11-3).

Viscosity Corrections

The techniques described here apply to nonviscous fluids and will suffice in all but a few situations. When the infrequent case of high-viscosity fluid is encountered, it should be handled as follows:

1. Calculate the Reynolds number (R)

$$R = \frac{(3,160)\,(\text{GPM})}{(D)\,(\text{Centistokes})}$$

D = inside diameter of pipe, inches-table

2. If R is greater than 2,000, then the following corrections to the tabulated ΔP's should be used.

Centistoke	Correction Factor
2	1.14
5	1.40
10	1.70
30	2.06
50	2.68
70	3.06
100	3.50

The conclusion is simply that the valve size will increase with viscosity, but the relationship is not linear. If very high-viscosity fluids are encountered, empirical test data are available from valve manufacturers to determine the C_v correction factor.

Now let us solve some sample problems to illustrate the techniques of valve selection and sizing.

VALVE SIZING: EXAMPLE 1 Let us assume that we have a control valve regulating liquid flow from a tank (Figure 11-7). The water level is to be controlled in this tank at a level of 25 feet by regulating the outflow. The measured inflow varies from 0 to 120 gallons per minute.

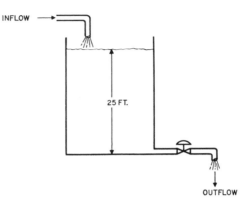

Fig. 11-7. Sizing valve to single tank.

The maximum outflow, then, must be equal to the maximum inflow, or 120 gpm. Since 1 foot of water develops a pressure of 0.433 psi, with the tank level at 25 feet; the pressure across the valve must be:

$$P = 25 \times 0.433 = 10.8 \text{ psi} \tag{11-1}$$

$$C_v = Q \sqrt{\frac{G}{\Delta P}}$$

when Q = flow rate, U.S. gallons per minute
        ΔP = differential pressure across valve in psi
        G = specific gravity (water = 1.0)
Hence $C_v = 120 \sqrt{1/10.8} = 36.5$

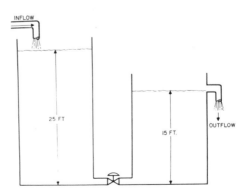

Fig. 11-8. Dual tank control system.

In Figure 11-11 (or Table 11-3), the smallest valve capable of providing a C_v of 36.5 is a 2-inch equal percentage valve. No valve less than a 2-inch provides the necessary C_v value.

VALVE SIZING: EXAMPLE 2 Instead of the valve discharging to atmosphere, let us assume that it discharges into a second tank with a head of 15 feet (Figure 11-8). If the maximum flow through is 120 gpm,

Maximum inflow = 120 gpm

Maximum outflow = 120 gpm

ΔP = (25 feet − 15 feet)(0.433) = 4.33 psi

$$C_v = Q \sqrt{\frac{G}{\Delta P}} \tag{11-1}$$

$$= 120 \sqrt{1/4.33} = 57.7$$

In Figure 11-11, the smallest valve that can satisfy the required C_v is 2½-inch. Note that the capacity of a valve depends on both the size and the pressure drop applied by the system.

VALVE SIZING: EXAMPLE 3 Now let us calculate the size of a control valve required to regulate the flow of natural gas through an 8-inch pipeline when the flowing conditions of the gas are as follows:

P_1	Upstream operating pressure	300 psi absolute
	Maximum upstream pressure	325 psi absolute
ΔP	Operating pressure drop across valve	50 psi
P_2	Downstream pressure $(P_1-\Delta P)$	250 psi
	Operating temperature of fluid (gas)	50°F
	Maximum temperature of fluid (gas)	90°F
Q	Maximum rate of flow	3,000,000 std cubic feet per hour
	Normal rate of flow	2,300,000 std cubic feet per hour
	Minimum rate of flow	2,000,000 std cubic feet per hour
G	Specific gravity at operating temperature (60°F) = 0.6	

$$C_v = \frac{Q}{1360} \sqrt{\frac{T_f G}{\Delta P(P_2)}} \tag{11-2}$$

$$T_f = 60 + 460 = 520$$

$$C_v = \frac{3,000,000}{1,360} \sqrt{\frac{520(0.6)}{50(250)}} = 348.5$$

Note that the upstream-operating pressure is used rather than maximum pressure. This produces a more conservative result. Determined from Figure 11-11, the smallest valve that satisfies the required C_v is a 6-inch wide-range equal percentage with a C_v of 449.

VALVE SIZING: EXAMPLE 4 Now let us determine the size of the control valve required for the heat exchanger described in connection with Figure 8-20. The following data are essential for the sizing calculation:

Maximum steam pressure	50 psi (65 psi absolute)
Normal steam pressure (P)	40 psi (55 psi absolute)
Maximum water pressure	65 psi
Normal water pressure	40 psi
Inlet water temperature	52°F
Outlet water temperature maximum	200°F
Maximum water flow rate	50 gpm
Normal water flow rate	20 gpm
Pipe size, steam header	4 inches

Btu per minute added to water $= (\Delta F)(\text{gpm})(\text{pounds per gallon})$

$$= (200 - 52)(50 \times 8.3)$$

$$= 148 \times 50 \times 8.3 = 61,600 \text{ Btu}$$
per minute

Btu per hour $= 61,600 \times 60 = 3,700,000$

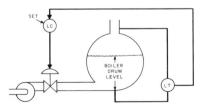

Fig. 11-9. Boiler drum level control.

Steam yields approximately 1,000 Btu per pound when condensed. Hence, to rounded out accuracy,

$$\frac{3,700,000}{1,000} = 3,700 \text{ pounds per hour steam}$$

Since the indicated downstream pressure (P_2) is that of low-pressure condensate, the actual pressure drop is more than half the upstream pressure. When the pressure drop reaches approximately half upstream pressure in a valve handling gas or vapor, the fluid velocity reaches the speed of sound and can increase no further. Further increase in pressure drop, then, will not increase the flow through the valve. This condition is known as "critical velocity," and $P/2$ is used instead of ΔP when downstream pressure is less than half that of upstream. Also, v is taken at the downstream pressure and is obtained from a steam table as the specific volume listed for 55/2 psi absolute, or 15.023. Hence,

$$C_v = \frac{W}{63.3} \sqrt{\frac{v}{\Delta P}} \tag{11-3}$$

$$= \frac{3700}{63.3} \sqrt{\frac{15.023}{55/2}}$$

$$= 47.76$$

Because more than half the total pressure drop (4-inch header) would appear across the valve, a 2-inch equal percentage globe valve would be a logical choice.

VALVE SIZING: EXAMPLE 5 Boiler feedwater preheated to 350°F enters the drum at a maximum feed rate of 200,000 pounds per hour. The feedwater pump produces a pressure of 800 psi and the pressure drop across the valve at full flow is 200 psi (see Figure 11-9). Determine the required valve type and size.

First step: From steam table (Appendix) find specific gravity. Interpolation

V_f at 358.43°F = 0.01809 ft^3/lbm

$\dfrac{V_f \text{ at } 341.27°F}{\text{Diff.: } \quad 17.16} = \dfrac{0.01789}{0.00020} \text{ ft}^3/\text{lbm}$

Increment °F = 358.43 − 350.00 = 8.43°F

$$\text{Increment } V_f = \frac{8.43}{17.16} \times 0.00020 = 0.00010 \text{ ft}^3/\text{lbm}$$

$$V_f \text{ at } 350°F = 0.01809 - 0.00010 = 0.01799 \text{ ft}^3/\text{lbm}$$

$$\text{Density} = 1/0.01799 = 55.59 \text{ lb/ft}^3$$

A cubic foot of water at standard conditions weighs 62.37 pounds

$$\therefore SP.GR = \frac{55.59}{62.37} = 0.89$$

The feed rate of 200,000 pounds per hour of steam is condensed to water. To convert to gallons per minute of water through the valve:

$$\frac{200,000 \text{ lb/h}}{55.59 \text{ ft}^3/\text{lb}} \times \frac{7.481 \text{ g/ft}^3}{60 \text{ min/h}} = 448.58 \text{ gpm}$$

Table 11-2

Valve	Characteristic and Rangeability	Use on Slurries, Dirty Solid-Bearing Fluids	Relative Cost	Maintenance Rating	Tight Shutoff	Rating as a Control Valve
Globe body with characterized plug or cage sizes from needle up to 24 inches (see Fig. 11-10a)	= % or linear Max. 50:1 Approx. 35:1 for needle	Very poor Can be constructed of corrosion resistant materials	High Very high in larger sizes	Good Easy to work on	Yes, if single soft seat is used	Excellent Any desired characteristic can be designed into this type
Ball valve Available up to 42 inches (see Fig. 11-10b)	Approx. = % with 50:1 ball can be characterized to modify	Reasonably good Can be constructed of corrosion resistant materials	Medium	Good Disassembly may be difficult	Yes, if lined	Excellent if characteristic is suitable
Butterfly valve Available up to 150 inches (see Fig. 11-10c)	Approx. = % with 30:1	Poor A variety of materials of const. available	Lowest cost for large sizes	Good	Yes, if lined type is used; No, if unlined	Good if characteristic is suitable
Saunders valve Available up to 20 inches (see Fig. 11-10d)	Approx. linear 3:1 – conventional 15:1 – dual range	Very good Available with liner to resist corrosion	Medium	Fair	Yes	Conventional is poor Dual range is fair. Generally used only when dirty flow dictates
Pinch valve Available up to 24 inches (see Fig. 11-10e)	Approx. linear 3:1—to 15:1 depending on type	Excellent Several materials available to resist corrosion	Low	Fair	Some leakage	Poor to fair Use only when ability to handle dirty flows dictates

Note: Globe, ball, and butterfly valves are available for high-temperature service (above 1,000°F). Saunders and pinch valves are limited by materials of construction to approx. 350°F with modest line pressures (less than 50 psi).

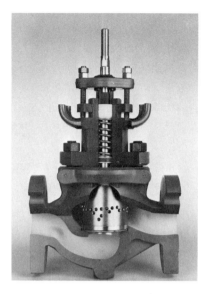

Fig. 11-10a

Fig. 11-10b

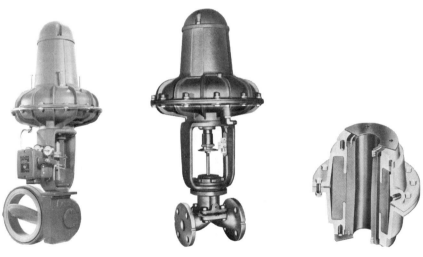

Fig. 11-10c   **Fig. 11-10d**   **Fig. 11-10e**

Fig. 11-10. See Table 11-2. *a*, Globe valve; *b*, Ball valve; *c*, Butterfly valve; *d*, Saunders valve; *e*, Pinch valve (photo courtesy of Clarkson Co.).

Then for water:

$$C_r = Q \sqrt{\frac{G}{\Delta P}} = 448.58 \sqrt{\frac{0.89}{200}} = 29.92$$

Referring to Tables 11-1 and 11-2, a 2-inch globe valve will satisfy the pressure and temperature conditions and is available with linear characteristics that satisfy all requirements.

VALVE SIZING: EXAMPLE 6 A 3-inch line carries 50,000 standard cubic feet per hour of air at an operating pressure of 75 psi. The downstream pressure is 50 psi and the specific gravity of air is by definition 1. Assume a valve is required for pressure control. Determine its size and type if the flowing temperature of the air is 500°F.

$P_1 = 75$ psi + 15 (atmos.) = 90 psi absolute

$P_2 = 50$ psi + 15 (atmos.) = 65 psi absolute

$\Delta P = 90 - 65 = 25$ psi

Since 65 psi is more than half the upstream pressure, $\Delta P = 25$ psi.

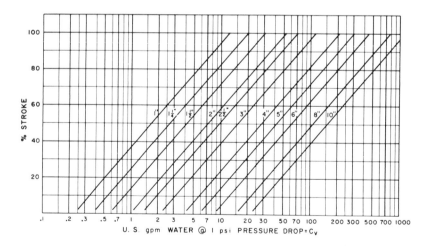

Fig. 11-11. Characteristic curves and C_r for typical globe, equal percentage valves.

$$C_v = \frac{Q}{1,360} \sqrt{\frac{Tf\,G}{P(P_2)}}$$

$$= \frac{50,000}{1,360} \sqrt{\frac{960 \times 1}{25(65)}}$$

$$= 28.25$$

This process will be satisfied by a 2-inch (Table 11-1 and 11-2) equal percentage globe valve.

VALVE SIZING: EXAMPLE 7 A ¾-inch chlorine line at 80°F carries 200 pounds per hour. The upstream pressure is 55 psi and it discharges into a tank at atmospheric pressure. Find the C_v of a control valve to be installed adjacent to the tank and recommend a valve type.

$P_1 = 55$ psi + 15 psi absolute = 70 psi absolute

$P_2 = 15$ psi absolute (0 psi)

Because 55 is greater than $\frac{70}{2}$ or one-half P_1, critical drop applies and $\Delta P = \frac{70}{2}$ or 35 psi.

Next, refer to a handbook that has a chlorine table, and, assuming adiabatic expansion is approximated, downstream temperature would be 60°F at 35 psi absolute and specific volume would be 2.18 cubic feet per pound.

$$C_v = \frac{W}{63.3} \sqrt{\frac{v}{\Delta P}}$$

$$= \frac{200}{63.30} \sqrt{\frac{2.18}{35.00}}$$

$$= 0.789$$

Because approximately 100 percent of the pressure drop occurs across valve A, linear characteristic indicated by the small C_v combined with a ¾-inch line would suggest a ¾-inch globe valve body with reduced ($C_v = 0.789$) trim. (Trim refers to the valve's internal working mechanism.) Materials would be carefully selected to stand up under the corrosive effects of chlorine; if the chlorine is wet, a stainless steel body and hastalloy C trim might be required.

Thus far only the maximum C_v value has been considered and often this is the only valve considered. However, it is better practice, when data are available, to calculate both maximum and minimum C_v values. The ratio of maximum to minimum then gives a required rangeability. If, for example, the rangeability worked out to be 125, it becomes evident that this cannot be accomplished by any single valve and that the sequencing technique would be required. On the other hand, the rangeability tells us if the selected valve is capable of acceptable control. This may be demonstrated with the solution of the following problems.

VALVE SIZING: EXAMPLE 8 A steam valve regulates the flow of saturated steam in a 10-inch header to a process. The maximum flow rate is 30,000 pounds per hour and the minimum is 7,200 pounds per hour. At maximum flow, the upstream pressure is 20 psi and the downstream pressure is 15 psi. At minimum flow the upstream is 25 psi and the downstream is 15 psi.

Maximum flow

$$P_i = 20 \text{ psi} + 15 \text{ (atmos.)} = 35 \text{ psi absolute}$$

$$P_2 = 15 \text{ psi} + 15 \text{ (atmos.)} = 30 \text{ psi absolute}$$

$$\Delta P = 35 - 30 = 5 \text{ psi}$$

$$C_v = \frac{W}{63.3} \sqrt{\frac{v}{\Delta P}}$$

Max.

$$= \frac{30,000}{63.30} \sqrt{\frac{13.744}{5}}$$

$$= 785.76$$

Minimum flow

$$P_1 = 25 \text{ psi} + 15 = 40 \text{ psi absolute}$$

$$P_2 = 15 \text{ psi} + 15 = 30 \text{ psi absolute}$$

$$\Delta P = 40 - 30 = 10 \text{ psi}$$

$$C_v = \frac{W}{63.3} \sqrt{\frac{v}{\Delta P}}$$

Min.

$$= \frac{7200}{63.30} \sqrt{\frac{13.744}{10}}$$

$$= 133.35$$

$$\frac{C_v \text{ max.} = 785.76}{C_v \text{ min.} = 133.35} = 5.89 \text{ rangeability}$$

This can easily be handled by a globe, ball, or butterfly valve. The most economical choice would be a 6-inch unlined butterfly valve. (See Table 11-3).

VALVE SIZING: EXAMPLE 9 Level is being controlled in a tank. The flow range is 100 to 1,000 gpm. The liquid is mineral oil and has a specific gravity of 0.88. Line pressure is 100 to 150 psi and the throttling pressure drop varies from 50 to 110 psi. The temperature may vary from 70 to 140°F.

Size the control valve and select a characteristic type that will satisfy the process.

$$C_v = Q \sqrt{\frac{G}{\Delta P}}$$

$$C_v \text{ max.} = 1,000 \sqrt{\frac{0.88}{50}} = 132.67$$

$$C_v \text{ min.} = 100 \sqrt{\frac{0.88}{100}} = 8.94$$

$$\text{Required rangeability} = \frac{C_v \text{ max.}}{C_v \text{ min.}} = \frac{132.67}{8.94} = 14.8$$

An equal percentage valve is indicated and a maximum C_v of 132.67 indicates a 4-inch globe or a 3-inch ball valve could be used.

Here, economy might be the deciding factor since either type could do the job. The 3-inch ball valve likely would be less expensive than a 4-inch globe valve, with equal percentage characteristics.

VALVE SIZING: EXAMPLE 10 A valve is used for flow control. The line size is 10 inches and the liquid is water. Maximum flow rate is 3,000 gpm and minimum 250 gpm. The line pressure varies between 40 and 50 psi. The downstream piping loss is 2 psi at 3,000 gpm and the downstream pressure is 25 psi.

Select a valve size and type that will satisfy this process.

$$C_v = Q \sqrt{\frac{G}{\Delta P}}$$

Maximum flow   *Minimum flow*

$P_1 = 40$ psi $P_1 = 50$ psi

$P_2 = 25$ psi (pipe loss) $= 23$ $P_2 = 25$ psi

$\Delta P = 17$ psi $\Delta P = 25$ psi

C_v max. $= 3,000 \sqrt{\frac{1}{17}}$ $C_v = 250 \sqrt{\frac{1}{25}}$

$\qquad\quad = 727.6$ $\qquad = 50$

Rangeability $= \dfrac{C_v \text{ max.}}{C_v \text{ min.}} = \dfrac{727.6}{50} = 14.55$

Table 11-3. Approximate C_v Values of Typical Valves

V1 SERIES CONTROL VALVE		
Valve Size	Trim Size	Rated C_v
1/2	1/8	0.25
3/4	3/16	0.50
1	1/4	1.0
	3/8	2.0
	1/2	5.0
3/4	3/4	10.
1		
1	1	17.
1 1/2	40%	15.
	FULL	34.
2	40%	24.
	FULL	60.
3	40%	48.
	FULL	120.
4	40%	80.
	FULL	200.
6	40%	160.
	FULL	400. *

* 360 for stem guided = % contoured inner valve

Butterfly Valves*		
Size	60° Open	90° Open
2	70	100
2 1/2	100	150
3	160	250
4	320	500
6	750	1500
8	1200	2500
10	1800	4000
12	2850	6000
14	3600	7500
16	4800	10000
18	5900	12500
20	7300	15000

* Representative C_v Values

BALL	TYPE	VALVE
Valve Size	C_v when line is valve size	C_v when line is twice valve
1	35	24
1 1/2	94	57
2	115	86
3	350	222
4	775	425
6	1000	755
8	2000	1430
10	3150	2240
12	5200	3450

Tables 11-1 and 11-2 indicate a linear valve is a good choice, but a valve with the C_v required would be rather large and very expensive. Therefore, the practical choice would be either a 6-inch butterfly or a 6-inch ball valve. The lined ball valve would provide shutoff, which could be advantageous. (See Table 11-3).

References

Valve engineering handbook. PUB 237B. Foxboro, MA., The Foxboro Company.

Liptak, B. G., *Instrument Engineers' Handbook,* Vol. II. Radnor, PA: Chilton Book Company, 1970.

Questions

Valve Specifications

1. Additive valve—Steel body, stainless steel trim, tight shutoff
   Flow range: 2 to 20 gpm
   Specific gravity: 1.2
   Line pressure: 45 to 60 psi (downstream piping loss 2 psi at 20 gpm)
   Temperature: 80 to 120°F
2. Level control valve—316 stainless steel body and trim
   Flow range: 200 to 2,000 gpm
   Specific gravity: 0.9

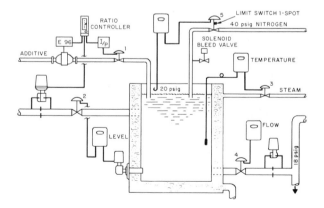

Fig. 11-12. Application data for Problems 11-1 through 11-5.

Line pressure: 75 to 125 psi (throttling pressure drop 40 to 100 psi)
Temperature: 70 to 140°F
3. Temperature control valve—iron or bronze body, 316 stainless steel trim
Flow range: 1,200 to 5,000 lb/h saturated steam
Line pressure: 65 to 90 psi (required valve outlet pressure 10 to 30 psi)
Temperature: 312-331°F
4. Flow control valve—Stainless steel body and trim
Flow range: 200 to 2,500 gpm
Specific gravity: 0.95
Line pressure: 22 to 30 psi (downstream piping loss 1 psi at 2,500 gpm)
5. Pressure control valve—Iron or steel body, 316 stainless steel trim, tight shutoff
Flow range: 0 to 5,000 scfh
Specific gravity: 1.0
Line pressure: 40 psi (required valve outlet pressure 22 psi at 5,000 scfh)
Temperature: 70 to 120°F

Directions:

11-1 through 11-5. Choose any or all of the five valves listed above and completely specify the proper type, size, etc., for each one.

11-6. Critical flow occurs in a gas or vapor valve when:
  a. Upstream pressure equals downstream pressure
  b. Downstream pressure equals one-half upstream pressure, both absolute
  c. Upstream pressure exceeds 15 psi absolute
  d. Downstream pressure is less than 3 psi absolute

11-7. C_v may be defined as:
  a. Critical velocity through a control valve
  b. The water flow in U.S. gpm discharged through a wide-open valve with a 1-psi drop across it
  c. Correction factor for the viscosity of the fluid
  d. Critical volume factor for gas flow

11-8. The liquid flow to be controlled contains suspended solids. A good choice of valve is:
  a. Globe   **c.** Saunders
  b. Wide-range V-port   **d.** Split body

11-9. If a control valve permits only minimum leakage when it is closed, the choice would be a:
  a. Single-seat   **c.** Butterfly
  b. Double-seat   **d.** Split body

Controller Adjustments

All controllers must be adjusted or tuned to accommodate the processes they control. There are several approaches to this task. Some of the techniques are mathematical. However, the most widely used approach is a combination of experience and trial-and-error.

The purpose of tuning a controller to a process is to match the gain and time functions of the controller with the rest of the elements in the control loop (process, transmitter, valve, and so on). The object of adjusting a controller's proportional band is to give the total loop a dynamic gain of less than 1 in order to force any induced upset to dampen out. If the dynamic gain of the loop equals 1, and in/phase feedback exists around the loop, cycling or oscillation will occur at its fundamental or natural frequency after an upset.

Proportional-Only Controller

In order to determine the dynamic gain of the loop exclusive of the controller, the following procedure may be followed with a proportional-only controller:

1. Adjust the proportional band to a maximum and the gain to a minimum.
2. Place the controller in automatic.

3. Make a step change in the controller set point.
4. Observe the resulting measurement cycle.
5. Reduce the proportional band, and repeat Steps 3 and 4.
6. Keep repeating Steps 3, 4, and 5 until constant amplitude cycle is observed. At this condition, the loop has a gain of 1.

The total loop gain is the product of the gain of the control loop, exclusive of the controller, multiplied by the gain of the controller. Assume that Step 6 gave a proportional band of 50%, i.e., a controller gain of 2. Since total loop gain is 1, the gain of the loop exclusive of the controller must be 1 divided by 2, or 0.5. To produce a damped response the total gain must be less than 1, and since the controller is the only part of the loop that can be adjusted, its gain must be less than 2 (i.e., a proportional band greater than 50%). The response that is usually desirable approximates Curve A for the proportional-only controller in Figure 12-1. This response, called quarter-amplitude dampening, causes each cycle to have an amplitude of one-quarter that of the previous cycle. A total loop gain of 0.5 will result in quarter-amplitude dampening. This, in the above example, the controller should have a gain of 1, or a proportional band setting of 100 percent. If the control loop has a much lower gain, the loop is said to be overdamped. Although the loop will be stable, excessive offset will result (see Curve C). A loop gain of 1 will yield Curve A. A loop gain greater than 1 will yield continuous cycling of increasing amplitude. This procedure should give satisfactory results in a proportional-only controller.

Proportional-Plus-Integral Controller

If a controller has an integral mode, then some measurement of time response of the control loop is necessary. By properly matching the

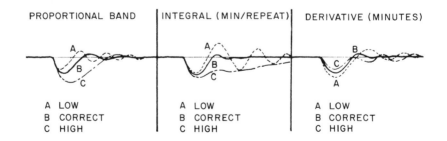

PROPORTIONAL BAND   INTEGRAL (MIN/REPEAT)   DERIVATIVE (MINUTES)

A LOW   A LOW   A LOW
B CORRECT   B CORRECT   B CORRECT
C HIGH   C HIGH   C HIGH

Fig. 12-1. Response to load depends on control mode and adjustment.

response of the loop (exclusive of the controller) to the response of the controller, corrective action will be applied as rapidly as the process will allow. In other words, with a properly tuned integral mode, the controller will allow an energy input to the process at the fastest rate at which the process can absorb it.

If a controller has too short an integral time, the process will not be able to keep up with the changes in energy input, and cycling will occur. On the other hand, if the controller has too long an integral time, sluggish response will occur. In this condition, the measured variable will not return to the set point after an upset as fast as possible, with optimum settings.

If a proportional-plus-integral controller is placed in manual and a step change is introduced into the output, a starting point for determining the integral time may be found. If it takes a short time for the full effect of the output change to be felt in the measurement signal, the controller's integral time will be correspondingly short.

An alternate procedure for tuning a proportional-plus-integral (two-mode) controller is:

1. Adjust the integral to a time setting (in minutes) that is equal to the natural period divided by 2.4. The natural period is the time in minutes between successive peaks.
2. Adjust the proportional band setting to an optimum value as described in the procedure for a proportional-only controller.

The effect of integral on process-response curves for a two-mode controller is also shown in Figure 12-1. Curve B represents the desired quarter-amplitude dampening; Curve C represents too long an integral time; and Curve A too short an integral time.

Adding Derivative Action

If the response time of the process is excessively long, derivative action may be necessary for optimum control. The adjustment procedure for a proportional-plus-integral-plus-derivative controller is:

1. Set the integral to maximum time.
2. Set the derivative to minimum time.
3. Adjust the proportional band to an optimum setting as before, except that a slight cycle should remain in the measurement.
4. Increase the derivative time until the cycle stops.
5. Narrow the proportional band until the cycle starts again.
6. Repeat Steps 4 and 5 until further increases in the derivative time fail to stop the cycle.

7. Widen the proportional band to stop the cycle.
8. Set the integral time equal to derivative time.

The effect of derivative action on process-response curves is shown in Figure 12-1 for the proportional-plus-integral-plus-derivative controller. Curve B represents optimum tuning; Curve C shows the effect of too long a derivative time; and Curve A shows too short a derivative time.

Tuning Maps

Assume the controller is on automatic and controlling the process. The adjustments have been made by either trial and error or experience. You now compare the record to the tuning map (Figures 12-2 and 12-3). This comparison should tell you what additional adjustment is necessary and what should be adjusted. The steps are as follows:

1. Determine which tuning map approximates the actual chart record.
2. Select the tuning map that, in your judgment, is most satisfactory for the desired response.
3. Adjust the proportional band, integral, and derivative in the directions indicated in order to obtain a more satisfactory chart record.

Only the proper direction of adjustment is provided because variations in actual processes dictate the magnitude of the change.

PROPORTIONAL PLUS INTEGRAL

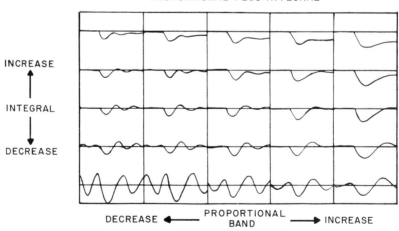

Fig. 12-2. Proportional-plus-integral controller.

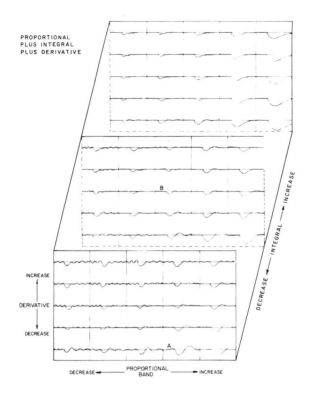

Fig. 12-3. Proportional-plus-integral-plus-derivative controller.

The tuning maps are shown with magnitude of deviation and a response time satisfactory for illustration. Measurement records will vary in deviation and response time from the tuning maps shown here. However, the relative effects can easily be compared.

Adjusting a Controller with the Proportional, Integral, and Derivative Mode

To tune a proportional-plus-integral-plus-derivative controller, let us assume that the tuning map marked A in Figure 12-3 approximates the actual chart record, although the deviations and the response times may be different. It is desirable to obtain a measurement response similar to the tuning map marked "B" in Figure 12-3.

Table 12-1. Controller Adjustment Equivalents

The following table enables an approximate conversion between different manufacturers' traditional "tuning" ' adjustments.*

PROPORTIONAL ACTION			INTEGRAL (RESET) ACTION	
Proportional Band (Per Cent)	*Gain*	*Sensitivity*	*Integral Rate (Repeats per Minute)*	*Integral Time (Minutes per Repeat)*
Foxboro	Bailey	Fisher	Bailey	Foxboro
Moore	Bristol		Bristol	F & P
Honeywell	Taylor		Honeywell	Moore
F & P			Taylor	
1	100	300	20	1.05
2	50	150	10	.1
5	20	60	5	.2
10	10	30	2	.5
20	5	15	1	1
50	2	6	.5	2
100	1	3	.2	5
200	.5	1.5	.1	10
500	.2	.6		
1000	.1	.3		

*Derivative action ("Rate" Action) given in minutes by the manufacturers above.

Note that the relative effects of proportional band, derivative, and integral can be easily determined from the tuning maps. Starting with tuning map A, we find:

Increasing the proportional band minimizes cycling but increases the deviation.

Decreasing the proportional band decreases the deviation, but causes prolonged cycling if too low.

Increasing the derivative decreases the deviation and minimizes cycling up to a critical value. Beyond the critical value, the deviation increases and prolonged cycling occurs.

Increasing the integral will minimize the amount by which the measurement crosses the set point and minimizes cycling, but a prolonged deviation occurs if the integral is too high. In this case, the proportional band must be decreased and the derivative and integral increased to obtain the desired results.

These tuning maps are designated to work with the dials labelled in current ISA and SAMA terms. If you are using a controller with labels in other terms refer to Table 12-1 for any necessary conversions.

Closed-Loop Cycling Method

The closed-loop cycling method uses a process-control loop that oscillates at a gain of 1 in order to predict all mode settings. A typical tuning procedure would be:

1. Set the integral to maximum time.
2. Set the derivative to minimum time.
3. Set the proportional band to maximum.
4. Place the controller in automatic.
5. Reduce the proportional band setting until the loop cycles with a gain equal to 1 (the ultimate proportional band, B_u) and as small an amplitude A as possible. Figure 12-4 shows these relationships.
6. Measure the loop's time period (t_u).
7. Adjust the settings as follows:
   a. For a proportional-only controller, $B = 2B_u$ for a 0.25 damping ratio. If the controller has gain units, the gain should be one-half the value that caused oscillation.
   b. In the three-mode controller the integral, which is phase lagging, and the derivative, which is phase leading, should have equal time values. This will produce minimum phase shift. The integral time = derivative time = $\dfrac{t_u}{2\pi}$ or approximately $\dfrac{t_u}{6}$
   c. After the integral and derivative dials are set, adjust the proportional band or gain to obtain the desired damping. If a 0.25 damping ratio is desired, B = approximately $1.77\,B_u$.

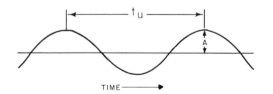

TIME ⟶

Fig. 12-4. Period of a cycle.

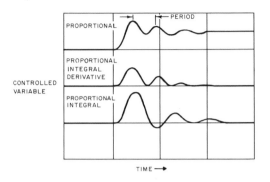

Fig. 12-5. Comparative load-response curves for three controllers.

Several factors must be kept in mind when making controller adjustments.

1. Load changes may vary the dynamic characteristics of the control loop, requiring either loose compromise operation that remains stable, or readjustment.

2. The dials used on controllers to designate mode values have poor accuracy due to the required wide rangeability. This makes precise adjustment impossible and requires that the final step be trial-and-error.

Once you have gained some practical experience in making controller adjustments, you will discover acceptable shortcuts. Table 12-2 gives some guidelines that should help with the typical loop:

Table 12-2. Guidelines of the Required Settings for Controller Modes

Control Loop	Proportional Band	Time Constant	Derivative
Flow	High (250%)	Fast (1 to 15 sec)	Never
Level	Low	Capacity dependent	Rarely
Temperature	Low	Capacity dependent	Usually
Analytical	High	Usually slow	Sometimes
Pressure	Low	Usually fast	Sometimes

In conclusion, it should be reemphasized that a controller must be properly adjusted to fit the loop if proper control action is to occur.

This chapter has presented the steps that may be employed to produce good controller adjustment.

Questions

12-1. A controller is adjusted to:
   a. Speed up the process
   b. Slow down the process
   c. Match its gain-and-time functions to the process
   d. Accommodate only process load changes

12-2. A proportional-only controller is:
   a. Very difficult to adjust
   b. Easily adjusted to accommodate load changes
   c. Slow to respond
   d. Going to cycle if the proportional band is too narrow

12-3. In a control loop using a three-mode controller that has been properly adjusted:
   a. Integral time will be equal to derivative time
   b. Integral rate will be equal to derivative time
   c. Derivative time will be one-half the integral time
   d. The proportional band should be the final adjustment

12-4. A limitation of calculating the controller settings to precise values is:
   a. The varying speed of process responses
   b. The poor accuracy of the dials on the controller
   c. Process error
   d. Unreliable data

12-5. The most common adjustment technique is:
   a. Closed-loop cycling method
   b. Tuning map method
   c. Reaction curve analysis method
   d. Trial-and-error

13

Step-Analysis Method of Finding Time Constant

Automatic control of a process has been defined in previous chapters as a technique whereby supply to the process is balanced against demand by the process over time. Dynamic analysis of the process clarifies how this balancing takes place. Knowledge of how the process functions with respect to time is provided by the *time constant,* or constants, of each component in the process. These time constants are useful in creating an analog of the system and in selecting the type of controller and its optimum control settings.

Block Diagrams

The system usually is represented by a block diagram, a simplified schematic, a mathematical description (mathematical analog), or an electrical analog. If, for example, the process is a tank with water flowing in and out, a simplified sketch of the process might be drawn as in Figure 13-1a. The system also can be shown in simple block form with a verbal description (Figure 13-1b), or the block can be labeled with a mathematical description of the component (Figure 13-1c). Figure 13-1d is an electrical analog of the system. The mathematical de-

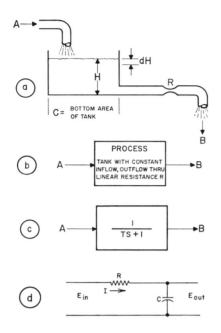

Fig. 13-1. Fluid process with single-time constant. Flow rate in is represented by A; flow rate out, by B. In the electrical circuit, E_{in} represents A and E_{out} is B.

scription is the most concise. For complex systems the mathematical approach is almost imperative because drawings are too complicated, and a verbal description is long and involved.

To describe a process, a mathematical expression must relate output to input (output/input) and show how one varies with respect to the other with the passing of time. Classically, this expression is a differential equation, differentiated with respect to time. For instance, the simple system shown in Figure 13-1a is described by

$$A - B = C(dH/dt) \qquad (13\text{-}1)$$

This expression says that if the volumetric outflow rate (B) is subtracted from inflow rate (A), the difference equals the area of the tank bottom (C) times the change in head (dH) during a change in time (dt). Or, stated differentially, rate A minus rate B times dt equals the volume change.

The hydraulic analog of ohm's law for flow through a linear resistance (R) can be used. Here (B), the outflow rate, is the equivalent of an electrical current produced by a difference of potential (in this case H) across a linear resistance (R). This assumes that there is no "backpressure," i.e., the pressure at the outlet is 0 psi, and that R is linear (actually rare in hydraulics). Since $E = IR$,

$$H = BR \qquad\qquad \textbf{(13-2)}$$

Substituting the value for H from Equation 13-2 into Equation 13-1,

$$A - B = Cd(BR)/dt \qquad\qquad \textbf{(13-3)}$$

Expressing d/dt as S (as is done in operational mathematics),

$$A - B = CSBR \qquad\qquad \textbf{(13-4)}$$

Solving for A,

$$A = B(CSR + 1) \qquad\qquad \textbf{(13-5)}$$

This value for A can be substituted in the basic expression for the process:

$$\frac{\text{output}}{\text{input}} = \frac{B}{A} = \frac{B}{B(CSR + 1)} = \frac{1}{CSR + 1} \qquad\qquad \textbf{(13-6)}$$

As T (the time constant) $= RC$, T can be inserted for RC in Equation 13-6:

$$\frac{\text{output}}{\text{input}} = \frac{1}{TS + 1} \qquad\qquad \textbf{(13-7)}$$

This basic equation 13-7 describes one type of first-order system (one with a single capacitance and a single resistance). This expression is often called a transfer function; it is simply the Laplace transform of the differential equation describing the system.

Any two systems that have the same transfer function have the same response with respect to time.

Figure 13-1d is the electrical equivalent of the same system and would be handled in the same way.

$$I = \frac{E_{in} - E_{out}}{R} \quad \text{and} \quad I = C\frac{dE_{out}}{dt} \tag{13-8}$$

Therefore,

$$\frac{E_{in} - E_{out}}{R} = C\frac{dE_{out}}{dt}$$

$$E_{in} - E_{out} = RC\frac{dE_{out}}{dt} \tag{13-9}$$

If

$$d/dt = S,$$

then

$$E_{in} - E_{out} = RCS\,E_{out} \tag{13-10}$$

Solving for

$$E_{out}/E_{in},$$

$$\frac{E_{out}}{E_{in}} = \frac{1}{RCS + 1} = \frac{1}{TS + 1} \tag{13-11}$$

Any single-capacitance, single-resistance system has this transfer function. Note that the first-order system can be described by one factor (T), the time constant.

A number of techniques have been developed and can be used for evaluating the time constant, including:

1. Step analysis
2. Frequency response
3. Ramp input
4. Correlation
5. Stastical analogies

Some of these techniques are easier to perform and require less equipment than others. For instance, the step-analysis method (which will be discussed here) requires only simple equipment and pencil-and-paper calculations; some of the others require computer evaluation. The best method in any given instance is dictated by the objectives of the test and the equipment or tools available.

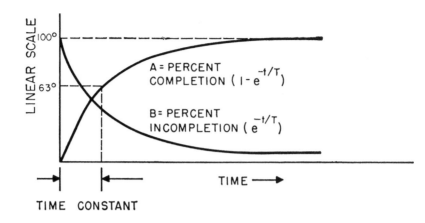

Fig. 13-2. Response of single-time-constant process on linear plot.

Step Analysis of Single-Time-Constant System

A step change in input to the single-time-constant process can comprise a sudden increase in flow rate (A) in the process shown in Figure 13-1a, or a step change in input voltage in Figure 13-1d. The output response curve is shown in Figure 13-2. The time constant of a single-time-constant process is found simply by finding the time required to reach 63.2 percent of the final change in the output following the step change. The equation for this response curve (A in Figure 13-2) is $1 - e^{-t/T}$. As will be shown, it is useful to plot the "percent-incomplete" response ($e^{-t/T}$) on semilogarithmic paper. Such a curve (B in Figure 13-3) repre-

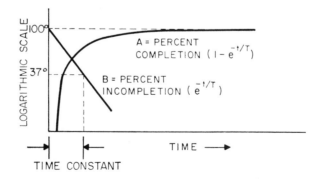

Fig. 13-3. Response of single-time-constant process on semilog paper, showing percent complete and percent incomplete response.

sents the percent of total rise that has not taken place at each point in time. Thus, the line starts at 100 percent nad drops down-scale.

In Figure 13-3, line B is straight because it is the plot of a pure exponential ($e^{-t/T}$ in Figure 13-2) on a logarithmic scale. Curve A (plot of $1 - e^{-t/T}$) is not straight. Therefore, the percent-incomplete (straight) curve ($e^{-t/T}$) is chosen and plotted on the logarithmic scale. When the percent-incomplete curve (B in Figure 10-3) is straight on logarithmic paper, it verifies that the process has a single time constant.

Time constant (T) is the time for 63.2 percent rise to final value (completion), which is the same as saying 36.8 percent incompletion, or 36.8 on the percent-incomplete scale. If a horizontal line is drawn on Figure 13-3 from 36.8 across to the intersection with curve B, the time found on the abscissa is T, the time constant for this process.

Step Analysis of Two-Time-Constant System

A two-time-constant process is assumed in Figure 13-4. With the switch in the manual position, all feedback or corrective action is eliminated.

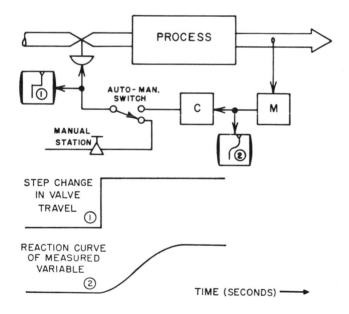

Fig. 13-4. Step test of a process with two time constants. The system can be applied to systems with one or more time constants.

Next a step change is applied to the valve and the resultant change in measured variable (reaction) is recorded (Figure 13-5). The magnitude of the step is not critical; the step should be large enough to produce the necessary data, but not large enough to disturb process operation or to cause the process to exceed its normal or linear limits. In general, the largest allowable upset is the best. If the upset must be kept small, the recorder can be recalibrated to expand the reading. From the resulting reaction curve, the first- and second-order time constants can be evaluated using the "percent-incomplete" method.

Percent-Incomplete Method

From Figure 13-5 (the reaction curve for a typical second-order, two-time-constant system) percent incompletion of the process (100 percent minus percent completion of the process) is calculated and tabulated. Note that if any dead time is known to exist, it should be subtracted from the time coordinate of each percent-incomplete point. Dead time (also called transportation time) is revealed by the reaction curve; it is the time it takes for a change in variable to "travel" to the measurement point. No such dead time is present in the Figure 13-5 curve—the output begins to change immediately. For example, dead time in a

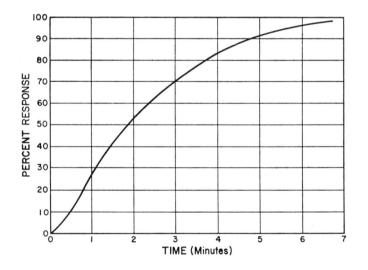

Fig. 13-5. Response of process with two time constants.

heating process could be the time required for the heated fluid to reach a measuring element located some distance downstream; then calculations would be made from a reaction curve whose time axis is shifted to eliminate this dead time.

The percent-incomplete curve is plotted on semilogarithmic graph paper (Figure 13-6); percent incompletion is plotted on the vertical axis and time on the horizontal axis. At time zero no change has taken place and the reaction is 100 percent *incomplete* (curve A in Figure 13-6).

In order to determine the first time constant (T_1), curve B is drawn by extending the linear portion of curve A back to the zero-time axis. The intercept of curve B at zero time is point P_1. In Figure 13-6 point P_1 is 133 percent. The first time constant (T_1) is the time it takes curve B to reach 36.8 percent of P_1. Time constant T_1 is found by marking a point (P_2) at 36.8 percent of P_1. In Figure 13-6, $P_2 = 0.368 \times 133 = 49$ percent incompletion. The time it takes for curve B to reach an ordinate (vertical height on the graph) equal to that of P_2 (49 percent incompletion) in Figure 13-6 is 2 minutes. Thus, T_1 is 2 minutes.

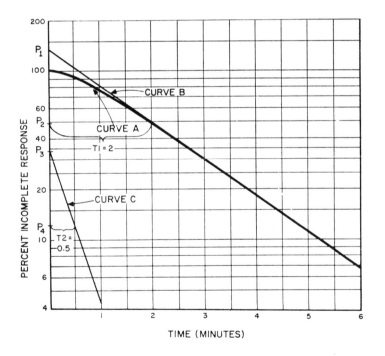

Fig. 13-6. Finding two time constants by the percent-incomplete method.

Curve C is plotted to determine the value of the second-order time constant (T_2). Curve C is the numerical difference between curves B and A; at any time the ordinate of curve C is equal to the ordinate of curve B minus the ordinate of curve A. If curve C is not a straight line, the system cannot be completely represented by two time constants.

Curve C has a zero-time intercept (point P_3) of 133 percent minus 100 percent, or 33 percent. Point P_4 is 36.8 percent of P_3, or 12.1 percent incomplete.

The second-order time constant (T_2) is equal to the time it takes for curve C to reach an ordinate of 12.1 percent incompletion. In Figure 13-6 this is 0.5 minute.

As an approximate check on the plotting, the zero-time intercepts of curves B and C may be found by using the time constants in the following expressions. For P_1,

$$P_1 = \frac{T_1}{T_1 - T_2} \tag{13-12}$$

From the graphical solution (Figure 13-6), T_1 equals 2 minutes and T_2 equals 0.5 minute. Therefore, point P_1, the vertical intercept of curve B, equals

$$P_1 = \frac{2}{2 - 0.5} = 1.33$$

or 133 percent, which is correct. For P_3,

$$P_3 = \frac{T_2}{T_1 - T_2} \tag{13-13}$$
$$= \frac{0.5}{2 - 0.5} = 0.33$$

or 33 percent, which is correct.

Thus, the computed values are consistent with the graphical values of Figure 13-6.

Using the time constants, the analog equivalent of the system can be constructed based on its transfer function. This transfer function for a first-order system is $1/(TS + 1)$. The percent-incomplete curve for such a system was shown to be a straight line. For the second-order system described above, the transfer function would be

$$K \left[\frac{1}{T_1 S + 1} \right] \left[\frac{1}{T_2 S + 1} \right]$$

where K = a constant.

The percent-incomplete method works well for second-order systems, but it is not generally practical for systems of higher order because data accuracy usually is limited. However, in the heat-exchanger (multicapacity system) analysis which will follow, the results are comparable to those obtained by more sophisticated methods. In general, the percent-incomplete method can be used with success for all second-order systems except those where curves A and B are too close to one another.

Multicapacity System

The heat-exchanger response curve shown in Figure 13-7 has a greater lag than the curve of Figure 13-5 for the two-time-constant system. Thus, the variation in the new curve A in Figure 13-8. The graphical solution for the time constants proceeds as follows:

1. Tabulate percent-incomplete data (since the measuring element is not located at any great distance, there is no predictable dead time).
2. Plot the percent-incomplete curve on semilog paper (curve A in Figure 13-8).

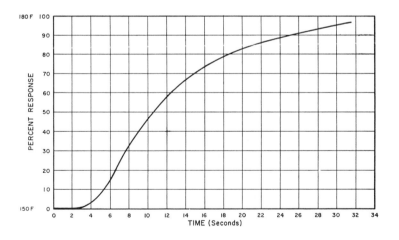

Fig. 13-7. Response of multiple-time-constant system.

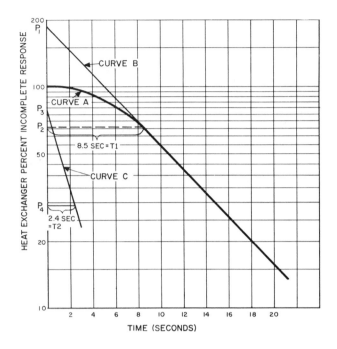

Fig. 13-8. Finding the two major time constants of the multiple-time-constant system.

3. Extend the linear portion of curve A to form curve B and establish point P_1 (176 percent).
4. Point P_2 equals 0.368 of P_1. From point P_2 (66 percent), draw a horizontal line until it intersects curve B. The length of this line represents the first time constant (8.5 seconds).
5. Plot curve C by subtracting curve A from curve B (i.e., $P_1 - P_2 = P_3$). P_3 is the point where curve C intersects the zero-time axis (76 percent).
6. Point $P_4 = 0.368 P_3$. From P_4 (28.5 percent), draw a horizontal line until it intersects curve C. The length of this line represents T_2 (2.4 seconds).
7. Using the check previously described in the text, verify the time-constant values.

Since the heat exchanger is a multicapacity, multiresistance process containing many time constants, the values of P_1 and P_3 do not check exactly. However, the time constants found by the percent-incomplete method are, in this case, nearly equal to those found by more exacting methods.

Finding Control Modes by Step Analysis

Step-change process-reaction curves not only reflect the dynamic behavior of a process, but also are a means for finding the first-order and second-order time constants. This is done by using the percent-incomplete method. Theoretically, this method could be used to find all of the time constants in a process, but usually it is limited to second-order systems in which the time constants are of different magnitudes. Equally important, the reaction curve reveals the best control action for the process, and additional data from the curve can be used in calculating the optimum controller adjustments.

Three criteria are of major importance in designing a process control system. One, continuous cycling of the controlled variable must be avoided (except where on/off control is acceptable). Two, offset should be kept to a minimum. Three, control should be capable of returning the variable to the set point as quickly as possible. For a given process, the simplest control system that still meets these requirements should be used. Information from the step-reaction curve can be used in the selection of best control mode.

Process controllability (or difficulty) is revealed by the shape of the reaction curve (Figure 13-9). The nature of the response can be appraised by two factors: the time (M) it takes the curve to reach its maximum rate of change, and the value of the maximum rate of change

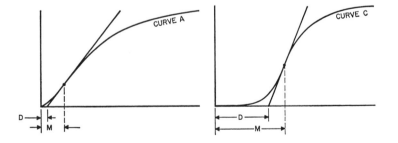

Fig. 13-9. Typical response curves of systems with multiple-time-constants. The tangent is constructed at the curve's maximum rate of rise. D is delay time; M is the time to maximum rate of rise.

(slope of the tangent line in Figure 13-9). In general, the larger the product of these two factors—that is, the more pronounced the "S" shape of the response curve—the more difficult the process. The process represented by the curve at the right in Figure 13-9 is thus more difficult to control than the process represented by the curve at the left.

On/Off Control Action

The on/off (two-position) controller operates only when the measurement crosses the set point. In on/off control, a valve would have only two positions—fully opened or fully closed; the energy or material supplied to the process is always either too much or not enough. As a result, the output measurement always cycles above and below the set point. However, when on/off control is applied to the right type of process, the amplitude of the oscillations of the controlled variable is so small that the output appears to be constant. However, small oscillations are always present.

On/off control is best applied to a *large-capacitance* process that has essentially *no dead time,* such as a large tank or bath. The rate of rise (or fall) of the output curve is small because the energy inflow is small compared with the large capacitance of the system. The response curve would be more like that of the left curve in Figure 13-9. *A system where dead time is short and rate of rise is slow can be handled adequately by on/off control.*

Proportional Control Action

Proportional action is required for control when the system has small capacitance and thus a faster response to input changes.

A proportional controller continuously throttles a control valve when the measurement is within the *proportional band* (Figure 13-10). A narrow band means full valve travel for a narrow range of variables; a wide band means that the variable moves over a wider range before full valve travel occurs. Unlike on/off control, the proportional controller can "meter" the energy input to counteract variations from the desired set point, and thus maintain one control point without oscillation around the set point.

With proportional control, however, offset occurs due to load changes. This is undesirable in some situations. Narrowing a proportional band can reduce the amount of offset, but too narrow a band leads to cycling. Accordingly, the proportional band setting for op-

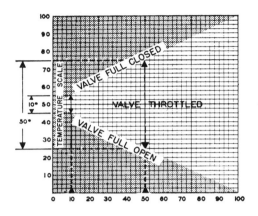

Fig. 13-10. Relation between valve position and measurement for different proportional bands. A narrow proportional band means control within narrow limits.

timum control is generally just above this critical oscillating point (Figure 13-11).

A bathroom shower is an example of a small-capacity process, that is, the capacitance of the system is small compared to the flow rate of energy to the system. The reaction curve for this system would show that the lag is relatively short and the response is rapid. In fact, most of

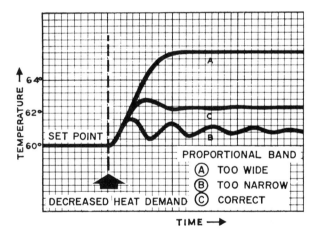

Fig. 13-11. Curve A results when the proportional band is too wide (insensitive). Oscillation results (B) when band is too narrow. Proportional band setting for optimum control (C) is just above the point of oscillation.

this is lag in the measuring system itself. The rate of rise of temperature in a shower is faster than that in a bath because the volume per minute of incoming hot water is larger in comparison with total volume of water in the system. Thus, a shower would need proportional control whereas the bathtub can use on/off control.

The response curve of a shower process would be similar to that illustrated in the curve at the right of Figure 13-9. Since lag is short and maximum rate of rise is relatively high, proportional control action is the proper selection.

Proportional-Plus-Integral Action

For some processes—for example, the measurement and control of flow—a wide proportional band is necessary to eliminate cycling. A wide band means that large amounts of offset can occur with changes in load, and it is necessary to introduce a control function that will reduce offset and thus return the measurement to the set point even if a load change occurs. This function was first known as *automatic reset* (Figure 13-12, but is now correctly called integral action.

The response curve of the flow loop shown in Figure 13-13 might be similar to that illustrated by the right curve in Figure 13-9. Note that the process takes a relatively long time to reach its maximum rate of rise, and that its maximum rate of change is higher than the larger capacity system represented by the left curve in Figure 13-9.

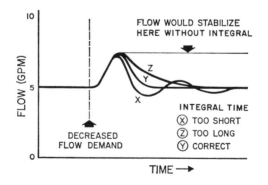

Fig. 13-12. Integral control action eliminates offset due to load changes.

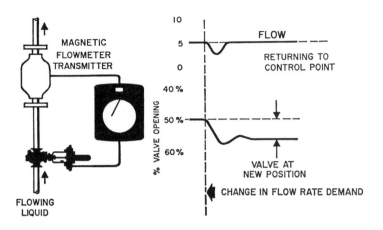

Fig. 13-13. Response of a flow loop to load change.

Proportional-Plus-Derivative Control Action

While it is possible to use proportional-plus-integral action on difficult (multicapacity) processes, recovery from process disturbances often is slow. In many such cases, *derivative action* can be added to hasten the recovery.

Derivative action of the controller is proportional to the rate of change of measurement. Therefore, the control valve reaches a given position sooner than it would with proportional action alone. Since the amount of "lead" that the valve has on the measurement is proportional to the rate of change of the measurement, derivative action can be described in terms of time. If derivative times is set in accordance with the requirements of the process, the proportional band may be narrowed without creating instability (Figure 13-14).

Proportional-plus-derivative control actions are generally employed on discontinuous processes (for example, batching operations involving periodic shut-down, emptying, and refilling). Here, the proportional and integral combination would not be applicable because time lags are long and operation is intermittent.

Proportional-plus-derivative control action also is recommended for processes employing duplex control action where one of the two control mechanisms always is inactive. In these situations, derivative control action (because it is *rate sensitive*) permits use of a narrower proportional band, thus reducing the deviation of measurement from

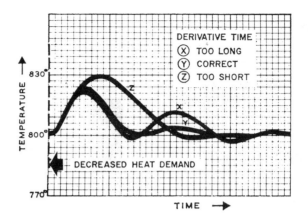

Fig. 13-14. Addition of derivative (rate) action can often permit a narrower proportional band without creating instability.

the set point. Derivative action also reduces the amount of overshoot at the start of a batch operation.

Proportional-Plus-Derivative-Plus-Integral Control Action

A gas-fired oil-heating furnace (Figure 13-15) presents a difficult control problem. The measurement is slow in reaching the maximum rate of rise (similar to right curve in Figure 13-1), but the maximum rate is high. Following a change in valve position, the measurement is slow to change because the furnace has large capacity and there is lag in the measuring system. However, once the process is under way, it reaches a high rate of change because the heat capacity of the furnace is large compared to the heat capacity of the fluid in the furnace tubes (oil). Therefore, all three control modes (proportional-plus integral-plus derivative) can be employed to advantage.

Using the Reaction Curve to Determine Controller Adjustments

A passive system is one which has no independent energy inputs; process gain can never exceed unity. For passive processes, there are empirical formulas which predict the optimum settings for proportional, integral, and derivative control actions. Four quantities are de-

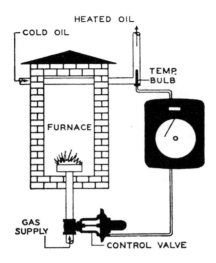

Fig. 13-15. Gas-fired oil-heating furnace is a "difficult-to-control" process that requires all three control modes.

termined from the process reaction curve for inclusion in the formulas: time delay, time period, a ratio of these two, and plant gain (Figure 13-16).

With a line drawn tangent to the curve at its point of maximum rise, the time delay (*D*) extends from zero on the horizontal axis to the point

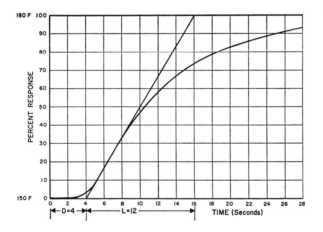

Fig. 13-16. Tangent to response curve at point of maximum slope gives basic time factors L and D.

where the tangent line intercepts the time axis. The time period (L) extends from the end of the delay period to the time at which the tangent intercepts the 100 percent completion (maximum measurement) line. The ratio (R) of the time period to the time delay describes the dynamic behavior of the system:

$$R = L/D$$

The last parameter to be used in the equations is the plant gain (C). Plant gain is defined as the percent change in the controlled variable divided by the percent change in the manipulated variable (valve stroke).

For the response of a typical heat exchanger (Figure 13-16), $D = 4$ seconds and $L = 12$ seconds. The ratio (R) is

$$R = L/D = 12/4 = 3$$

Since the span of the measuring transmitter is 200°F, and the temperature change is 30°F, the percent temperature change is

Table 13-1. Control Action Relationships for the Idealized Controller

Type of Control	QUICKEST RESPONSE WITHOUT OVERSHOOT		QUICKEST RESPONSE WITH 20% OVERSHOOT	
	Step Change in Manipulated Variable	Step Change in Load	Step Change in Manipulated Variable	Step Change in Load
STRAIGHT PROPORTIONAL				
Proportional band (percent)=	$\dfrac{333\ C}{R}$	$\dfrac{333\ C}{R}$	$\dfrac{143\ C}{R}$	$\dfrac{143\ C}{R}$
PROPORTIONAL & INTEGRAL				
Proportional band (percent)=	$\dfrac{286\ C}{R}$	$\dfrac{167\ C}{R}$	$\dfrac{167\ C}{R}$	$\dfrac{143\ C}{R}$
Integral (time units of D)=	$3.33\ DC$	$\dfrac{6.67\ DC}{R}$	$1.67\ DC$	$3.33\ \dfrac{D}{R}$
PROPORTIONAL & INTEGRAL & DERIVATIVE				
Proportional band (percent)=	$\dfrac{167\ C}{R}$	$\dfrac{105.2\ C}{R}$	$\dfrac{105.2\ C}{R}$	$\dfrac{83.3\ C}{R}$
Integral (time units of D)=	$1.67\ DC$	$2.5\ \dfrac{DC}{R}$	$1.43\ DC$	$1.67\ \dfrac{DC}{R}$
Derivative (time units of D)=	$0.3\ \dfrac{RD}{C}$	$0.4\ \dfrac{RD}{C}$	$0.45\ \dfrac{RD}{C}$	$0.5\ \dfrac{RD}{C}$

$30°F/200°F = 0.15 = 15$ percent. The percent change in valve travel which produced this temperature rise was 5 percent. The plant gain is, therefore,

$C = 15/5 = 3$

The empirical equations in Figure 13-9 describe the optimum process control settings for each of three types of control actions. They are defined for the idealized controller and, therefore, require slight corrections, depending on actual controller configurations. They do, however, provide a close approximation to optimum settings. The equations are based on the work of Chien, Hrones, and Reswick (Ref. 3).

For control without overshoot, one of the first two columns of Table 13-1 is used. If 20 percent overshoot can be tolerated, then the formulas in the third or fourth column are applicable. Overshoot, as shown in Figure 13-17, reduces the time it takes for the measurement to reach the set point.

Since a heat exchanger is classed as "difficult to control," it probably will require all three control actions—proportional, integral, and derivative. In the reaction test, the step change was in the manipulated variable and, assuming no overshoot, the formulas in column 1 would be used. Again, $R = 3$ (dimensionless), $C = 3$ (dimensionless), $D = 4$ seconds, and $L = 12$ seconds.

Substituting into the equations in Figure 11-9,

Proportional band $= 167 \, C/R = 167 \times 3/3 = 167$ percent

Integral time $= 1.67 \, DC = 1.67 \times 4 \times 3 = 20$ seconds

Derivative time $= 0.3 \, RD/C = 0.3 \times 3 \times 4/3 = 1.2$ seconds

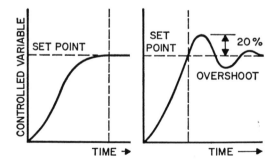

Fig. 13-17. Response from a step change in manipulated variable for quickest recovery with no overshoot (*left*) and quickest recovery with 20 percent overshoot (*right*).

Controllers manufactured by The Foxboro Company are of the series-connected type. For these controllers, the proportional band needs no correction factor. However, the integral time should be multiplied by 100/percent *PB* and the derivative multiplied by percent *PB*/100. The corrected values therefore, are

Corrected integral time $= 20 \times 100/167 = 12$ seconds

Corrected derivative time $= 1.2 \times 167/100 = 2$ seconds

Many other controllers are parallel connected. If a parallel-connected controller is employed, the correction becomes considerably more complex.

These values have been confirmed by actual tests on a heat exchanger. The resulting recovery curve is similar to the left curve in Figure 13-10.

If 20 percent overshoot is permissable, the formulas in column 3 would give

Proportional band $= 105.2C/R = 105.2 \times 3/3 = 105.2$ percent

Integral time $= 1.43DC = 1.43 \times 4 \times 3 = 17.6$ seconds

Derivative time $= 0.45\ RD/C = 0.45 \times 3 \times 4/3 = 1.8$ seconds

The corrected values of integral and derivative are:

Corrected integral time $= 17.6 \times 100/105.2 = 16.9$ seconds

Corrected derivative time $= 1.8 \times 105.2/100 = 1.89$ seconds

These settings also have been confirmed on a heat exchanger, and results are similar to the right curve in Figure 13-17.

The step analysis of a reaction curve is not exact but is an empirical approximation. However, it requires little or no special equipment and serves as a reasonably complete identification of dynamic behavior.

References

Anderson, N., Step analysis method of finding time constant. *Instruments and Control Systems,* Vol. 36, No. 11 (Nov. 1963) 130-136.

Catheron, A. R., S. H. Goodhue, and P. D. Hansen. Control of shell and tube heat exchangers. ASME Paper No. 59-IRD-14, presented at the Instrument and Regulators Conference of ASME, Cleveland, OH, March 1959.

Chien, Kun Li, J. A. Hrones, and J. B. Reswick. On the automatic control of generalized passive systems. *Transactions of the ASME,* Vol. 74, No. 2 Feb. 1952 175–185.

Questions

13-1. A transfer function relates:
   a. Input to output attenuation
   b. Input potential to output potential
   c. Input dead time to output dead time
   d. Behavior of the output to the behavior of the input

13-2. When used in block diagram the transfer function is:
   a. A shorthand notation
   b. An algebraic formula
   c. A static expression describing the relationship
   d. Useless if more than one time constant is involved

13-3. The percent-incomplete method enables you to determine the following:
   a. First-, second-, and third-order time constants
   b. First- and second-order time constants
   c. Second- and third-order time constants
   d. Any number of time constants

13-4. The percent completion curve is $1 - (e^{-t}/T)$, where e is the natural logarithm base 2.718, t is actual time, and T is time-constant time. When the actual time equals the time-constant time, the curve will be at the following percentage of completion:
   a. 36.8 percent
   b. 63.2 percent
   c. 100 percent
   d. 0 percent

13-5. Pure dead time such as transportation lag;
   a. Should be subtracted from the time coordinate for each percent-incomplete point
   b. Makes the percent-incomplete method useless
   c. Makes the percent-incomplete method more accurate and very useful
   d. Should be added to the time coordinate for each percent-incomplete point

13-6. The shape of a process reaction curve indicates:
   a. Process productivity
   b. Process safety
   c. The type of control valve required
   d. Process controllability

13-7. On/off control can be effectively applied to a process where:
   a. Dead time is short and rate of rise is slow
   b. Lag is short and maximum rate of rise is high
   c. The operation is continuous
   d. Dead time lag is less than 5 minutes

13-8. A passive system is one that:
   a. Has 0 gain
   b. Has no independent energy inputs
   c. Will never explode
   d. Is entirely safe for anyone to operate

13-9. An inspection of the relationships for the idealized controller:
   a. Reveals that integral time always equals derivative time
   b. Proves that, properly applied, derivative tends to stabilize the system
   c. Proves that, properly applied, integral tends to stabilize the system
   d. Establishes the values of the process time constants

13-10. A control valve is moved 5 percent and the resultant reaction curve has a D value of 3 seconds and an L value of 9 seconds. The measuring transmitter has a span of 400°F, and after the process has restablized, the temperature is 40°F above the initial value. A three-mode controller has been selected to control the process. The approximate controller settings for no overshoot would be:
   a. PB = 167 percent, Integral = 20 seconds, Derivative = 1.2 seconds
   b. PB = 100 percent, Integral = 44 seconds, Derivative = 18.5 seconds
   c. PB = 74 percent, Integral = 4.4 seconds, Derivative = 1.85 seconds
   d. PB = 110 percent, Integral = 10 seconds, Derivative = 1.35 seconds

13-11. Having calculated the controller adjustments and dialed them into the controller we should then:
   a. Never again touch the controller
   b. Trim the adjustments in small increments, carefully observing the results until the optimum results are achieved
   c. Apply these results to every other process in the plant
   d. Apply these results only to the process from which the data were obtained or to similar systems in the area

13-12. A heat exchanger is successfully controlled with a three-mode controller. Then, due to an accident, the measuring transmitter with a span of

200°F must be replaced. The replacement transmitter is very similar, has the same center scale reading, but has a span of 300°F. All the controller adjustments:

    a. Will remain the same

    b. Must be altered by a factor of 1.5

    c. Must be altered by a factor of 0.66

    d. Will remain the same except for the proportional band, which will become approximately two-thirds of its previous value

14

Frequency Response Analysis

Open-loop frequency response analysis of a control system or its components provides dynamic characteristics of the system. Corner frequencies reveal the time constants.

The step-response method described in Chapter 13 provides a means for obtaining the process time constants and for selecting the type of controller and its optimum settings. Although the method is simple and usually sufficient, there are limitations. For example, a step upset could disrupt process operation in a plant.

Frequency response is another method for analyzing the characteristics of the process. Since it usually is less upsetting to a process than step response, it is more widely applicable. The data derived from a frequency test can be used to indicate the type of controller, optimum settings, and other factors influencing the design of the process and the control system.

The basic technique consists of changing a variable (such as the input signal to a valve) in sinusoidal fashion and measuring its effect on the system. For example (Figure 14-1), a sine-wave generator and a transducer can be used to apply a pneumatic sinusoidal input signal to a valve actuator. The resulting variation in process output pressure is recorded and is a measure of the system response (valve, surge tank, piping, and pressure device) to the input signal.

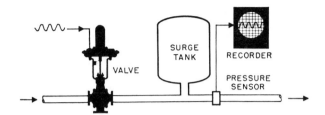

Fig. 14-1. Frequency response test of a pressure system.

What happens to the sinusoidal signal that is transmitted? The frequency of the output signal is always the same as that of the input, but the amplitude of the pressure variation will change. As the frequency of the input signal (valve pressure signal) increases, the amplitude of the output (pressure) signal will probably diminish. The ratio of output signal amplitude to input signal amplitude is the *gain* of the system, sometimes called *amplitude ratio,* or *attenuation* (when gain is less than unity).

The second basic consideration is the change in *phase* which the signal undergoes. Since the surge tank and piping contribute both capacitance (C) and resistance (R) to the system, there will be a lag that produces a change in phase—that is, the output fluctuations will not occur simultaneously with the input fluctuations but will be offset in time. Since each RC component in any system produces a lag, and since the lag depends on the frequency, a certain combination of frequency and gain will induce cycling in any closed-loop operation. The control system designer must see that this combination does not occur.

Gain and phase shift should be noted over a range of frequencies from very low (where attenuation and phase shift are minimal) to high (at which attenuation and phase shift are marked). Typical input and output signals are shown in Figure 14-2. Each chart shows the gain and phase change at a given frequency. A series of such charts gives the amplitude ratio and phase shift for many frequencies. Gain and phase versus frequency then can be plotted on a Bode diagram (Figure 14-3). If special Bode graph paper is available, the two graphs can be plotted on the same piece of paper; otherwise separate sheets will suffice.

The gain can be expressed in terms of amplitude ratio on a log scale, or decibels on a linear scale as shown in Figure 14-3. (The decibel rating is equal to 20 times logarithm to the base of 10 of the amplitude ratio.) Using either of these scales, the gains of individual system com-

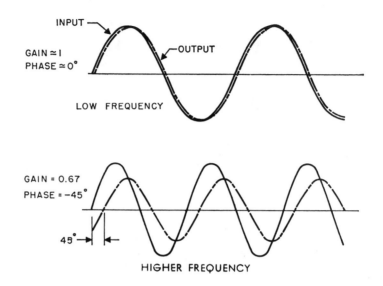

Fig. 14-2. Input/output signal relationship. As frequency increases, gain decreases and lag (phase) increases.

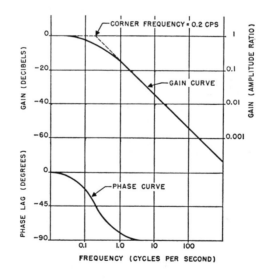

Fig. 14-3. Bode diagram of a valve actuator (single-time-constant system). Corner frequency gives the time constant and also the 45 degree phase-shift point. A Bode diagram can be drawn from time constants alone.

ponents can be added graphically to produce the gain of the entire system. This is the same as multiplying the individual gains (since both scales are based on logarithms), and the product of these gains is the gain of the entire system.

Although the frequency response technique is theoretically restricted to linear systems, nonlinear systems also can be tested if the amplitude of the input signal is kept sufficiently small, and provided there are no discontinuities.

Finding the Time Constant from the Bode Diagram

The Bode diagram of the valve shown in Figure 14-3 is similar to that of a single-capacity single-resistance system. The time constant of a system is the time it takes to develop one radian (one cycle divided by 2π) at the so-called corner or break frequency. Gain $(G) = 1/(1 + j\omega T)$; at high ω, $G \simeq 1/j\omega T$; where $G = 1$, $\omega T = 1$; $T = 1/\omega$. The corner frequency lies at the intersection of the unity-amplitude-ratio line (gain $= 1$, db $= 0$) and the line which is asymptotic to the downward slope of the curve. It is a measure of system response; the higher the break frequency, the faster the response. In other words, as the flat (horizontal) portion of the curve extends farther to the right, higher frequency disturbances can be handled. In Figure 14-3, the corner frequency f_c is 0.2 cycle per second; therefore, the time constant is

$$T = \frac{1}{2\pi f_c} = \frac{1}{2\pi \times 0.2} = 0.8 \text{ second} \tag{14-1}$$

Testing a System

Figure 14-4 is a block diagram of a heat-exchanger control system. If the transfer function of each component or block is known, actual tests are unnecessary because a mathematical approach can be used to obtain the response curves. In some cases, curves for components such as

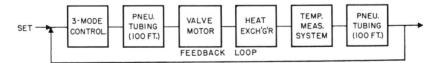

Fig. 14-4. Block diagram of heat exchanger control system.

transmitters, receivers, and valve motors are available from the instrument manufacturer or from the technical literature. An entire process and its control system can sometimes be simulated wholly on paper with no equipment necessary.

If a step-response test has been performed, and the time constants determined, the frequency response curves can be constructed by using the time constants—because the response curves for a first-order system have a characteristic shape. Since the time constant is known, the corner frequency can be calculated from Equation 14-1 and the decibel curve can be fitted to the corner frequency valve, using a slope of -6 db/octave (gain decreases by a factor of 2 as frequency increases by a factor of 2).

Location of the phase curve for a single time constant system also depends on the corner frequency—this is where the 45-degree phase shift occurs. At corner frequency ($\omega T = 1$), $G = 1/(1 + j\omega T) = 1/(1 + j)$, representing a vector at an angle of 45 degrees, with length $1/\sqrt{2}$, or 3 db down. This curve is shown at bottom in Figure 14-3.

However, a frequency test must be run if response data are not available for all components. Although each system component could be tested individually, it is easier to group several components and test the group. For instance, the heat-exchanger, valve motor, and temperature-measuring system can be tested simultaneously to yield the frequency response curves (gain and phase) for the three components. The test could be expanded to include the two lengths of pneumatic tubing. However, since tubing length might be changed at some later date, it is best to obtain tubing data separately. If the length does change, new response curves can be graphically added to the curves for the other components.

Once the response data have been gathered, the gain and phase curves are plotted (Figures 14-5 and 14-6).

In the heat exchanger example, it will be assumed that the steady-state gain of each component, and thus the entire process, is unity. In Figure 14-5, the gain curve for the three components (heat exchanger, valve, and temperature-measuring system) is added to the curve for the pneumatic tubing to produce the curve for the entire process. Since there are two 100-foot lengths of tubing, the tubing curve is added twice. (For simplification, this curve is an average of two curves, one for a pneumatic line terminating in a valve motor and the other for a line terminating at the controller.) Measurements are made from the zero-db line, and points on the same side of the line are additive.

A controller has a response curve for each combination of control-

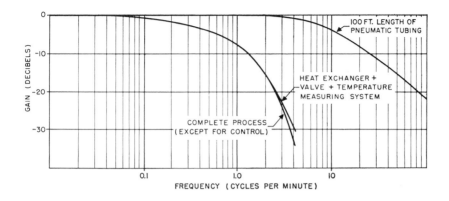

Fig. 14-5. Gain curves.

ler settings. However, this family of curves can be simplified into a single curve for each control mode because the integral action affects the low-frequency response, the derivative action affects the high-frequency response, and the proportional action is simply a horizontal line. A set of the simplified curves is shown in Figure 14-7. All curves are shifted up or down as the proportional band is changed; the integral line is shifted to the left as the integral time is increased, and the derivative curves also are shifted to the left as the rate action is increased.

For continuous cycling to occur, the system must simultaneously have a gain equal to or greater than zero db (gain of unity or more) at a

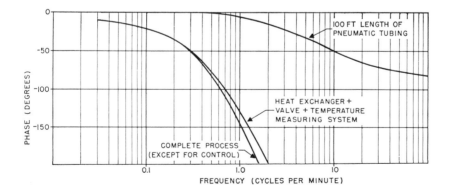

Fig. 14-6. Phase curves.

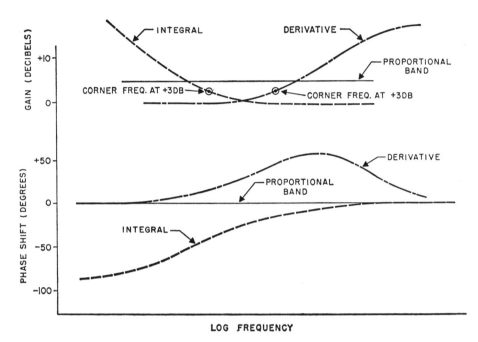

Fig. 14-7. Simplified control-mode response curves.

phase shift of −180 degrees. The proportional band that results in a gain of unity at this −180-degree point is called the *ultimate proportional band*. The ultimate proportional band is the one which, when added to the process curve, produces a gain of unity at the critical −180-degree phase point.

In Figure 14-6, the −180 degree phase shift occurs at a frequency of 1.4 cpm. At this frequency, the gain is −11 db, or an amplitude ratio of 0.28. To raise the process curve to zero db at 1.4 cpm, the proportional band (*PB*) curve would be a horizontal line at +11 db or 3.55 amplitude ratio. Since *PB* is (1/amplitude ratio) × 100, the ultimate proportional band (*PB$_u$*) is, therefore,

$$PB_u = \frac{1}{3.55} \times 100 = 28\%$$

A second factor of interest, called *frequency ratio,* is a measure of the system response beyond the −180-degree phase point. The ratio of the frequency at −270-degree phase change to the frequency at −180° phase change gives this. For the heat exchanger, this ratio (*f$_r$*) is 2.8 cpm/1.4 cpm, or 2.0.

Dead time is another factor to consider. Dead time increases phase lag without affecting the gain curve. If the total dead time for the system is greater than about 0.1 of the largest system time constant, there will tend to be overshoot with three-mode control.

The PB_u tells much about the process. A rough estimate of offset (with straight proportional action) is $PB_u/2$. In the heat exchanger this would mean an offset of approximately 28/2, or 14 percent. Therefore, if the span of the temperature measuring instrument were 100 degrees, offset would be in the neighborhood of 14 degrees. If this amount of offset proved to be troublesome, the integral mode should be considered, keeping in mind that the addition of integral action can cause overshoot on startup.

Another rough indicator can be used to point out the necessity for derivative action. If the 270 degree/180 degree frequency ratio (f_r) is two or greater, derivative can be used to advantage. Derivative action raises the high-frequency end of the process curve to increase the -180-degree frequency.

For the heat exchanger, all three control modes were selected.

Control Objectives

Since cycling occurs when a process plus its control has a gain of one or greater at the same time that its phase is -180 degrees, the major control objective is to separate this phase and gain by as large a frequency spread as practical. To ensure this spread, two quantities are defined—*gain margin* and *phase margin* (Figure 14-8). Gain margin is the amount that the gain curve differs from unity (0 db) at the -180-degree frequency. Phase margin is the amount by which the phase curve differs from -180 degrees at the 0-db frequency.

When one of the margins is set at a desired value, the other margin is fixed. Widening either the phase or gain margin reduces the settling-out time (after a change in set point) until, in the ultimate, all overshoot is eliminated. In general, the gain margin does not fall below -5 db nor exceed $+10$ db. When the phase margin is used as the controlling factor, it usually lies between 40 degrees and 60 degrees, with 50 degrees as a good compromise.

Control mode curves must be added in sequence—derivative first, followed by integral and proportional band. The proportional band must be added last because it adjusts the process gain without affecting the phase; the other modes affect both gain and phase simultaneously.

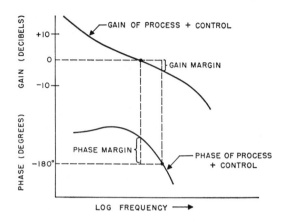

Fig. 14-8. Gain and phase margin described graphically.

Adding Derivative

Although placement of the derivative curve is somewhat arbitrary, one successful method is to align the +50-degree point of the derivative curve with the −180-degree point of the process curve (Figure 14-9). When these two curves are added graphically, the −180-degree fre-

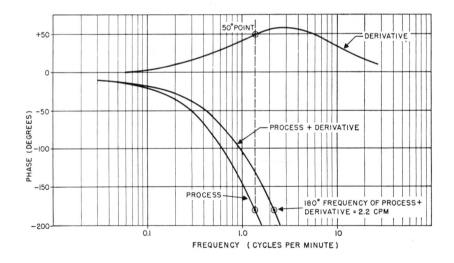

Fig. 14-9. Phase curve of process plus derivative.

quency shifts to the right. The position of the derivative phase curve determines the position of the derivative gain curve.

Adding the derivative gain curve to the process curve (Figure 14-10) raises the process gain curve. At the new − 180-degree frequency of 2.2 cpm, the gain of process plus derivative becomes − 11 db. The derivative time is also a function of the corner frequency of the derivative gain curve; it is the frequency at which the gain of the curve is + 3 db.

$$\text{derivative time} = \frac{1}{2\pi \times \text{derivative-curve corner frequency}}$$

From Figure 12-10, the corner frequency is 1.1 cpm; therefore,

$$\text{derivative time} = \frac{1}{2\pi \times 1.1} = 0.142 \text{ minute}$$

Adding Integral

The next control mode to be added to the process-plus-derivative curve is the integral (Figure 14-11). Here the − 10-degree point of the integral

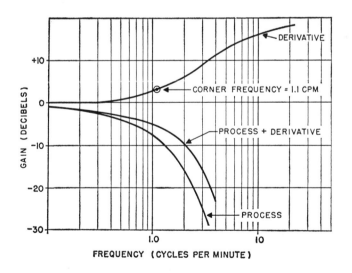

Fig. 14-10. Gain of process plus derivative.

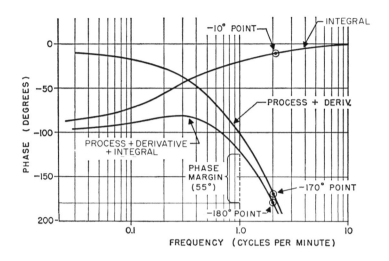

Fig. 14-11. Phase of process plus derivative plus integral showing phase margin.

curve is aligned with the −170 degree point of the process-plus-derivative curve. This shifts the −180 degree slightly left to a frequency of 2.0 cpm. In general, adding integral lowers the total phase curve since the integral phase curve lies entirely below zero.

Adding the integral gain to the process-plus-derivative curve (Figure 14-12) raises the low-frequency response. Note that the high-frequency end of the curve does not change.

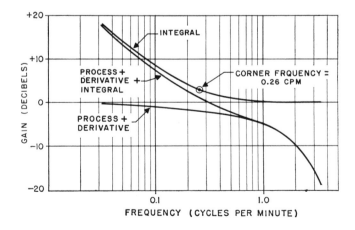

Fig. 14-12. Gain of process plus derivative plus integral.

$$\text{integral time} = \frac{1}{2\pi \times \text{integral-curve corner frequency}}$$

The corner frequency here is also at the 3-db point and is equal to 0.26 cpm. Therefore,

$$\text{integral time} = \frac{1}{2\pi \times 0.26} = 0.615 \text{ minute/repeat}$$

If the controller is calibrated in repeats per minute, the reciprocal of this number is used.

Adjusting the Proportional Band

Adjusting the proportional band moves the gain curve vertically but has no effect on phase. From Figure 14-11, the -180-degree frequency of process-plus-derivative-plus-integral curve is 2.0 cpm.

From Figure 14-12, the gain of process plus two control modes (corresponding to 2.0 cpm) is -10 db, or the gain margin without proportional band is 10 db. To obtain a gain margin of 5 db, the curve can be raised 5 db. This is accomplished by adding a proportional band of $+5$ db to the process-plus-derivative-plus-integral curve. Since $+5$ db $= 1.78$ amplitude ratio,

$$PB = \frac{1}{1.78} \times 100$$
$$= 56.2\%$$

The final gain curve is shown in Figure 14-13.

The phase margin is found by first noting the frequency of the gain curve of the process plus the three control modes at 0-db gain. From Figure 14-13 this is 1.0 cpm. Since at 1.0 cpm the phase is -125-degrees, the phase margin (Figure 14-11) is 55 degrees (180 degrees $- 125$ degrees). The curve of process plus derivative plus integral ($PB = 0$) in Figure 14-11, and the gain curve of the process plus all three control modes in Figure 14-13, are the open-loop response curves of the process plus its control.

Closed-Loop Response

Using data from the open-loop response curves, closed-loop response can be found from a *Nichols diagram*. The Nichols diagram, or phase

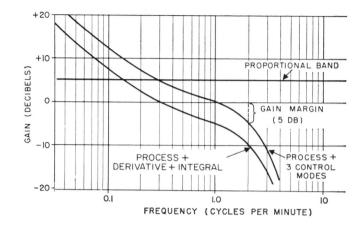

Fig. 14-13. Gain of process plus all three control modes. A controlling gain margin of 5 db determines the value of the proportional band.

margin plot, may be used to define system performance. It is a plot of magnitude versus phase. The magnitude is plotted vertically, and the phase is plotted horizontally. For a more detailed description, refer to an advanced text on frequency response techniques. Closed-loop response is automatic control; manual control is open loop. If the open- and closed-loop responses are plotted on a single Bode diagram, much interesting information can be obtained. For instance, a plot of the ratio of closed-loop to open-loop magnitude ratios reveals the effectiveness of the controller in handling various disturbance frequencies. Also, if the process plus its control resembles that of a second-order system (this is valid in many cases), such quantities as the natural frequency of the system, damping factor, and settling time can be estimated. Also calculable is the peak deviation after a set-point change.

Conclusion

The frequency response method is both a test and a graphical technique to obtain information about the control loop. The instrument manufacturer can put these tests to good advantage in evaluating instruments. However, few plants would allow the process disruptions caused by frequency response testing. Perhaps the greatest benefits are the graphical solutions of rather complex problems and the simultaneous understanding of the system that is developed.

Questions

14-1. The frequency response technique is theoretically restricted to:
  a. Nonlinear systems   **c.** Closed-loop systems
  b. Resonant systems   **d.** Linear systems

14-2. The frequency response analysis technique is:
  a. A test that can be quickly conducted without special equipment
  b. A test that does not upset the process in any way
  c. A test that will reveal controller settings, process time constants, etc.
  d. Not suitable for large-capacity systems

14-3. The first-order time constant or single time constant occurs at a phase shift of:
  a. 90.0 degrees   **c.** 45.0 degrees
  b. 63.2 degrees   **d.** 36.8 degrees

14-4. In constructing a Bode diagram on linear graph paper the gain (vertical) axis should be expressed in terms of:
  a. Decibels   **c.** Linear gain
  b. Gain ratio   **d.** Maximum gain

14-5. In using the frequency response method as a graphical analysis technique, the only component that usually requires actual testing is the:
  a. Controller   **c.** Closed-loop combination
  b. Process   **d.** Control valve operator

14-6. The frequency response of a three-mode controller:
  a. Displays the quality of the controller
  b. Varies according to the manufacturer
  c. Can be manipulated to a desired curve shape through mode adjustment
  d. Is fixed and always remains rigid

14-7. The frequency response gain curve is *not* affected by:
  a. Proportional band   **c.** Integral
  b. Dead time   **d.** Derivative

14-8. When control mode curves are added, the following order must be observed:
  a. Proportional, integral, and derivative
  b. Derivative, proportional, and integral
  c. Integral, proportional, and derivative
  d. Derivative, integral, and proportional

14-9. When the frequency response curves of all loop components are added, the resultant curve becomes:
  a. The closed-loop curve
  b. Useful in that the Nichols diagram will then reveal closed-loop response
  c. Interesting from an academic standpoint only
  d. A measure of controller effectiveness

14-10. Perhaps the greatest benefit of the frequency response technique is that it:
  a. Provides a method of controller adjustment
  b. Makes it possible to predict stability
  c. Provides a method of measuring time constants
  d. Provides the student of automatic control with a clearer insight as to what occurs within the control loop

Split-Range, Auto-Selector, Ratio, and Cascade Systems

Although conventional controllers satisfy most process control requirements, improvements can be achieved in some situations by combining automatic controllers (duplex, cascade, ratio). All of the combination controller actions to be described can be achieved with either pneumatic or electronic equipment.

Duplex or Split-Range Control

A duplex controller has one input and two outputs. It may have two control mechanisms, each with an output, or a single-control mechanism operating two control valves by means of relays or positioners.

The need for such a system is apparent in a process such as electroplating. For example, the quality of bright chromium plating depends largely on proper temperature control of the bath. With a given current density and bath composition, variations in temperature not only affect the final appearance, but also the rate of chromium deposition.

Temperatures a few degrees too high produce dull, lusterless plating. When tools are being plated with a hard chromium finish, low temperatures can cause hydrogen embrittlement of the steel, resulting in cracking and tool failure.

Desired thicknesses and finishes can only be produced on successive jobs if the operating conditions are duplicated exactly (Figure 15-1). Plating current flowing through the electrolyte generates heat that is normally dissipated by a controlled flow of cooling water through the tank coils. Frequently, however, when large, cold metal pieces are introduced, the solution becomes too cool for good plating. Also, on startup, the temperature must be brought up to the operating level or the initial batch will be below specifications. The problem is solved by a duplex controller that adds cooling water when the temperature is too high, and steam to the heating coils when the temperature is too low. If the temperature is within acceptable limits, neither cooling water nor steam is admitted.

The control action is pneumatically produced by a conventional controller with proportional action (left in Figure 15-1). Controller output is simultaneously fed to (1) the receiver bellows in a valve positioner, located on the water (cooling) valve, and (2) a similar bellows in a second positioner located on the steam (heating) valve. If the controller proportional band is adjusted to 10 percent, the response will be as shown in Figure 15-2. The positioner on the steam valve is adjusted to stroke the valve through its full travel when the air signal on the valve positioner receiver bellows goes from 3 to 9 psi (20 to 60 kPa); the valve is closed with 9 psi (60 kPa) applied to the positioner and wide open with 3 psi (20 kPa) applied. The positioner on the water valve is adjusted to operate in the opposite direction; it opens the valve fully with

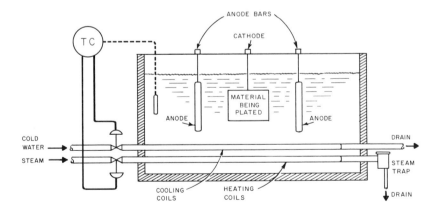

Fig. 15-1. Plating process requires a duplex controller, one with one input but two outputs.

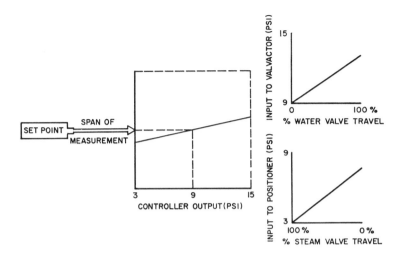

Fig. 15-2. Response of proportional controller with a proportional band setting of 10 percent.

15 psi (100 kPa) applied and closes it when the signal pressure is 9 psi (20 kPa).

A properly aligned pneumatic proportional controller produces an output of 9 psi (60 kPa) when measurement and set point agree. In this example, both valves would be closed with a signal pressure of 9 psi (60 kPa). If measured temperature rises above, or falls below, the set point, water or steam would be circulated in proportion to the measurement's deviation from the set point.

Integral and derivative action could be incorporated in the electroplating control system. However, neither one is required normally because a system of this type usually is required to control only within a narrow band of temperature (say 10 percent band) between heating and cooling. Thus, integral and derivative modes are unnecessary.

On/off control sometimes proves to be the most economical and satisfactory solution to the same problem—provided the process can tolerate the cyclic operation that will occur.

Auto-Selector or Cutback Control

Auto-selector control is the opposite of duplex in that it allows the automatic selection between two or more measurement inputs and provides a single output to control a single valve. The auto-selector-

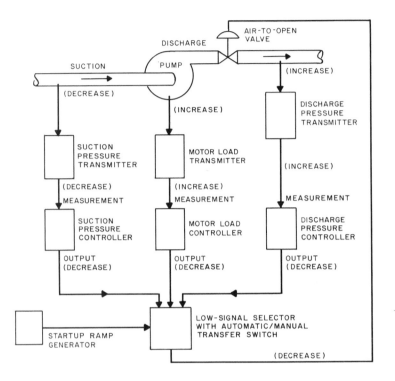

Fig. 15-3. Pipeline operates safely on auto-select/cutback control.

control system continuously senses all measurements applied to it and provides control action to the valve based on the value of the measurement closest to its particular set point. All measurements must be interdependent, that is, directly related to each other, for successful auto-selector application.

While control may be either pneumatic or electronic, the electronic type is ideal for this technique. A simple selector system, often found in pipeline control, will be used as an illustration (Figure 15-3).

In attempting to operate a pumping station efficiently on a pipeline, it is desirable to operate the control valve wide open at all stations except one. One station will then become the limiting or throttling unit and will pace the line based on the delivery requirements. However, if any of the following conditions occur, it is necessary to cut back on the position of the control valve:

1. Suction pressure drops too low.
2. Motor load rises too high.
3. Discharge pressure rises too high.

If the suction pressure drops below a predetermined level, there is danger of causing the pump to cavitate. If motor load rises above a predetermined level, there is danger of overloading the motor and burning it out. If discharge pressure rises too high, there is danger of causing problems downstream. Cutting back on the position of the final operator will cause each of these variables to return to a safe operating level.

Consider the operating parameters in the following example for a pumping station:
1. All variables are on the "safe" side of their respective set points. Suction is sufficiently high; motor load and discharge pressure are sufficiently low.
2. A decreasing suction pressure will cause the output of the suction-pressure controller to decrease.
3. An increasing motor load will cause the output of the motor-load controller to decrease.
4. An increasing discharge pressure will cause the output of the discharge-pressure controller to decrease.
5. A low-signal selector is used. Its output will be equal to the lowest input.
6. A decreasing input to the low-signal selector will cause its output to decrease.
7. The control valve opens as its input increases (air-to-open).

Since all variables are operating in their "safe" zones, all controllers will have a maximum output, that is, 100 percent, and the control valve will be wide open.

Assume that the suction pressure drops below the predetermined low limit, as will be indicated on the suction-pressure controller's set-point dial. Output of the suction-pressure controller will start to decrease. This decreasing output will be selected by the low-signal selector and transmitted to the control valve. The control valve will close until the suction pressure is brought equal to the preset limit. Assume this condition is met with a 50 percent valve position.

If all of the controllers in a cutback system contain proportional-plus-integral action, integral windup should be anticipated. In order to prevent this condition, a special type of "external feedback" is employed. The selected output is fed back to each controller's integral circuit. This feedback (rather than conventional feedback, where a controller's output is fed to its own integral circuit) conditions all of the controllers whose outputs are not being selected.

Assume one of the other variables, for example motor load, approaches its limit. As soon as the load exceeds the limit, the output of the motor-load controller will begin immediately to decrease from 50 percent (the output of the low selector) and not from 100 percent as would be the case if external feedback had not been provided. If the control valve must start to move before the motor load reaches the limit, a batch-cutback (auto-selector) controller should be used.

A slow, ramping-open of the control valve is a common requirement in starting up a pumping-station pipeline-control system. A ramp generator providing a constantly increasing signal (commencing on operator command) will smoothly open the valve at a controlled rate. This generator simply becomes another input to the low-signal selector.

Provisions for switching the entire system to the manual mode of control are normally incorporated into the system as an integral part of the signal selector, or sometimes as a separate unit.

If an air-to-close valve is required, a high-signal selector should be specified. The switch on each controller that determines whether an increasing measurement will cause an increasing or decreasing output should be changed to the action opposite to that used in the preceding example.

Nonpipeline Application of Auto-Select/Cutback Control Systems

A nonpipeline application of a cutback system is shown in Figure 15-4. In this system, the outlet temperature of the product is sacrificed if the oil pressure drops below a preset low limit. In normal operation, the temperature controller will regulate the oil pressure, and thus the heat input, in order to maintain the product temperature at the desired control level.

If the demand for oil by the exchanger should start to diminish the main oil supply (indicated by a falling oil-header pressure), the oil-pressure controller will cut back the control valve, reducing the product temperature and, at the same time, the consumption of oil. The header pressure will now be allowed to recover and remain at the predetermined low limit. Should other process demands for oil be reduced, the oil-pressure controller will gradually open the oil valve until complete control is returned to the temperature controller.

The one major requirement for auto-selector application is that all of the measured variables must be interdependent, that is, regulated by the single manipulated variable.

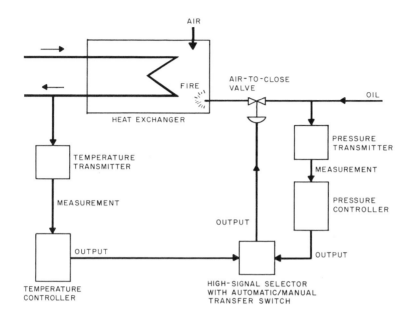

Fig. 15-4. Temperature adjusts to reduce oil pressure.

Flow-Ratio Control

Assume a fixed ratio between two flow rates is to be controlled. This would suggest a flow-ratio control. Acid-to-water ratio for a continuous pickling process found in certain plants working on metal would be typical. An acid-water-ratio control system for continous pickling automatically controls the addition of fresh acid in correct proportion to the water used. The accurate flow control and proportioning of the acid results in a stricter adherence to pickling specifications, and, therefore, a minimum consumption of acid. The system records and integrates the consumption of both water and acid. Control of solution level by the operator is also simplified since he need adjust only the water valve. An installation is schematically shown in Figure 15-5.

Normally, variation of the flow of water, either to fill the pickling tank or maintain the operating solution level, is accomplished by pneumatically positioning a control valve in the water line by means of a pneumatic manual setting station.

This is a typical ratio control system; however, it does have a limitation. It is assumed that the instrument components are selected—sized to provide the desired ratio of acid to water, and that this ratio will never need to change. Unfortunately, this is seldom the case, making a method of ratio adjustment necessary.

The ratio factor is set by a ratio relay or multiplying unit. This unit would be located between the wild flow transmitter and the flow-controller set point (Figure 15-5). It is desired to control Flow B in a preset ratio to Flow A. Flow transmitter A senses the wild flow. The

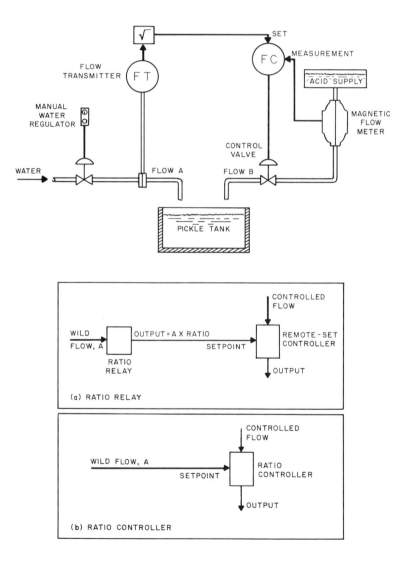

Fig. 15-5. Acid-water ratio control system.

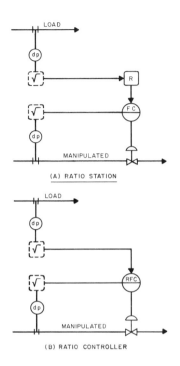

(A) RATIO STATION

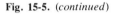

(B) RATIO CONTROLLER

Fig. 15-5. (*continued*)

ratio relay multiplies the output (0 to 100 percent) of the flow transmitter by a manually set factor:

(Flow output A) × (Preset factor) = Output of ratio relay

This output becomes the set point of the controller that regulates Flow B. At equilibrium, Flow B equals the set point of the controller, or:

Flow B = (Flow A) × (Ratio factor)

Ratio factor = Flow B/Flow A

The following example may help in understanding a ratio-control system. Assume that the range of Flow A is 0 to 100 gpm, the range of Flow B is 0 to 100; the output of transmitters A and B changes linearly from 0 to 100 percent as the flow changes from 0 to 100 percent. The ratio relay has a factor adjustment ranging from 0.3 : 1 to 3 : 1.

If the ratio factor is set at 1 (meaning a 1 : 1 ratio), for every measured gallon of A, the controller will allow 1 gallon of B to flow. If 50 gpm of A flow, then 50 gpm of B will flow. If the ratio factor changes to 2, then 2 gallons of B will flow for every gallon of A. If 50 gpm of A flow, 100 gpm of B will flow.

If the range of flow transmitter B is changed to range from 0 to 10 gpm, while A remains 0 to 100 gpm, a 1 : 10 ratio factor is built into the system. In other words, if the factor set on the ratio relay is 1, then for every gallon of A that flows, 0.1 gallon of B will flow. Hence, the range of the transmitters, as well as the multiplying factor set into the ratio relay, determines the ratio factor for the system. Therefore, care should be taken in selecting transmitter range in order to allow maximum system flexibility. Try to choose a range so that the ratio factor is normally in the middle. However, make sure that the transmitter range will cover all process conditions.

Flowmeters used in a ratio control system are often orifice meters or other constriction meters with differential-pressure transmitters. Thus, transmitter output is proportional to the square of flow, rather than being proportional to a linear relationship as in the above example. (The transmitters in the example could be magnetic flowmeters, turbine flowmeters, rotameters, and so on.) In this case, the ratio relay provides the same 0.3 : 1 to 3.0 : 1 ratios between transmitter outputs, that is between differential pressures. Since flow is proportional to the square root of differential pressure, the actual flow ratios will be $(0.3 : 1)^{1/2}$ to $(3.0 : 1)^{1/2}$, or about 0.6 : 1 to 1.7 : 1. The ratio dial will have a square root layout and indicate flow ratios of 0.6 : 1 to 1.7 : 1. This ratio range is the square root of the gain range.

A transmitter with a linear output cannot be used with another having a square root output unless one or the other output is passed through a converter to obtain its square or square root, as required.

More complicated systems may involve ratioing two or more flows to the wild flow.

A ratio control system may be implemented in several different ways with pneumatic or electronic hardware. The ratio delay may be incorporated in the controller, and a ratio dial replaces the normal set-point dial. This is normally called a single-station ratio controller (Figure 15-5).

Cascade Controller

A cascade-control system consists of one controller (primary, or master) controlling the variable that is to be kept at a constant value, and a second controller (the secondary, or slave) controlling another variable that can cause fluctuations in the first variable. The primary controller positions the set point of the secondary, and it, in turn, manipulates the control valve.

The objective of the cascade-control system is the same as that of any single-loop controller. Its function is simply to achieve a balance between supply and demand and thereby maintain the controlled variable at its required constant value. However, the secondary loop is introduced to reduce lags, thus stabilizing inflow to make the whole operation more accurate.

The cascade-control technique is shown in schematic form in Figure 15-6. Two feedback controllers are used but only one process variable (m) is manipulated. The primary controller maintains the primary variable (C_1) at its set point (r_1) by automatically adjusting (r_2) the set point of the secondary controller. The secondary controller controls the secondary loop responding to both its set point (r_2) and the secondary measurement (C_2).

The secondary controller may be regarded as an elaborate final control element, positioned by the primary controller in the same way as a single controller would ordinarily position the control valve. The secondary variable is not controlled in the same sense as the primary; it is manipulated just like any control medium. If, for example, the secondary controller is a flow controller, then the primary controller will not be dictating a valve position, but will, instead, be dictating the prescribed flow.

A simple, single-loop temperature-control system is shown in Figure 15-7, wherein the temperature of the liquid in the vessel is controlled by regulating the steam pressure in the jacket around the vessel. Since such a process normally involves a longtime constant, a three-mode controller having a long integral time is required. This will provide satisfactory control as long as the supply of steam is constant, that is, the upstream pressure does not change.

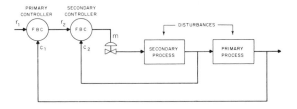

Fig. 15-6. Cascade control technique.

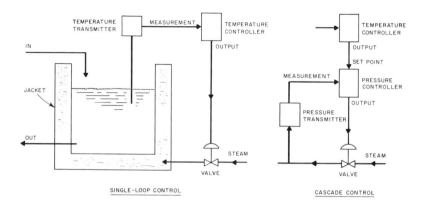

Fig. 15-7. Temperature controller provides set point for steam valve or steam-pressure controller.

However, if the supply steam is subject to upsets, a different control scheme is needed. The temperature controller will not know that the heat input (steam pressure) has changed until the temperature of the liquid in the vessel starts to change. Since the time constant of the temperature loop is long, the temperature controller will take a long time to return the process to an equilibrium (a new steam-valve position corresponding to the new level of heat input). Therefore, it is desirable to correct for the change in heat input before it affects the temperature of the liquid.

The cascade control system shown in Figure 15-7 will both control the temperature in the vessel and correct for changes in the supply steam pressure. The steam-pressure controller, called the secondary or slave controller, monitors the jacket inlet steam pressure. Any changes in supply pressure will be quickly corrected for by readjusting the valve because this loop has a fast time constant, and, hence, a short integral time. The temperature controller, normally called the primary or master controller, adjusts the set point of the pressure controller as dictated by the heat requirements of the incoming feed. Thus, the secondary controller has its set point set by another controller rather than by a manually adjusted dial. These secondary devices are termed remote-set controllers.

For the cascade control system in the above example, the time constant of the secondary control loop must be significantly faster than that of the primary loop. If the time constant of the secondary loop is too long—equal to or even greater than the primary loop—cascade control will provide no benefits. It may, in fact, cause stability problems. An example of this occurs when a valve positioner, described in Chapter 10, is used for flow control. The flow process is faster than the valve positioner and this produces a continuous cycle. Stating it another way, the slave controller will not accept orders from the master controller faster than it can carry them out.

Typical secondary loops or inner loops and primary loops may be generally classified as follows:

Types of Inner Loops	*Set By*
	Anything but flow
Valve position	Anything
Flow	Temperature (with longer time constant)
Temperature	

Inner (Secondary) Loops	*Control Modes*
Valve position	Proportional only
Flow	Proportional plus integral
Temperature	Proportional only

Mathematically, it could be demonstrated that a cascade system will increase the natural frequency and reduce the magnitude of some of the time constants in the system, both beneficial results. However, the more evident benefits are reduced effect from disturbances, more exact adjustment in the presence of disturbances, and the possibility of incorporating high and low limits into the secondary control.

Saturation in Cascaded Loops

Integral windup or saturation, mentioned in Chapter 11, can become a problem in cascade control. Assume, for example, the process load demands more than the control valve can deliver. This would result in a sustained deviation of the secondary measurement from its set point. The primary measurement will then change and result in a change of the primary controller's output. This will readjust the set point of the secondary controller.

In practice we do not want the primary controller to change its output. A simple way to accomplish this in a pneumatic system is to use the secondary measurement as the integral to the primary controller

(Figure 15-8). In a conventional controller, as in Chapter 11, the integral input is the output. In the cascade system this is the secondary set point. As long as the secondary measurement and set point are equal, integral action of the primary controller proceeds in normal fashion. If the secondary measurement can not follow its set point, integral action in the primary controller will stop. It will resume only when the secondary measurement once again can follow set point. To accomplish this, both primary and secondary controllers must have integral action and the primary controller must be equipped with an external integral connection.

This arrangement also simplifies transfer from automatic to manual. With conventional integral action, the primary controller would saturate if the secondary were placed in the manual position. With the arrangement described, this cannot happen as the primary controller's integral bellows will track the process (secondary) measurement.

In the electronic system (SPEC 200), a similar arrangement may be employed to prevent saturation of the primary controller. Just as in the pneumatic system described, secondary measurement is used as the integral input to the primary controller resulting in operation similar to the pneumatic system.

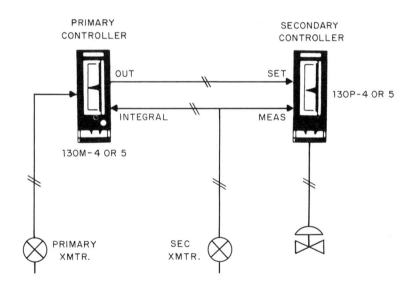

Fig. 15-8. Primary and secondary controllers.

Logical use of cascade control systems requires nothing more than standard control instruments. A standard unit can provide indication or recording and control of both the primary and the secondary variables, plus output indication and manual-automatic transfer. That is all that is needed for normal operation.

To tune a cascade-control system, the following procedure is suggested:

1. Place both controllers in manual.
2. Adjust proportional and integral settings to conservative values (wide proportional band and long integral time) on the secondary controller.
3. Place secondary controller in automatic and then tune, using any of the procedures outlined in Chapter 12.
4. Adjust the proportional, integral, and derivative settings on the primary controller to conservative starting values.
5. Place the primary controller in automatic and then tune, using any of the procedures in Chapter 12.

Always tune the secondary controller first (with the primary controller in manual), and then the primary controller.

Summarizing the major advantages of cascade control:

1. Disturbances affecting the secondary loop are corrected by the secondary controller before they are sensed in the primary loop.
2. Closing the secondary loop reduces the lags sensed by the primary controller, thus increasing the speed of response.

Questions

15-1. A duplex controller has the following number of inputs:
a. One
b. Two
c. One for each output
d. As many as required

15-2. The duplex controller provides the following number of control valve signals:
a. One
b. Two
c. One for each input
d. As many as required

15-3. An auto-selector control system should be considered when:
a. The process has more controlled variables than manipulated variables
b. The process has more manipulated variables than controlled variables
c. Two independent control systems are not economical
d. Two or more measured variables must be isolated

15-4. The auto-selector controller will function only when the measurement inputs are:

a. Interdependent

b. Independent

c. Isolated from one another

d. Flow measurements

15-5. The ratio controller:

a. Can be used with any combination of related process variables

b. Has one measurement input and two outputs

c. Can be used for even-numbered ratios

d. Must always employ the derivative mode in the controller

15-6. Two flows are to be ratio controlled, the first is measured with an orifice and the second with a shedding vortex flowmeter. The signals must be:

a. Both made linear by using a square root extractor on the orifice signal

b. Used as if a 1 percent error can be tolerated

c. Both made linear by using a square root extractor on the shedding vortex transmitter

d. Flow-calibrated

15-7. A cascade control system will increase the natural frequency of the control loop along with having the following influence on the magnitudes of the associated time constants:

a. Reduce them

b. Increase them

c. Not affect them

d. Have an unpredictable effect

15-8. In a cascade control system the secondary may be regarded as:

a. A valve actuator

b. An elaborate final control element

c. A means of slowing down regulation of the manipulated variable

d. A sophisticated noise filter

15-9. If two square root flow signals are to be ratio-controlled and the ratio factor adjusted, the maximum practical factor is:

a. 0.6 to 1.7

b. 0.3 to 3.0

c. 1.0 to 10

d. 1.0 to 1.0

15-10. A very common secondary loop that may be used with almost any process except flow is:

a. Flow

b. Temperature

c. Pressure

d. Valve position

15-11. A cascade control system is to be adjusted. You should first:

a. Place the primary controller on manual and adjust the secondary controller

b. Place the secondary controller on manual and adjust the primary controller

c. Place both controllers on automatic and go through the conventional adjustment routine

   d. Bypass the secondary controller and adjust the primary controller by
   the conventional method

15-12. If the time constant of the secondary loop is greater than the time
constant of the primary loop, cascade control will:
   a. Shorten the period of the total system
   b. Lengthen the period of the total system
   c. Improve the operation of the system
   d. Provide no benefits

16

Feedforward Control

Definition

The term feedforward control applies to a system in which a balance between supply and demand is achieved by measuring both demand potential and demand load and using this information to govern supply.

History

The first feedforward control system was used in about 1925 to control boiler drum level. In this type of feedforward control, which is described in Chapter 9, water inflow is directly regulated by steam outflow. Feedback in the form of bias, or trim, corrects for any cumulative measurement or other error in the system. This system, which is still in everyday use, met an urgent need for better level control. It was applied to a process that was completely understood and required no special instrumentation. Many years passed before feedforward techniques were applied to other process applications.

Advantages

The potential applications of feedforward techniques are virtually endless, but they usually involve a distillation process. Several improvements are provided by feedforward control. One is reduced measurement lag. For example, in a distillation column, the lag between input and output may vary from a few minutes to many minutes, depending on the size of the column and other variables. A feedback control system cannot make any required corrections until an error is sensed. With the long lags inherent in such a process, good control becomes difficult. However, with the feedforward technique, an exact balance between supply-and-demand potential is achieved before the process is influenced, and improved control results.

Technique

Before feedforward control can be applied effectively to any process, the process must be completely understood. It must be possible to write one equation that accurately states the material balance, and another equation, if required, that shows the energy balance. If the process creates interactions between energy and material balances, this also must be understood. The development of feedforward control was delayed because of the lack of such understanding.

When the equations that completely describe the process are written, the next step is to manipulate the equations into a form that can be solved by available instrumentation. The absence of this type of analog-computing instrument was another reason for the delay in developing feedforward techniques. This type of instrumentation has become generally available only in the last few decades. Other instrumentation that will provide dynamic compensation must also be incorporated into the system to overcome the unavoidable leads and lags contributed by other process components. If all elements were perfect, the straight feedforward system would result in ideal control of the process. However, in practice, very small errors can accumulate over a period of time, and small nonlinearities can add up to difficulties. They can be overcome by adding feedback to trim the feedforward system. This results in the optimum feedforward control system.

Application to Heat Exchanger

Figure 16-1 represents a typical heat exchanger set up for conventional feedback control. Figure 16-2 shows the recovery curves that result from typical load upsets under feedback control using a three-mode

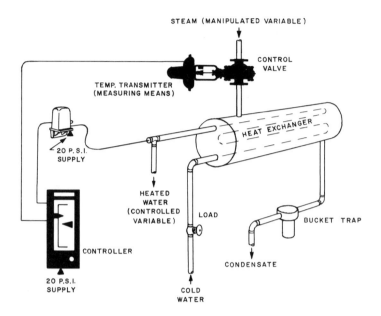

Fig. 16-1. Heat exchanger with conventional feedback control.

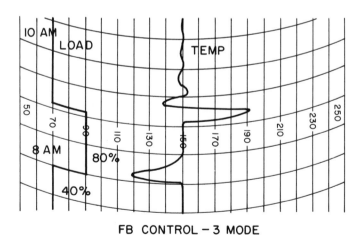

Fig. 16-2. Curve or steam response with feedback.

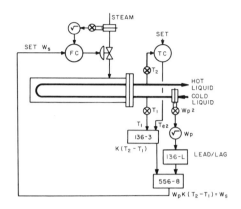

Fig. 16-3. Heat exchanger with feedforward control.

controller. The recovery curve in Figure 16-2 is a good example of ideal feedback control, but because no correction can occur before an error develops, the overshoot reaches a substantial value.

Figure 16-3 shows the same heat exchanger, controlled with a feed-forward system. The energy balance that the control system must maintain is:

heat in = heat out

or

$$W_s H_s = W_p C_p (T_2 - T_1)$$

where:

W_s = steam flow (pounds per hour)
H_s = heat of vaporization (Btu per pound)
W_p = water flow (pounds per hour)
C_p = specific heat of the liquid (Btu per pound [F])
$T_2 - T_1$ = temperature rise

When solved for the manipulated variable W_s, the equation becomes

$$W_s = W_p \frac{C_p}{H_s} (T_2 - T_1) \tag{16-1}$$

If the specific heat of the liquid (C_p) and the heat of vaporization of the steam (H_s) remain constant, the equation may be written:

$$W_s = W_p K (T_2 - T_1) \tag{16-2}$$

This equation is easily handled by readily available analog computers. One device subtracts T_1 from T_2 and multiplies the result by a constant. An example of this is the Foxboro Model 56-3.

$$T_1 \rightarrow \boxed{-\Sigma \ +} \leftarrow T_2$$
$$K(T_2 - T_1)$$

Water flow is measured by an orifice plate and a differential pressure transmitter. Thus, the signal represents $W_p{}^2$. The Foxboro Model 557 may be used to extract the square root.

$$W_p{}^2 \rightarrow \boxed{\sqrt{}} \rightarrow W_p$$

These two signals $K(T_2 - T_1)$ and W_p must be multiplied to complete the equation for the manipulated variable W_s (steam flow):

$$K(T_2 - T_1)$$
$$\downarrow$$
$$W_p \rightarrow \boxed{X} \rightarrow W_p K(T_2 - T_1) = W_s$$

This signal is then fed to the steam flow control loop.

If typical values are now assigned to the heat exchanger process, the techniques of scaling the system can be demonstrated.

Assume the following conditions:

$W_s = $ 0 to 3,000 pounds per hour
$H_s = $ latent heat (steam saturated at 100 psi condensing at 0 psi), 1,010 Btu per pound from steam table
$W_p = $ 8.33 (pounds per gallon) (60 gpm) (V_w)
       $V_w = $ 0 to 50 gpm
       and $W_p = $ 8.33 (pounds per gallon) (60 minutes per hour)
                  $= $ 8.33 (60) (V_w) pounds per hour
$C_p = $ specific heat of water $= $ one Btu per pound
$T_2 = $ outlet temperature $= $ 50 to 250°F
$T_1 = $ inlet temperature $= $ 0 to 100°F

Applying these conditions to Equation 16-1

$$W_s = 8.33 \ (60) \ \frac{(1)}{1,010} \ V_w(T_2 - T_1)$$

$$= 0.495 \ V_w(T_2 - T_1) \tag{16-3}$$

From this point, scaling depends on the transmission method. Assume a pneumatic system:

When pneumatic devices are to be used, the subtraction of T_1 from T_2 must be made separately prior to multiplication.

The first problem to be resolved is the discrepancy between the ranges of T_1 and T_2. The most direct way to resolve this is to change the scale of T_1 from 0 to 100°F to 50 to 250°F. Now

$$T_1 = T_1$$

If the input signal is given as T_1', then

$$T_1 = 100 \ T_1' \ \text{(where } 100 = \text{span of input)}$$

If the output signal is T_1'' then

$$T_1 = 50 + 200 \ T_1'' \ (50 = \text{base and } 200 = \text{span of output)}$$

Combining

$$50 + 200 \ T_1'' = 100 \ T_1'$$

The scaled equation is

$$T_1'' = 0.5 \ T_1' - 0.25 \ \text{(this solves for output in terms of } T_1'' \text{ which is}$$
$$\text{desired scaled output)}$$

The pneumatic computer has a common multiplier for both positive and negative inputs or bias terms. This makes it necessary to rearrange the equation

$$T_1'' = 0.5 \ (T_1' - 0.5)$$

This equation is solved by a Foxboro M/56-693. Its gain is adjusted to a value of 0.5 and two bellows are replaced with springs. Now a scaled equation may be written for the system:

$$W_s = 3,000 \ W_s'$$
$$V_w = 50 \ V_w'$$
$$T_2 = 50 + 200 \ T_2'$$
$$T_1 = 50 + 200 \ T_1''$$

Substituting (Equation 16–3):

$$3,000 \ W_s' = 0.495 \ (50 \ V_w') \ (200 \ T_2' - 200 \ T_1'')$$
$$W_s' = 1.65 \ V_w' \ (T_2' - T_1'')$$

The only scaling factor appearing is the gain 1.65. This can be distributed between the multiplier and the subtracting computer. For convenience, the element factor for the multiplier can be selected at 2.00, leaving 0.825 for the subtracting relay. Or it may be selected at 1.65 with 1.00 for the gain of the subtractor. However, in no case should the factor for the multiplier be less than 1.65, because the output of the subtractor then may have to exceed full scale for a given set of conditions, since its gain would exceed 1.00.

Now let us review the steps essential to scaling. Bear in mind that analog computers operate on signals that vary across a standard range, for example, 3 to 15 psi (20 to 100 kPa) or 4 to 20 mA dc. Within the nest of the Foxboro SPEC 200 system, the signal is 0 to 10 V, and this simplifies scaling. In order to make a correct calculation, information about the scales of all inputs and output signals, as well as conversion factors, must be considered. There is a straightforward procedure which simplifies this task of "scaling," but it must be followed rigorously:

1. Write the equation to be solved in its standard dimensional form.
2. Write equations giving each input and output in terms of its analog signal range of 0 to 1.0 (0 to 100 percent of scale). It is convenient to use a prime to indicate the analog signal in contrast to the variable itself, with its dimensions.
3. Substitute the analog signals into the original dimensional equation. Then solve for the output signal in terms of all the input signals.

The resulting equation contains entirely pure numbers and is called a "scaled equation." It may be used directly in electronic computers,

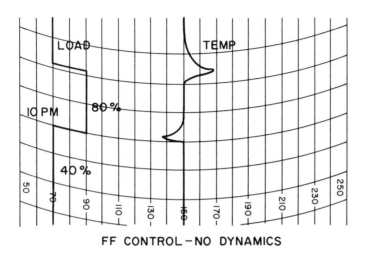

FF CONTROL – NO DYNAMICS

Fig. 16-4. Curve or system response with steady-state model.

but will have to be converted to a standard form to fit the mechanical properties of the pneumatic computers. Occasionally two or more instruments will be required to solve a given equation. Scaling requires some judgment, then, because the constants may be distributed among the various instruments; improper distribution can result in an off-scale output from one of the intermediate instruments.

The system thus has its energy balance equation solved at all times. Figure 16-4 shows a response to a load change using the system shown in Figure 16-3. Curve 4 shows improvement over the feedback system shown in Figure 16-2, but the dynamic balance between the load and the manipulated variable prevents the result from being ideal. Figure 16-5 compares a step change in water flow with a step change in steam

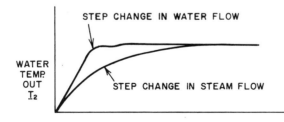

Fig. 16-5. Step changes in water and steam flow.

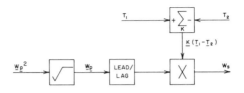

Fig. 16-6. Computing system with dynamic compensation.

flow. It can be seen that the dynamic response of the load variable (water flow) is faster than the dynamic response of the manipulated variable (steam flow).

A device capable of correcting these dynamic differences must be added to the system. This dynamic compensator is called lead/lag unit. The analog lead/lag unit is similar to a controller. Integral and derivative adjustments in conjunction with gain become the lead/lag adjustments. The unit is inserted in the flow signal (W_p) line as shown in Figure 16-6. The response of the system with dynamic compensation is shown in Figure 16-7. This is as close to perfect control as is possible at present and far better than the results obtained with the feedback system. The complexity of the feedforward control system makes it a great deal more expensive than the conventional feedback system. The feedback system is also easier to adjust. The practical conclusion is generally to use the feedforward approach only when the process requires

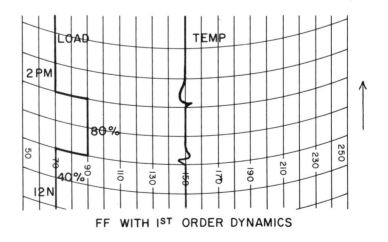

Fig. 16-7. System response with lead/lag.

the accuracy it provides, and this would seldom occur with a simple heat exchanger.

Distillation Control

Although feedforward control can be applied to almost any process, many of the most successful systems designed and applied over the past several years have been in the complex area of distillation control. Control of the distillation process has needed improvement since the column became a key part of most chemical and petrochemical processes. The problems associated with the control of a distillation process are slow response and long dead times. Distillation is affected by many variables. On-line analysis is often not available, and varying feed rates subject the column to constant upsets.

Figure 16-8 indicates that both an energy balance and a material balance are involved in the operation of a distillation column. This operation is further complicated by the fact that considerable interaction occurs between energy and material. The relationships between energy and material have been thoroughly investigated and reasonably well defined. It is possible, therefore, to measure and analyze the feed input and manipulate the energy input along with distillation (reflux rate) to keep the column under efficient control. Figure 16-9 shows a column with feedforward control on two loops arranged to control product composition. Thorough explanations of this process and the

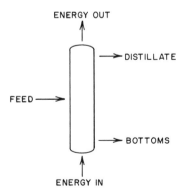

Fig. 16-8. Both energy and material balances are involved in a distillation column.

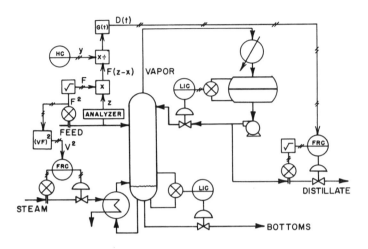

Fig. 16-9. Feedforward loops on two variables set product composition.

relationships that exist within it are given in the references listed at the end of the chapter.

Conclusion

Many feedforward systems have produced increased efficiencies in distillation processes and have justified their cost in a very short time. However, to achieve this, the process generally must be one that can demonstrate economic advantages through closer control. Each application requires considerable engineering effort, since the control system must be designed for the particular process. The characteristics of any distillation column are unique, and the control system must be both carefully selected and carefully adjusted if it is to produce good results. The feedforward system has unlimited possibilities in many process applications. The only limitation of feedforward control is that it can only be applied where complexity justifies its expense.

References

Shinskey, F. G., *Process Control Systems*. New York: McGraw-Hill Book Company, 1979.

Shinskey, F. G., *Distillation Control for Productivity and Energy Conservation*. New York: McGraw-Hill Book Company, 1977.

Questions

16-1. The feedback control system:
   a. Cannot make corrections until a measurable error exists
   b. Makes a change in output which is the differentiated error
   c. Is always superior to a feedforward system in operation
   d. Is theoretically capable of perfect control

16-2. The feedforward control system:
   a. Cannot make corrections until a measurable error exists
   b. Makes a change in output that is the integrated error
   c. Requires little knowledge of the process before installation
   d. Is theoretically capable of perfect control

16-3. The original boiler drum level application of feedforward control, developed in 1925:
   a. Is still being applied every day
   b. Was abandoned because of poor measuring techniques
   c. Worked well only with the particular boiler for which it was originally designed
   d. Was immediately recognized as the ultimate control system

16-4. A properly designed feedforward control system:
   a. Should be applied to every process
   b. Should be employed when its use can be justified economically and technologically
   c. Is always easier to adjust than a feedback system
   d. Will always result in more economical process operation

16-5. The most dramatic application of feedforward techniques has occurred in their application to:
   a. Heat exchangers
   b. Level processes
   c. Flow processes
   d. Distillation columns

Computer Interface and Hardware

For many years the slide rule, an analog device, was a universally popular and useful calculator. In recent years, the slide rule has been replaced by the pocket calculator, or electronic slide rule—a digital device. The advantages of the calculator are obvious. Similar advantages, proportional to the required investment, are possible from properly applied digital control systems.

A digital computer cannot reason for itself. Therefore, it must be provided with all the relevant data and told exactly how to solve the problem. This is called the program, or "software." Essential to a basic understanding of software and its operation is the identification of its functional components—the hardware. This chapter will be devoted to a discussion of hardware.

Basic Elements of the Computer

Computers are designed to perform the particular varied tasks assigned to them. For example, a computer designed for machine control would not make a very good filing system. Nevertheless, there are several basic elements common to all computers. These elements are shown in Figure 17-1.

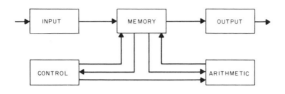

Fig. 17-1. Computer elements.

Input

In its simplest form, the input system translates the input information prepared by people into a form to which the machine can respond. Generally, the source of computer input information includes punched cards, punched or magnetic tape, or a special typewriter.

The input system converts information into a series of signals. Each signal is merely the presence or absence of a voltage or electrical current. A signal is either there or it is not there at any particular time. Just as the 26 symbols of the English alphabet can be combined into many different words to convey information, so can the two symbols of the computer alphabet (signal and no signal) be combined to form words, phrases, and sentences in computer language. When the information is put into this form, the computer is able to work with the information and process it through a series of logical operations.

The input information, now in the form of signals, is next sent to the computer memory.

Central Processing Unit (CPU)

The combination of the working memory, control, and arithmetic elements is often called the CPU. The CPU has a bus for transmitting and receiving signals, and the controllers, or modules, which are tied directly to that bus. At the "head" of the bus is the central processor, which decodes and executes instructions, a series of which are stored in memory as programs. The central processor controls the activity of its bus, performs the computations with the data it obtains from various devices connected to its bus, and stores the results of these computations in its memory, alternatively transmitting them to the appropriate bulk storage device or controller.

Memory

This element stores information until it is needed. Just as people store in their memories multiplication tables, phone numbers, addresses, schedules of what they plan to do, the computer memory stores information for future reference. The memory element is passive in that it merely receives data, stores them, and gives them up on demand. Devices used as memory elements include magnetic cores, magnetic drums, tape recorders, disks, and solid-state devices.

Control
The next element, the control unit, is active. It selects information from the memory in the proper sequence and sends the information to other elements to be used. In addition, the control unit conveys commands so that the next element down the line will perform the proper operations on the information when it arrives. Thus, the control element makes decisions.

Arithmetic
The arithmetic unit receives information and commands from the control unit. Here, the information, still in the form of symbolic words, is analyzed, broken down, combined, and rearranged in accordance with the basic rules of logic designed into the machine and according to the commands received from the control unit.

The astonishing phenomenon is that there are so few rules of logic designed into the machine. The power of the digital computer lies in the fact that it can perform complex operations rapidly by breaking them down into a few simple operations that are repeated many, many times. By performing these few simple operations over and over again in many different combinations, immense problems can be solved a bit at a time, but at fantastic speed.

Output

When the signals have passed completely through the arithmetic unit, they no longer take the form of a problem, but now assume the form of an answer. This answer is passed back to the memory unit and along to an output element. The output element reverses the process, converting the new train of signals back into a form that can be undestood by the operator or by other machines.

Types of Memory

Core

This is the main memory associated with the computer. The basic unit of storage within main memory is called a word. Process computer systems typically have main memories with storage capacities between 24,000 and 64,000 words. Memory sizes are usually expressed as multiples of 1,024 words, for example, 24K, 64K, and so on, where K equals 1,024. (A 24K memory actually contains 24,576 words). Since main memory is essentially part of the basic computer, access to it is extremely rapid.

Drum

A drum is essentially a continuous loop of magnetic tape. A cylinder is coated with a ferrous material similar to that used on tape. The cylinder is then kept in rotation and read/write heads are positioned around its circumference. Each head writes only on that track (band) of the drum with which it is aligned.

Heads are energized by addressing circuits that select the desired track and are synchronized by the output of a prerecorded timing track, which generates pulses at bit and word time intervals. In this fashion, computer words can be written on any track and at any word position of that track.

In either writing or reading a word into a given position, the longest wait encountered is the time of one complete drum revolution. Drums typically rotate at speeds of up to 6,000 RPM. Figure 17-2 shows the basic elements of drum memory.

Disk

A disk is similar to a drum. Data are stored on the magnetic surface of a flat circular plate, like a phonograph record. The disk material is somewhat similar to that used in magnetic tape. Stored information is divided into tracks, or sectors, as with a drum. The primary difference is that the disk has only one read/write head that moves from track to track. The advantage to this design is that it is less expensive than

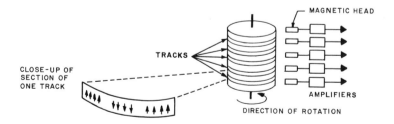

CLOSE-UP OF SECTION OF ONE TRACK

TRACKS

MAGNETIC HEAD

AMPLIFIERS

DIRECTION OF ROTATION

Fig. 17-2. Magnetic drum memory.

multiple heads. However, the mechanical movement of the head causes data access to be slower than with drums (Figure 17-3).

Semiconductor (Solid-State)

The semiconductor memory, like the others, is simply an information storage device. These memories are classified into a number of types according to such variables as manufacturing technique, how information is put in and taken out, permanence of storage, and other characteristics. In any memory, the binary bit storage location must be addressable. Further, it must be possible to read the state of every binary word. If the binary words can be read but not changed, the memory is called a read-only memory (ROM). If the contents of the memory can be changed as well as read, it is called a read/write memory. Read/write memories are commonly called random-access memories (RAM). A memory is said to be randomly accessible if individual binary words within the memory can be accessed directly. Solid-state memories are also divided into two categories—static and dynamic. Static memories are bistable. Information is stored in either of the two stable states. It remains in a state until changed by an external signal. A flip flop is a common form of static memory. A dynamic memory circuit uses logic absence or presence, zero or one, in the form of a capacitor charge, as a storage element. Because a capacitor charge can leak off, it must be periodically recharged or refreshed.

The specifications for the memory used in the FOX 3 computer are as follows: N-channel MOS, 4K by 1 bit RAM. This translates into N-type material, manufactured by metal oxide semiconductor (MOS) technology, 4096 bits on each chip, and 16 chips per array for a total of 4096 words with 16 bits per word. One logic board contains two arrays, which provides 8192 addresses per board. The unit can accommodate from three to eight boards or 24 to 64K words capacity of random-access memory. The FOX 3 also contains a diskette (small disk) which is used

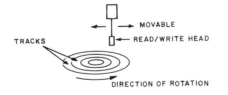

Fig. 17-3. Magnetic disk memory.

diskette (small disk) which is used for both bulk storage and for initiating programs into the working solid-state memory.

Operator Communication Devices

The process computer system supports various operator communication devices that enable the process operator to observe process measurements, enter control set points and tuning parameters, request logs and summaries, and trend critical process variables. These devices include operator's consoles, auxiliary special-function keyboards, alarm and message typers, trend recorders, Teletype® and CRT terminals, and remote CRT monitors.

Process Interface Equipment

Process parameters that are monitored by the control system are called inputs. Voltage or current signals from measurement sensors are classified as analog inputs, since the voltage, or current value, is analogous to the value of the measured process parameter over the entire parameter range. Some process parameters, such as flow rates measured using turbine flowmeters, are monitored by pulse-generating devices and are classified as pulse inputs. Limit switches and other contact sense devices indicate which of two possible states a parameter is in, such as pump (on or off) and pressure (high or low). This type of process measurement is called a digital input.

Since the FOX 3 is a digital system, there must be some means of converting the analog signals from the field devices into equivalent form before it can be used in the computer. This conversion is done by an analog to digital convertor which, when activated, converts the input voltage to some equivalent digital number. The key phrase is "when activated." Since there could be a large number of inputs to any computer system, it would be wasteful to have one A/D convertor for each input. Therefore, one convertor will function for a number of inputs, and each time any of these inputs is sampled the A/D convertor is also activated.

The input/output (I/O) module reads the current process signal, converts it to a number that can be read and processed by the control computer, and notifies the computer that the value is available by sending it an interrupt. The computer reads the converted parameter value and stores it in its memory for further processing.

There are several types of computer systems. The most common type is the business or data-processing system. Next is the scientific type, and last—but most important to the process industry—is the process control computer system.

A process control computer system is an on-line computer system. Measurements and other indications from the process are sent directly to the computer system. The computer system changes the analog signals to digital words, and then performs the proper control equations. Results of control calculations are sent to the computer system's output circuits, which change digital words to analog signals, which then return to the process. Many loops are controlled by this means. Since data for each loop reside somewhere in the computer system, occasionally these systems may also be used simply to collect information for the operator. However, this is simply an extension of the conventional analog system. The true process computer, when used for control, is part of the closed loop.

The outputs from the computer connect to the controller via the analog output subsystem. In this closed-loop configuration, the computer will take the measurement and perform calculations based on the measurement and other process parameters. The computed results, or control strategy, will be sent to the selected controller to update the controller set point or to change the controller output directly, thereby driving the valve.

If the computer determines the set point of an analog controller, it is called a set-point control system. Set-point, or supervisory control systems, use the computer to determine new set points. Measurements are sent to the computer and the analog controllers. The computer partially replaces the operator by changing set points in response to process changes, production schedules, and supervisory control programs while control is by conventional analog means. This frees the operator for more important tasks.

Direct digital control is a system in which the manipulated variables are controlled directly by the computer. A valve is driven according to the result of computer calculation of the control equation or algorithm for that loop. (An algorithm is an equation that is repeated many times. Algorithms are named for the control action they provide, i.e., Lead-Lag, Proportional-plus-Integral, Ratio, etc. These algorithms make up most of the control actuating program of the control software. Thus, the actual operation of the closed-loop computer control system can be subdivided into DDC (direct-digital control) or SPC (set-point control). The typical process computer has the capability of operating in either fashion. At times, a combination of these techniques may be employed.

Having explored the functions that are carried out by a computer, let us look at an actual computer. Computer systems are available in many sizes with varying abilities. Since this presentation is simply an introduction, the computer selected is of modest size, but is used widely for monitoring and process control of plants such as the one shown in Figure 17-4.

This computer system has a multiprogrammed operating system and utilizes modular equipment. The process interface is through SPEC 200/INTERSPEC (Foxboro multiplexing system). The INTERSPEC port plugs into the FOX 3 the same as any other peripheral device and accepts, converts, directs, and transmits the many signals that flow between a SPEC 200/INTERSPEC system and the FOX 3, allowing the

Fig. 17-4. Typical process plant.

Fig. 17-4. (*continued*) Analog/digital instruments.

FOX 3 to perform process monitoring, supervisory set-point control, or direct-digital control of even the most complex loops. Standard software packages have been developed and are available for process monitoring and control applications. Those discussed in the next chapter may be employed to create video displays, as well as other support facilities.

The FOX 3 system equipment provides the speed, storage, and I/O capabilities needed for real-time, on-line data acquisition and control

Fig. 17-4. (*continued*) Fox 3 computer.

applications. It combines the reliability essential to process applications and the functional modularity necessary for economical growth and ease of maintenance.

Functional modularity means that the system is composed of basic building blocks, such as a completely independent, intrinsically safe, I/O subsystem.

Fig. 17-4. (*continued*) Parts of this figure show major components in a computer system: process plant, instruments, computer, and operator.

The major hardware components of a FOX 3 system include:

Processor cabinet (Figure 17-5)

Peripheral equipment (Figure 17-6)

Process I/O interface (Figure 17-7)

All of the electronics necessary for system operation are housed in this single cabinet, which contains the following elements:

Microprocessor

The major functional elements are four microcontroller chips chosen for their high-speed characteristics and application flexibility. The microcontrollers form a 16-bit processor that interprets the language statements stored in memory that directs the FOX 3 to perform its control functions. It can perform arithmetic calculations in both fixed and floating-point formats.

Fig. 17-5. Processor cabinet.

Fig. 17-6. Peripheral equipment.

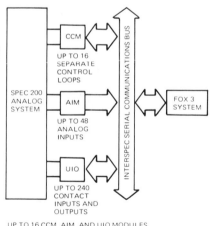

UP TO 16 CCM AIM AND UIO MODULES
IN ANY COMBINATION

Fig. 17-7. Process interface through
Foxboro SPEC 200/INTERSPEC.

383

Diskette
The diskette provides a simple, fast, safe, and convenient method for the user to save a permanent record of tasks, schemes, or the entire contents of memory (Figure 17-8). Data can be reloaded from diskette to memory at any time.

Processor Service Unit
This unit is an integral part of the processor and performs off-line functional testing of critical processor operating parameters. It is used in conjunction with a diagnostic diskette and eliminates the need of procuring separate processor test equipment. Just as the patient tells the doctor where it hurts, the diagnostic program tells the technician where the trouble lies.

System Clock
The line frequency clock provides a time reference that schedules many functions vital to accurate and effective control applications. It

Fig. 17-8. Diskette.

can be used to generate interrupts, provide references for measuring elapsed time, control program sequencing, and so on.

Peripheral Interface Logic
This logic provides a standard interface channel for easy interconnection with all keyboard/printers, alarm loggers, video consoles, and so on. Each interface channel supports full-duplex operation. The FOX 3 provides up to eight channels for devices that may be located up to 8 kilometres (5 miles) from the processor cabinet if optional models are used.

Other components located within the cabinet include a battery backup unit that provides power to system memory for up to 30 minutes (optionally for 24 hours) during power outages, power fail/restart circuits that assure continuous process control during power dips or outages, a service panel that monitors cabinet temperature and power, as well as the mode (normal/backup) of the process controllers and the system power supplies.

Conclusion

We should note that the physical size of a computer in no way relates to its power or performance. Computers cannot be compared in terms of the sizes of memory banks or cycle times, since these factors do not indicate the power available to perform the required tasks. A reliable method for selecting a computer involves running the same higher language program on several types and sizes. Note the ease of development, the results in time of execution, the amount of storage consumed by each computer, and the potential ability to handle development of programs while the process is under computer control.

A digital computer system may or may not be more economical than the equivalent analog system, depending upon the complexity of the system, performance requirements, projected long-term changes, and other relevant factors. For many applications, the digital system will prove to be more economical than an equivalent analog system.

Questions

17-1. A digital computer
   a. Has the ability to solve problems by reasoning
   b. Can only do the things it has been taught to do

    c. Has all the answers on file in its memory
    d. Makes assumptions and estimates the answer

17-2. The computer's memory element is
    a. Passive   **c.** Always a magnetic device
    b. Aggressive   **d.** Always a capacitive device

17-3. The arithmetic unit receives information and commands from
    a. The operator's console   **c.** The control unit
    b. The memory   **d.** A punch tape reader

17-4. The longest delay in reading information from a drum is
    a. One drum revolution
    b. 1 second
    c. $1/60$ second
    d. Dependent on track configuration

17-5. A disk recording surface is made of
    a. Aluminum
    b. Cast iron
    c. Fiber
    d. Plastic containing magnetic material

17-6. A process control computer system is
    a. A real-time system
    b. An off-line system
    c. Independent of the control loop
    d. On-line until the operator presses the release key

17-7. A process control computer system
    a. Always costs more than an equivalent analog system
    b. Always costs less than an equivalent analog system
    c. May prove cost effective if future changes and process efficiency are considered
    d. Will always provide superior control

17-8. Most process inputs to the computer control system are
    a. Digital   **c.** Pulses
    b. Analog   **d.** On/off

17-9. The input/output module reads the process signal and notifies the computer it is available by an
    a. Interrupt   **c.** Activating pulse
    b. Alarm   **d.** Equivalent digital number

17-10. A diagnostic program is useful for
    a. Locating information in memory
    b. Locating process malfunctions

    c. Off-line functional testing of operating parameters

    d. Indicating when routine maintenance is needed

17-11. A system clock

    a. Keeps the operator aware of the time

    b. Schedules functions vital to control applications

    c. Programs the rate at which memory is searched

    d. Places the alarms on a periodic basis

17-12. The power of a process control computer

    a. Is a direct function of physical size

    b. Depends only on available memory

    c. Can be evaluated only by applying a higher language to the computers being compared

    d. Is measured by the total number of control loops it can handle

Computer Software and Operation

A Brief Introduction

By definition, software is the collection of programs and routines associated with a computer. The first working process computer became operational in 1958. These routines involve other programs such as compilers, library routines, input/output (I/O) device handlers, plus various other devices shown in Figure 18-1. This figure is termed a software configuration. The computer performs many tasks in decimal parts of one thousandth of a second. While it is extremely fast, it can do only one thing at a time; therefore, an executive program is provided to control the order in which tasks will be accomplished. This is called a priority schedule. Under the control of the executive are process I/O service handler programs, terminal I/O handler programs, and application programs such as scanning, alarming, controlling, and logging. Thus, the executive is both a manager and a traffic director.

Computer systems also have on-line compilers and off-line compilers. The example shown in Figure 18-1 is an on-line system. This means that, while the system is actively monitoring or controlling process routines, the engineer or programmer may at the same time modify his control scheme or compile new scan-and-alarm-logging programs.

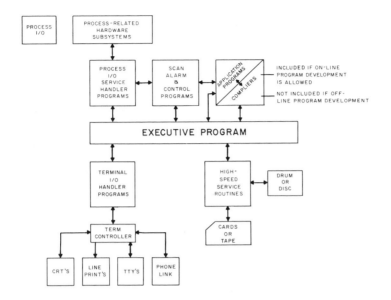

Fig. 18-1. On-line system.

Thus, changes in the various application programs or the overall control scheme may be made without shutting down the computer or removing the plant from the control of the computer. The on-line capabilities of some computer systems have been used to develop maintenance programs that allow the computer to diagnose itself and to generate alarms when preventative maintenance is due.

On the other hand, some computer systems require that compilers be loaded off-line, which means that the computer must be shut down for modification of programs. No process control is carried out during the modifications. With certain types of computers, program development must be carried out on a different machine altogether. To compare on-line with off-line operation, let us assume that a particular solenoid valve fails to energize as a result of a minor computer I/O failure. If the computer has on-line capabilities, a test program can be compiled immediately to determine the location of the problem. The maintenance man, using the test program, can isolate the specific point of failure and determine the proper action to take without affecting the rest of the process. However, with an off-line system, the computer would have to be shut down to develop or load a test program into the compiler and isolate the problem. Some off-line computers will not even allow the tuning of control loops without shutting down.

Basic Computer Operation

The steps the computer negotiates when it follows the program are shown in a general way in the following discussion. First, a program is written and converted into a language the machine can understand. This is called machine language.

For our illustration, let us: calculate output No. 1 by solving the equation $\dfrac{A \times B}{C}$ where $A = 25$, $B = 4$ and $C = 5$. The program will be stored in memory, beginning at location 200. For the purpose of simplicity, these instructions have been shown here symbolically.

Instructions enter the machine through some type of input equipment (Figure 18-2). The input equipment would normally be used to convert these instructions into electrical pulses that would represent the data. The control logic sends command signals to the input equipment and the memory, causing data to be transferred to memory for storage. The control logic probably would be initiated by an operator function. Once the program and its data have been stored in the memory, the program can be run. Again, this likely would be initiated by an operator function.

The operator would set the starting address where the program has been stored on a set of computer console switches—in this case, ad-

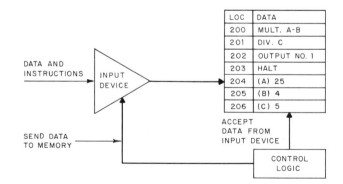

Fig. 18-2. Instructions entering memory.

dress 200—then press a start button. As a result, the control logic would obtain or "fetch" the first instruction from the address that was provided by the operator through the console (Figure 18-3).

After fetching this first instruction, the control logic would have to interpret and determine the type of instruction. In this case, the control logic would establish this as a MULTIPLY instruction. The control logic is capable of determining the type of instruction because of the way in which an instruction is coded. Each type of instruction has a particular operation code associated with it. Therefore, the control logic decodes that portion or field of the instruction that contains the operation code and, as a result, provides other pieces of logic that will be needed for the execution of that instruction. To execute the MULTIPLY instruction, the control logic must allow the two factors (A and B) to be obtained from memory and brought to the arithmetic unit. The addresses of these two factors (204 and 205) would also have been provided in this hypothetical instruction.

Upon entry into the arithmetic unit, the two numbers will be multiplied, resulting in a product of 100, which would reside in the arithmetic unit's temporary storage register or accumulator.

After executing the multiply instruction, the control logic sets itself up to fetch the next instruction, which is located at Location 201 (Figure 18-4). After fetching this instruction from its memory location, the control logic will interpret it. This time, the control logic establishes the operation code as a DIVIDE instruction. The control logic will be set up to perform the division using data from the address specified by the

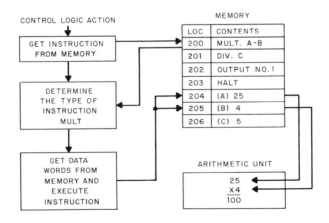

Fig. 18-3. Fetching the first instruction.

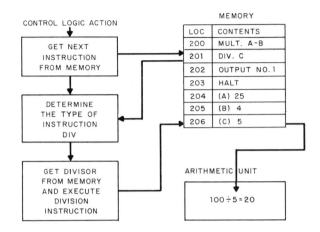

Fig. 18-4. Performing the divide operation.

DIVIDE instruction—Location 206. The divisor value (5) will then be brought to the arithmetic unit and the division process will be performed upon the accumulator contents and the data from Location 206. The result of this operation is the quotient of 20, which becomes the new accumulator contents.

The control logic will now fetch the next instruction from Location 202, this time finding an output instruction (Figure 18-5). Since the

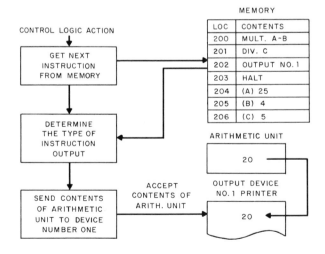

Fig. 18-5. Output of data.

output instruction is not an arithmetic instruction, the control logic will handle it somewhat differently than for the arithmetic MULTIPLY and DIVIDE instructions.

Instead of looking for an address where it can find other data words, the control logic will look for a code that will specify the device to which the contents of the arithmetic unit should be sent. In this case, the output instruction is specifying device No. 1, which we will call a line printer. The contents of the arithmetic unit will be sent to the line printer as a result of control logic signals sent during the execution of the instruction.

Finally, after the output instruction has been completed, the control logic will return to fetch the next instruction, this time from Location 203 (Figure 18-6). When it determines that it has fetched a HALT instruction, it will bring the memory and control unit timing circuits to an orderly shutdown.

Many steps are necessary to perform even an elementary problem. In some of the most complicated problems, the number of instructions necessary to define the problem completely may require hundreds of instructions. However, it must be kept in mind that the computer can execute these programs at lightning speeds, since cycle or operating

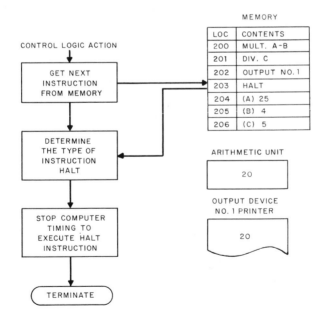

Fig. 18-6. Halting the program.

time of these instructions is in the order of microseconds (millionths of a second). Any problem or condition that can be represented logically or numerically, can be solved with the digital computer. In addition, once a program has been written, the computer is able to solve that problem repeatedly, using original or new data.

Real-Time Clock and Power Fail/Restart Logic

The *real-time* (line frequency) *clock* interrupts the system at a high priority level every $1/60$ of a second so that the system can time events to this degree of accuracy. The clock "tick" is accumulated into seconds, minutes, and days so that user programs can respond to time-oriented events.

The *power fail/restart module* continually senses system ac power. If power drops below a predetermined threshold, the module interrupts the system, forcing it to come to an orderly shutdown (before insufficient power is available to read and write memory). On the return of suitable operating power, a second interrupt forces the system to resume operations.

The term "control system software" describes all programs that go into the making of a complete control system. On large systems, programs are placed in bulk storage (drum, disk, tape, etc.) and the operating system controls the transfer into cone or semiconductor memory for program execution. These programs are grouped into three basic categories: the operating system, control software, and support software. The operating system consists of programs that control system resources and do internal housekeeping. The control package consists of programs that interface with the process to accomplish some level of control. The support programs consist of programs that minimize the user's effort in using the system.

Control Software

The primary approach to process and plant control is to use a building block concept to complete control loops. Each block in this control loop represents a mathematical algorithm that is performed on some piece of information to achieve some other piece of useful information in the control scheme. These algorithms are performed at fixed periods to achieve the proper control. These periods, called scan periods, generally range from one second to several minutes. Every control loop in a plant will be represented by a number of these control blocks. To minimize the user's efforts in creating a control scheme, most control systems employ the standard algorithms currently used in analog control. These blocks are assembled in a file called the data base. The data base contains all the information necessary to define each control loop, both internal to the computer and external to the user.

Building the data base is the first task that the user must perform with the control system. This building effort is often accomplished by using fill-in-the-blank forms to describe the type of control and the type of control scheme that is to be accomplished. These forms ask such questions as: What is the name of the control loop? What is the range of the measurement and the engineering units? It asks for such things as set-point and measurement limits. It also asks for the hardware address of the measurement and the control valve. After all the questions are answered, these forms are converted into punch cards, paper tape, or other means and fed into one of the control programs called a data-base generator. The data-base generator converts this information into especially formatted words that are filed away and utilized by the control program.

Having created the data base and, of course, having the operating system and the control program, the user can then load the data base into the system and start to control the process. Since data base modifications are frequently required, provision must be made for on-line data-base-modifying routines. This is a part of the control package. Data-base modifications would include simple set-point changes through more sophisticated modifications.

As the level of control increases, the building block approach soon becomes inadequate to handle the special needs of individual plants. Therefore, rather than try to create standard algorithms that apply to all control needs, an interface is normally provided within the control package to allow user-written programs to assume some of the control. With the capability of user programs interfacing to the control package, either the engineer or the programmer, or both, can accomplish various sophisticated functions in terms of plant level control. Production scheduling, inventory control, and material tracking are just a few of the many powerful things that can be accomplished. The only limitation is the capability of the engineer/programmer and the time available to develop the control scheme. An important component is the support software, called the compiler. The compiler converts the language employed by the engineer/programmer into machine language the computer can understand.

Now let us demonstrate how a typical process computer, the FOX 3, is programmed to carry out its control assignments. The system language is called Foxboro Process Basic (FPB). This language has evolved from the American National Standard Basic with some modifications provided by Foxboro Process Language (FPL), which was developed originally as a programming tool for small computers.

In FPB, tasks (programs) are written like operating instructions, using expressions (statements) familiar to process people. This FPB enables process engineers and operators to build, operate, and modify process monitoring and control tasks, using conventional process terminology.

Within the computer's memory there exists a program that will convert the FPB into so-called machine language that the computer can utilize. This function is called a compiler.

Achieving Process Control

The purpose of a language like FPB is to allow the user to tell the computer what to do and how to do it, in a concise and orderly way. To attain a control objective, the user first defines the problem in functional terms, then designs an appropriate solution. The next step is to express the solution as a structured set of statements. The supervisory portions of the solution are written as FPB tasks.

These tasks are groups of FPB statements arranged so that they describe a set of procedures for the FOX 3 system to follow. Typically, these tasks describe supervisory functions, such as generating reports and controlling the sequence of process operations. To define a task, the user types FPB statements at the interactive device currently serving as the system terminal. Once entered, a task is ready to run.

The illustrations below show a simple batch process. The task called CHARGE involves mixing two materials in a tank, heating the mixture, and holding it at a constant temperature for 20 minutes before draining the tank. Figure 18-7 shows the process diagram. Figure 18-8 is the flow chart depicting the steps to be taken to control the process. Figure 18-9 is the FPB task that performs those steps.

When the user types the command: RUN CHARGE, the task will begin executing; that is, the materials will be mixed, heated, and held at the specified temperature, and, after the designated interval, the tank will be drained.

An important part of the FPB software is the control package, which is the means of describing the process interface and control structure to the FOX 3 system.

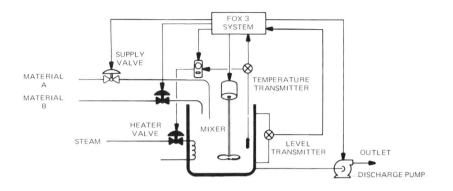

Fig. 18-7. Process for charge task.

In a question-and-answer format, the user enters the specific characteristics of the process control and computational elements to be used. This description is called a "block." Blocks can be connected to form a variety of control strategies, called "schemes," which tell the system how to control the process.

A block is comparable to an analog device. It is a conceptual entity, identified by a name, for which certain monitoring and control functions (called "algorithms") are defined. By logically connecting blocks of various functions, the user can create a block scheme in the computer which can replace or enhance an analog control scheme—

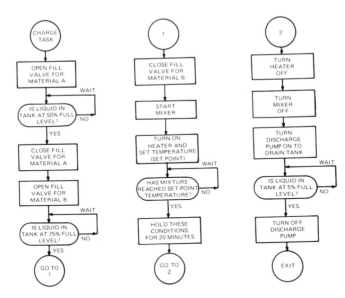

Fig. 18-8. Flow chart for charge task.

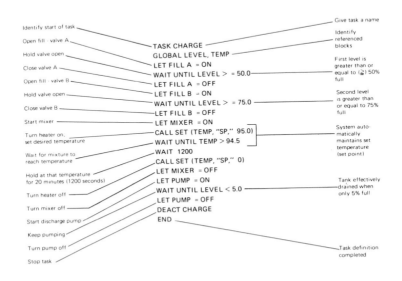

Fig. 18-9. Charge task written in FPB.

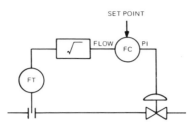

Fig. 18-10. Analog flow loop.

receiving measurements from the plant, processing these signals, performing control or computational functions, and returning control signals to the plant.

In a simple analog flow loop (Figure 18-10), the signal from a differential flow loop is conditioned and transmitted by the square root element ($\sqrt{\ }$). The resulting flow signal is passed to a controller, where it is subtracted from a set point to produce an error value to be used in the control function.

The control function, or algorithm, is two-mode proportional-integral (PI) control, which calculates the new valve position. This is a building block approach to process control, since the process engineer selects standard "black boxes" that perform defined functions and connects them into loops.

The Foxboro Control Package (FCP) refers to the computer equivalent of an individual analog device as a "block." For every analog "black box," there is an equivalent computer "block." Using the control package, the process engineer logically connects blocks to achieve control schemes exactly as with analog devices. Figure 18-11 depicts a computer-based control scheme equivalent to that in Figure 18-10. This

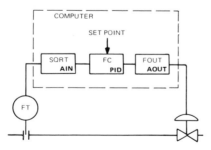

Fig. 18-11. Digital flow loop.

is a direct digital control (DDC) scheme because the computer is producing the valve position signal. In certain situations, flow control would be superior with an analog controller because of its speed. When this is necessary, the computer would set the analog controller's set point.

FCP consists of standard control blocks that provide a wide range of process monitoring, control, and calculation functions. Each FCP Control Package block is a standard skeletal definition of one process function. The user completes the definition of a block by participating in a question-and-answer dialogue with the FOX 3 system. That is, the user signifies the kind of block to be defined, and the system asks questions about the alarm limits, engineering units, and so on. The user's answers, which define the characteristics of the block, are called the block parameters. As in the comparable analog control scheme, the block itself is called an algorithm. A wide variety of control block algorithms is available. These software blocks now replace analog components in the system.

The user can define each block in the system by specifying all of the control parameter values that pertain to each block type. For example, all analog input blocks start with the same basic block structure, yet each one can have its own unique set of block parameter values such as the frequency of block execution, high and low limits, engineering units conversion, and so on.

In addition to control functions, blocks are available to perform alarm functions, displays, monitoring calculations, and virtually every function an analog device would accomplish.

The software for the modern process computer system has progressed considerably in recent years. When the first computer was applied to the control of a refinery in 1958, it was necessary to program the computer in machine language. This was a long and extremely tedious job. Modern technology, combined with controller languages like FPB, makes the job much easier and infinitely more pleasant. Significant improvements in computer technology continue to occur at a rate that staggers the imagination.

Questions

18-1. A compiler

   a. Stores bulk data

   b. May be on-line or off-line

   c. Establishes priorities

   d. Is an input/output handler

18-2. In computer operation even a simple operation
  a. Requires a sophisticated CPU
  b. Takes many steps
  c. Must be scheduled by the real-time clock
  d. Must be reprogrammed each time it is run

18-3. An algorithm is
  a. A sequence of calculations performed to obtain a given result
  b. A formula
  c. Always a differential equation
  d. Determined from a table

18-4. Every analog function
  a. Is duplicated by digital hardware
  b. Should be reviewed to confirm its usefulness
  c. Can be duplicated by digital software
  d. Can be improved upon in a digital system

18-5. A major advantage of the digital system is
  a. Energy efficiency
  b. The ease of making changes
  c. Economy
  d. Its ability to control without feedback

18-6. The first working process computer became operational in
  a. 1976   **c.** 1968
  b. 1948   **d.** 1958

18-7. The number of tasks that a computer system can carry out depends mostly on
  a. Real-time clock   **c.** Its restart module
  b. Capacity of its memory   **d.** Its data base

18-8. If the computer is examined as a process control tool, we conclude
  a. It has reached its full potential
  b. It has the potential to become virtually anything you wish to make it
  c. Economically it cannot be justified
  d. It is a more economical method of control for any process.

19

Programmed Control Systems

In a programmed control system, the control point is automatically adjusted to follow a predetermined pattern with respect to time and process conditions. For example, assume that a food product is to be cooked and the cooking process requires the following sequence: 1. Raise the batch temperature uniformly up to the cooking temperature over a 1-hour period. 2. Cook the batch for 1 hour at the cooking temperature. 3. Allow the batch to cool to room temperature gradually and uniformly over a 2-hour period.

This type of process can be satisfied by a programmed control station. Originally, most programmed controllers were cam-operated, and this technique is still widely used. In the cam-operated controller, the control setting movement is replaced by a follower, riding on a cam that has been cut to produce the desired schedule. The cam is friction-driven by a clock. The cam can be laid out to provide any predetermined schedule of time-and-process requirements. The cam-operated, programmed controller offers the advantages of simplicity and ease of operation. The limitations are that it is only a single loop; and a program change requires a new cam layout, and, in some cases, a change in cam speed (motor). These limitations made the elapsed-time controller popular.

The elapsed-time controller is designed to bring the measurement up to a set point at the start of a preset period (holding period) and control it at that point for the remainder of the period. At the end of the period, control ceases and the process shuts down until another cycle is started. Elapsed-time controllers may be either electronic or mechanical instruments for application on any process variable. On/off or proportional control may be used.

The timer periods are adjustable and available to provide holding periods from minutes to hours. Figure 19-1 shows a typical process curve controlled by an elapsed time controller.

Description of Circuit

The Model 40E-A elapsed-time controller consists of a conventional pneumatic controller with an additional flapper-nozzle operated by an electrical circuit (Figure 19-2). With the elapsed time switch in the manual (M) position, the flapper is held against the nozzle irrespective of the solenoid plunger position, and the pneumatic circuit functions in the usual manner. However, with the elapsed-time switch in the A (automatic) position, the flapper is free to be operated by the solenoid, which, in turn, is controlled by the timer.

When the timer is set to the desired value of the holding period, and the timer start button (14) is pressed, the timer circuits (1) and (3) are completed, as shown in Figure 19-2. The solenoid is energized, moving the plunger to the right in the diagram, covering the nozzle, and thereby starting pneumatic control. The control valve opens to raise the measurement to the holding value and maintain it there for the duration of

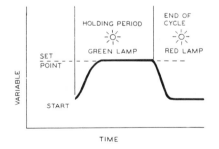

Fig. 19-1. Typical process curve controlled by elapsed time controller.

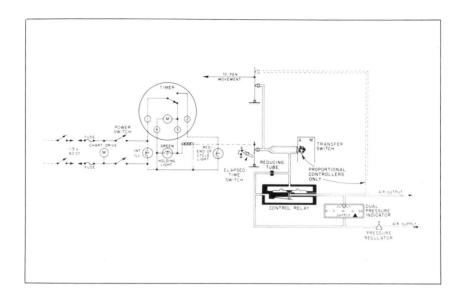

Fig. 19-2. Foxboro 40E-A elapsed time controllers.

the holding period. Electric circuits are also completed that operate the timer motor and the green holding light.

At the end of the holding period, timer circuits 1 and 2 are connected, all previous circuits are broken, and the solenoid is deenergized. Air bleeds from the pneumatic circuit and the control valve (air-to-open) close to shut down the process. At the same time, the electric circuit through the "End-of-Cycle" light is completed.

This type of elapsed-time controller is, like the cam-operated controller previously described, a single-control loop device.

If more than one control loop is involved and additional functions are required, the control systems described are inadequate. The initial solution to the problem might be to use a cam-operated controller for each loop, which has been done on occasion. This approach would have limited success, since any required change in the program would be difficult to incorporate into the system. This type of problem can be resolved quite readily by the digital computer.

A digital computer-oriented system provides versatile, preprogrammed sequential logic control functions. It is used to automate adjustments of analog set points and the check and inhibit actions of interlock logic. Also, the synchronized starts and stops of process equipment are completely automated with the use of a computer system. This computer-compatible logic system bridges the gap between mechanical programmers and process computers. The sequential control program is designed with an easily understood ladder logic language. The logic operation is created by pushbutton, without wires or wiring, and is stored in the memory of a programmable controller. This programmable controller is capable of operating multiple, completely independent process units at the same time.

Sequential Control Systems

In addition to startups and shutdowns of continuous systems, batch processes are potential applications for computer-controlled systems. A stored control program is used to repeat all the events in an operation, step-by-step, throughout a production cycle. The program elements consist of on/off relay logic, time-delay and elapsed-time periods, predetermined counts of pulses, and pulse outputs for set-point control.

Elementary mathematical manipulations (addition, subtraction, multiplication, division) can be performed within a step of the logic program. Each of the stepped phases of all operational events is controlled by input contacts and feedback. These contact inputs are from pushbuttons for manually initiated action or from limit switches for positional action. Also, contact inputs can be derived from analog inputs for signal comparison or from other permissive interlock or inhibit actions. The memory and logic program, shown in Figure 19-3, controls

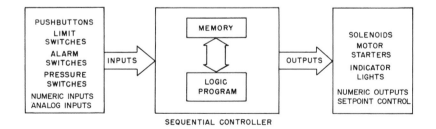

Fig. 19-3. Functional diagram of sequential controller.

the conditions when a field output is turned ON or OFF for any of the following operations:

Opening or closing individual valves

Starting or stopping pumps or agitators

Operating the lights of communication displays

Programmable Controllers

The programmable controller is a solid-state, sequential logic control system. It replaces mechanical logic equipment, such as relays, stepping switches, and drum programmers. It is designed and packaged for operation in harsh industrial process environments and for high reliability and long-operating life. The logic sequence program, implemented by pushbuttons, simplifies the development, checkout, and maintenance of the control system design.

The programmable controller consists of four functional sections: a core memory, a logic processor, storage registers, and input/output signal conditioners. The size of the memory determines the logic storage capacity for individual programs. These programs are executed by the control routines of the logic processor and are used to generate signals to turn on or off operating field outputs or to adjust set points of controllers. To communicate these input/output signals, storage registers act as bridges between the signals at field levels and the high-speed circuit of the logic processor. The input/output signal conditioners convert input signals at field levels for compatibility with the logic processor and control the output power for operation of field circuits.

Programming Language

The language used to describe sequential on/off logic is an electrical ladder diagram with logic line numbers and contact symbols. A logic control program closely resembles the format used for mechanical relay systems. A typical ladder diagram is illustrated in Figure 19-4. In this diagram, the *source* of power is indicated by the vertical uprights of the ladder. The horizontal rungs indicate switch contacts to control the flow of power. The four contact symbols used in logic sequences are shown in Table 19-1.

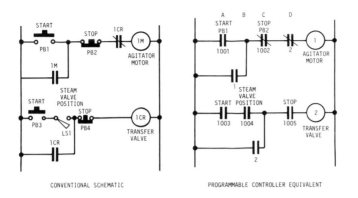

Fig. 19-4. Logic language (ladder diagram) format.

Generally, the logic program diagram elements use the same number designations as the conventional relay diagram. However, an individual rung is limited to a maximum of four contact elements. If a logic sequence requires more than the four elements, it may be handled as follows: The original line output number is used as the first contact input reference to the next line. For step-controlled events, a single-input contact is used to control many contacts by assigning the same reference number as if it were a relay with unlimited poles.

Table 19-1. Logic Sequence Symbols

TYPE OF CIRCUIT	NORMALLY OPEN (NO)	NORMALLY CLOSED (NC)
SERIES	─┤ ├─	─┤╱├─
PARALLEL	─┤ ├┘	─┤╱├┘

Programming

The logic control program is created in memory by the use of a programming panel. A typical panel is shown in Figure 19-5. Logic is entered line by line. The programming panel is plugged into the service port of the programmable controller. A new program is manually entered into the memory by pushbuttons marked with the same symbols as the four relay contact types. The line number and the individual contact reference number are dialed on a thumbwheel; one of the pushbuttons, marked with the logic symbol, is then selected and pushed. The programming panel contains other types of logic functions, including time delay or elapsed time, event counting, and calculation (addition, subtraction, multiplication, division, and comparison of two numbers) for control of a logic line output.

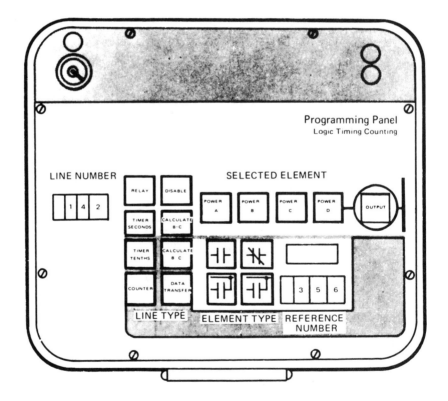

Fig. 19-5. Programming panel.

The logic program lines are designed with seal circuits for momentary contact inputs or with latch relay circuits for retentive output action in case of power failure.

Logic System Automation

Proper application of a programmable control system begins with an economic justification analysis. Savings are developed principally from reduced cycle time and scheduling. Cycle automation provides rigid control enforcement to eliminate human errors and to minimize necessary manual interventions. Increased efficiency in scheduling is to be expected with maximum utilization of equipment and reduction of fluctuating demands on critical equipment.

Guidelines should be established to determine future plant goals and the direction for the analysis. For instance, in a limited production

Table 19-2. Operational Displays

PROGRAM READY	THE CONTROL CYCLE PROGRAM IS ENTERED AND THE SYSTEM IS PREPARED TO RUN A NEW CYCLE.
RESET CYCLE	TO REPEAT THE SAME RECIPE, THE OPERATOR RESETS THE CYCLE TO THE STARTING POINT.
CYCLE START	THE OPERATOR CAN INITIATE THE CYCLE.
PROGRAM HOLD	THE CYCLE OPERATING CONDITION IS FROZEN AT THE (LAST) OPERATING LEVEL.
CHECK/RUN	ALL FIELD OPERATIONS ARE INHIBITED, AND THE PROGRAM SEQUENCE CAN BE MANUALLY ADVANCED FOR A PREVIEW OF EACH STEP AND OPERATING VALUE.
MANUAL ADVANCE	(OPERATED IN THE CHECK POSITION) STEP-BY-STEP SEQUENCING AHEAD OR BACK THROUGH THE PROCEDURE OR RETURNING TO A PRIOR STEP CAN BE ACHIEVED.
EMERGENCY SHUTDOWN (OPTIONAL)	COMPLETE AND SUDDEN SHUTDOWN OF THE PROCESSING OPERATION HAS BEEN INITIATED.
RATE INTERRUPT (OPTIONAL)	THE CONTROLLER SET POINT RAMPING ACTION IS STOPPED AND WILL REMAIN IN THE LAST POSITION UNTIL THE MEASUREMENT IS WITHIN THE CONTROLLER THROTTLING RANGE.

plant, the economic incentives are immediate from increased throughput. In a plant that produces for a limited market, quality control, cost reductions, and future flexibility for change are more important. With this type of system, the many possible control schemes are as variable as the individual processing applications.

System Control Displays

The effectiveness of the programmed logic control system depends on cycle status communication. This, in turn, depends on the proper selection of compatible display hardware and an understanding of the needs of the operator for efficient performance of his job in varying circumstances. The display task is to advise not only the *correct* information, but the most *useful* information. Table 19-2 lists typical displays that might be incorporated into the system.

A good display technique simplifies the communication of the operation tasks. There should be no time lost while the operator waits for operating status and instructions.

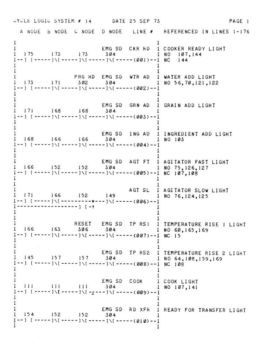

Fig. 19-6. Computer printed ladder diagram.

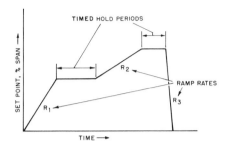

Fig. 19-7. Typical program cycle.

The outputs controlled by this program adjust the correct processing values by moving the set points of the analog controllers.

The logic system can program the final set-point position by comparing the programmed set-point signal to the measurement signal from the process. These signals are interlocked to prevent the sequence from advancing to the next step until the measurement has reached the preprogrammed set point. A typical program cycle is shown graphically in Figure 19-7.

Conclusion

The batch processes in the food, chemical, paper, and cement industries are sequential in nature, requiring time-or-event-based decisions. Programmable controllers are being used more and more as total solutions to a batch problem rather than just a tool. A process control computer system generally has the ability to perform a programmable controller function.

Industry's goal is to improve quality and cut costs. This can be done by increasing throughput by minimizing cycle time per batch, and consistent production of on-spec products.

Questions

19-1. A programmable control system is
    a. A single-cam set controller
    b. A digital computer programmed to carry out a multiloop control sequence

    c. A data logger attached to a multiloop system
    d. Any digital computer that performs set-point control

19-2. Programmable control systems are generally used with
    a. Flow loops   **c.** Batch processes
    b. Pressure loops   **d.** Large-capacity systems

19-3. Cam-operated controllers are programmed by
    a. The layout of the cam   **c.** Varying the timer settings
    b. Using the programming panel   **d.** The operator

19-4. The effectiveness of the programmed logic control system depends on
    a. Cycle status communication
    b. Economic justification analysis
    c. The mathematical manipulations required
    d. The computer it interfaces with

19-5. Proper application of a programmable control system should begin with
    a. A difficult control problem
    b. A passive process
    c. Proper selection of compatible hardware
    d. An economic justification analysis

20

Constructing and Instrumenting Real and Simulated Processes

A process control loop can best be understood when its components are combined in a loop and studied in operation. This, of course, requires both the instrumentation and a process to be controlled. The process may be real, or it may be an analogy or a simulation of a real process. This chapter describes how both simulated and real processes can be constructed. In terms of operation, there is little difference between the real thing and the simulation, but in terms of being able to visualize the operation, the real process definitely has the advantage. First, let us consider the simulated process.

Simulated Processes

The axiom in plane geometry that "two things equal to the same thing are equal to each other" applies to the technique of simulating a process and its control. The axiom is applied by constructing a system, a simulator, which simulates the behavior of the physical system under study. The information from the simulator is applied to the actual system.

The technique is convenient because it is generally much easier to manipulate the components of the simulator than it is to work with the actual system being investigated. It also is a great time-saver, because the simulator time base can be manipulated. For example, if a dynamic study of a process with a 30-minute time constant were made, it would take many hours to collect the necessary data. However, an equivalent simulator with a time constant of a few seconds, or even milliseconds, could provide the same data in a few minutes.

Still another area of effectiveness for the simulator is its application to the training of instrument engineers and operators. Frequently, it is desirable not only to demonstrate control action, but also to let the student make adjustments and observe the results. An actual process may be expensive and unwieldy, and may contain time constants that would make the demonstration too long to be practical. Both of these obstacles can be overcome by a suitable simulator.

A process simulator may be analog or digital; that is, it may use voltages or pressures, currents or flows, or other continuous variables to represent the process. Or a digital computer may be used to make a digital simulation of the process. Only the analog simulator will be considered here. The process simulator may be electric or pneumatic; the choice usually is dictated by the type of instrument mechanisms being investigated.

The response of many process elements is similar to that of a single resistance-capacitance (*RC*) electric circuit. More complicated process elements usually have a response similar to several *RC* elements in series. Thus, three *RC* groups can simulate a three-element process.

Constructing an Electric Process Simulator

A three-element electric analog simulator for a process containing multiple capacitances and resistances, including an adjustment for gain and load changes, is shown in schematic form in Figure 20-1 and pictured in Figure 20-2. This simulator consists of three *RC* networks whose resistance value—and hence the *RC* (time-constant) value—is adjustable. The middle capacitor in the series has an adjustable resistor in parallel with it to simulate process load changes.

The output of the *RC* network *P* has an adjustable gain control, making it possible to simulate various process gains. Output *P* is amplified, recorded, and brought to the controller input to close the loop. With the adjustments available, this analog simulator has the capability of simulating a wide variety of process types and operating conditions. It has the further advantage of being relatively inexpensive and easy to construct.

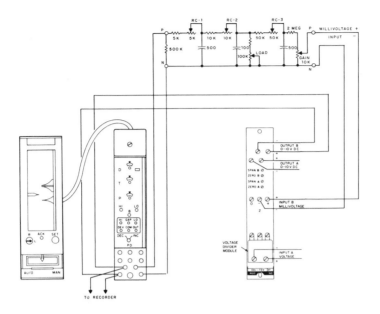

Fig. 20-1. A three-element electronic simulator for a process containing multiple capacitances and resistances and used with Foxboro SPEC 200 equipment.

Now let us examine a commercially available analog simulator and the features it makes available.

The Foxboro Electronic Process Simulator

The Foxboro Company electronic process simulator is a lightweight, portable, self-contained real-time simulator (Figure 20-3). A wide range of industrial processes can be simulated with this device.

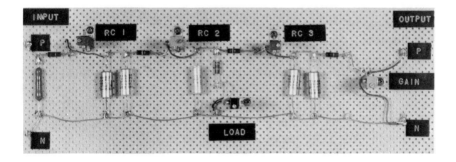

Fig. 20-2. Electronic analog.

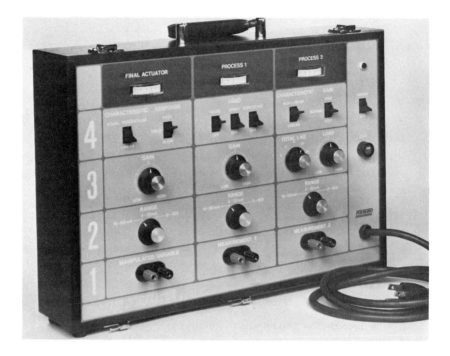

Fig. 20-3. Process simulator.

The simulator consists of three serially connected special electronic circuits. One circuit simulates a final actuator—commonly valves or dampers of various sizes and characteristics. A second circuit (Process 1), simulates a fast-flow process. The third circuit (Process 2), simulates a thermal process.

Simulated final actuators (valves, dampers, and so on) may be scaled to a range of sizes. The simulated final actuator characteristics may be linear or nonlinear, with either a slow or fast response time. A hold position is provided to allow display of the controller output without the dynamic effect of measurement feedback.

Process 1, the fast-flow or materials feed process, is adjustable for process gain. Single or repeated load upsets, with or without noise, can be superimposed on the input to Process 1.

Process 2, the slow heating, tank, and reactor process, is adjustable for total-process time lag and process load. In addition, the circuit allows the simulation of linear or nonlinear processes with three different sensor gains.

Fig. 20-4. A student and the author at electronic process simulator.

When using pneumatic instrumentation, controllers, and recorders, it is necessary to add only pneumatic-to-electric and electric-to-pneumatic converters between the simulator and the instruments. With such converters, useful and direct comparisons can be made between electric and pneumatic instruments.

The process simulator input, the manipulated variable, is an analog signal either 10 to 50 mA dc, 4 to 20 mA dc, or 0 to 10 volts dc.

The process simulator outputs (MEASUREMENT 1 and MEASUREMENT 2) from, respectively, Process 1 and Process 2, appear as the same type of signals.

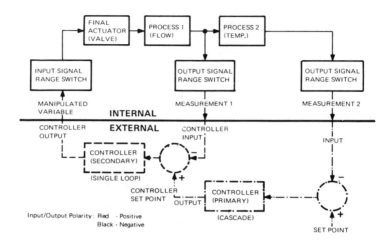

Fig. 20-5. Connection block diagram.

Inside the simulator, special purpose solid-state electronic circuitry provides the features and characteristics to provide realistic simulation of a variety of processes.

Circuit block diagram Figure 20-6 shows the internal arrangement.

Assume that we wish to simulate a heat exchanger—a single-element temperature control as shown in Figure 20-7.

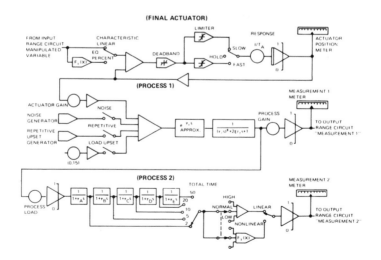

Fig. 20-6. Circuit block diagram.

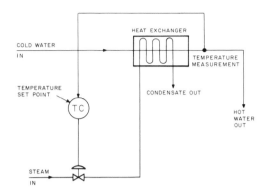

Fig. 20-7. Heat exchanger—single-element temperature control.

In this single-element temperature control system, the controller compares a measurement of hot water temperature with its manually adjusted set point to develop an output signal. This output signal feeds to and controls a fast-acting equal percentage steam valve.

The size of the heat exchanger, hot water load, steam pressure variations, and many other factors determine the success of the controller in its efforts to control the hot water temperature.

As stated in Chapter 11, the final actuator or valve must be properly sized. To accomplish this, several steps must be taken. The valve characteristic linear or equal percentage also must be selected.

The final actuator should be sized to bring the simulated process (1 or 2) measurement to 100 percent when the controller output (manipulated variable) is 100 percent. To do this:

1. Adjust controller output to 100 percent.
2. Adjust Process 2 load to the highest anticipated load. Low - 0.2, High - 5.0.
3. When using Process 1 but not Process 2, adjust to normal (1), center (up) position.
4. Adjust final actuator gain to bring Process 2 measurement to 100 percent.
5. Adjust Process 1 gain to bring Process 1 measurement to 100 percent.

Now we are ready to tune the controller to control the simulated heat exchanger process, using the procedures outlined in Chapter 12. The results would be similar to those shown in Figure 20-9. Once the simulator has been set, it should not be adjusted to get the desired results. The process is not tuned to match the controller, but rather the controller is adjusted to accommodate the process.

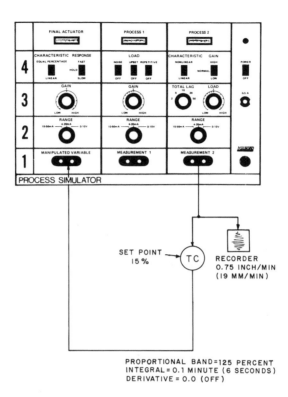

PROPORTIONAL BAND = 125 PERCENT
INTEGRAL = 0.1 MINUTE (6 SECONDS)
DERIVATIVE = 0.0 (OFF)

Fig. 20-8. Simulator setup for single-element temperature control.

The variety of processes that can be simulated is virtually endless. The limitations are time-related. For example, a simulation of a distillation column with time constants of an hour could not be done in real-time with this simulator. Generally, this creates no problem, since shorter time periods are much more practical for learning.

Pneumatic Process Simulator

The pneumatic-simulated process shown in schematic form in Figure 20-10 and pictured in Figure 20-11 consists of three capacitances (tanks) and three resistances (variable restrictions) alternately in series, and a method of simulating a change in process load. Components above the dotted line in the schematic provide simulated load changes. In order to simulate load changes, a simple proportional controller is used.

HEAT EXCHANGER WITH FAST-ACTING
EQUAL PERCENTAGE VALVE

LOAD NORMAL
SET POINT 85 PERCENT

LOAD NORMAL
SET POINT 55 PERCENT

LOAD NORMAL
SET POINT 15 PERCENT

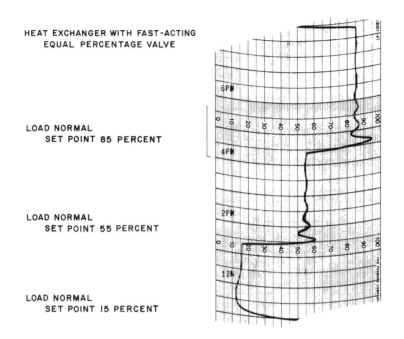

Fig. 20-9. Heat Exchanger—normal load, equal percentage valve.

The Foxboro Model 56-3 relay is basically a proportional controller, that is, the output (bellows P, called the feedback bellows) has a signal proportional to the difference in signals in bellows A and C. Bellows C is normally called the "measurement bellows," and A is called the "set bellows." In this case bellows C is used as the measurement input, but bellows A is used as a "simulated process."

This analog demonstrator uses three standard Foxboro instruments (a Model 122F [two-pen] recorder, a Model 130-M-N5 [PR&D controller], a three-unit pneumatic control shelf #100-3N-R1-C1, and a Model 56-3-9 computing relay).

The set-point transmitter, controller, and transfer switch are all in the same unit. The controlled variable (pressure) is recorded by the one pen (red in the Foxboro Model 122F). Another pen (green) may record any of several possible inputs, such as controller set point, controller output, or air pressure applied to the C bellows of the relay. The ability to record these factors provides flexibility for demonstration.

The set-point transmitter delivers a signal to the controller corresponding to the desired operating level. The transfer switch permits

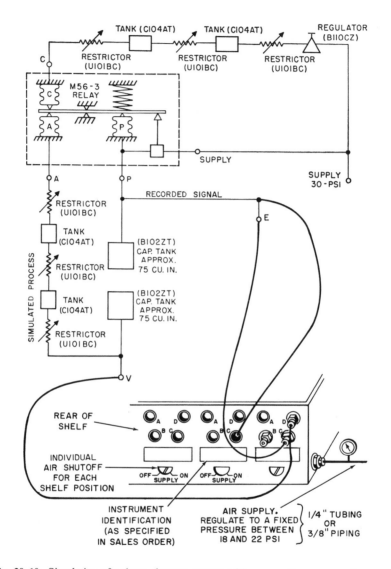

Fig. 20-10. Simulation of a three-element system with pneumatic components.

switching control from manual to automatic, or from automatic to manual.

The three-mode pneumatic controller has proportional, integral, and derivative functions. The recorder should have a ¾-inch per minute chart drive.

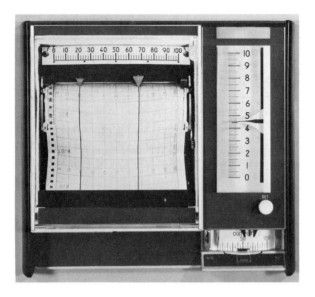

Fig. 20-11. Pneumatic Consotrol-100 Line, three-unit shelf, and two-pen recorder.

Operation of the Pneumatic Simulator

Operation of the control loop can be traced on the schematic of Figure 20-10. First assume that a stable condition has been reached with the recorder pen on scale. A step change then is created by moving the set point. This results in a change in set-bellows pressure, and in the moment of force exerted by this bellows on the controller's floating disk.

The controller immediately sets out to rebalance itself. The result is a change of controller output pressure, which is applied through the resistance-capacitance network to the A bellows of the relay. The three resistance-capacitance networks and their time constants establish the process time characteristics, or phase shift. It is by altering these values that the time constants of the process may be varied. The pressure change in the A bellows unbalances the moments of force in the computing relay. This unbalance is sensed by the flapper-nozzle detector, and the relay rebalances itself by changing the pressure in bellows P. This pressure in bellows P becomes the controlled variable, and is fed back to the measurement bellows of the controller and the recorder. Within the controller, the moment of force created by the pressure in this bellows is concerned with that created by the set-bellows pressure.

When these moments are again in equilibrium, the controller has rebalanced and the system has returned to a steady state.

The C bellows is the "measurement" bellows of a conventional transmitter. Load changes are simulated by feeding changes through a three-element RC circuit into the "measurement" bellows C of the Model 56-3 relay.

Load changes may be simulated by changing the pressure in the C bellows. The knob on the pressure regulator varies the process load. The resistance-capacitance network in series with the regulator adds delay time to the load-change signal and makes the operation more realistic.

Many processes have the characteristic of pure dead time. Pure dead time is the lapse between the instant a change occurs in the process and the time it is sensed by the control system. For example, assume that the outlet temperature of a heat exchanger is to be measured and controlled. If the temperature sensor is placed in the outlet line 200 feet from the exchanger, and if the flow velocity through the line is 5 feet per second, then 40 seconds (200 feet per 5 feet per second) must elapse before any temperature change can be sensed by the sensor. This 40-second time delay would be pure dead time. Pure dead time is very difficult to simulate in any process analog. In the pneumatic process simulator, the pure delay is approached by substituting a lag for dead time. This is accomplished by introducing a resistance in series with the P signal. Several feet of 0.007-inch bore capillary in this line will provide the lag. A bypass valve will eliminate it as desired. For demonstration purposes, the simulated process should be operated without dead time, and then dead time can be introduced as an additional complication or a greater challenge.

A list of adjustments that will provide a starting point might be as follows: VR_1 = 60 percent; VR_2 = 60 percent; VR_3 = Min; C = 25; PB = 100 percent; integral = 0.5; derivative = 0.5; no dead time resistance (bypass open).

Time Base

One advantage of the process simulator is its ability to operate on a faster time base. For example, assume that an actual open-loop process is subjected to a step upset and output is recorded on a chart having a ¾-inch per hour chart drive.

If the simulator recorder chart moves at ¾ inch per minute and we manipulate the RC values, we discover that the resulting curve has the

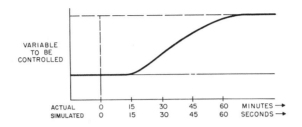

Fig. 20-12. (*Top*) The response curve, the "signature" of the system, can have its time axis changed by simulation. Only the time base is different. One curve is the real system; the other is simulated.

same shape, though our time base is only $1/60$ of that of the real process (Figure 20-12).

If the control loop is now closed and the controller is manipulated, the general problems encountered with the simulated process are essentially the same as for the real process, but the speed of action has been increased by a factor of 60.

If, after a few trial-and-error adjustments, the controller modes are set to produce the recovery curve desired, these same settings can be used in the controller attached to the actual process, but those factors

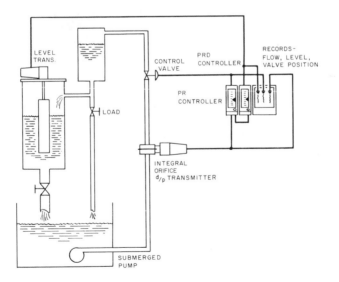

Fig. 20-13. Liquid-level process.

Fig. 20-14. Level/flow process.

having time elements (derivative and integral) must be changed by a factor of 60.

Time scaling has perhaps been oversimplified in this illustration. If the analog is used in the solution of a differential equation, the time factor must be introduced into the equation so that the resultant solution will be in proper form.

Proportional band has no time factor and would remain the same for either case. An open-loop step upset can be simulated by switching the controller to manual control. Using this technique, and the faster time base, the step upset or signature curve of the simulated process can be adjusted to correspond to the process under investigation.

Real Processes

Simplicity and economics are major considerations in the construction of a process for instructional purposes. After investigating many possibilities, the logical choice became a simple level process. The process and instrumentation are shown schematically in Figure 20-13 and pictured in Figure 20-14.

The process consists of a tall cylindrical main tank which empties into a short tank large enough in diameter to contain the contents of the tanks above it. A small tank is placed in the flow line between the bottom tank and the main tank to provide a realistic lag. Liquid (colored water) is pumped by a submersible pump from the bottom tank through an orifice plate—mounted within a differential pressure transmitter—a control valve, the small-capacity tank, and then back into the main tank. Level in the main tank is measured by a Foxboro Model 17B buoyancy transmitter. The list of material required to construct the system follows. The materials employed cost approximately $7,000 in 1980.

1 Foxboro Model 130-M-N5 PR&D controller

1 Foxboro Model 130F-N4 P&R controller

1 Foxboro Model 123, three variables

1 Foxboro Model 102-4N, four-unit mounting shelf

1 Foxboro Model V4A ¾-inch body, H needle, $C_v = 0.56$ control valve

1 Foxboro Model 17B4-K41 pneumatic buoyancy transmitter

1 Foxboro Model 17B displacer A0100WM, 14-inch, rod = 8 inches

1 Foxboro Model 13A d/p Cell transmitter

1 Foxboro Model 13A d/p Cell transmitter with integral orifice, 0.159-inch-diameter

1 Little Giant submersible pump; Little Giant Corp., Oklahoma City, OK

20 feet of Imperial plastic tubing, ⅜-inch

50 feet of Imperial plastic tubing, ¼-inch

Assorted Imperial fittings

1 pressure regulator and manifold

Wire and junction boxes

1 plywood 2 × 6 feet mounting panel and 2 × 6 feet instrument panel

1 plastic tank, 4-inch inside diameter × 42 inches high

1 plastic tank, 4-inch inside diameter × 13 inches high

1 plastic pail

Use all nonferrous materials to prevent rusty water.

The combinations that may be arranged with this equipment are many and can demonstrate a variety of control actions. Figures 20-15 through 20-20 show some of these arrangements.

If you plan to do the feedforward loops Figure 20-19 and Figure 20-20, two square root extractors, such as Foxboro Model 557 and an analog computer, Foxboro Model 136-1, will be required.

Now let us examine a more deluxe real-time process designed for training. Figure 20-21 presents a picture of a process designed for use in educational activity at Foxboro. It is shown in schematic form in Figure 20-22. This process unit features a graphic panel to which all instrument inputs and outputs are terminated. It is possible, then, with pneumatic patch cords, to make an almost endless number of loop configurations.

The dead time unit is a form of hydraulic screw. The colored water

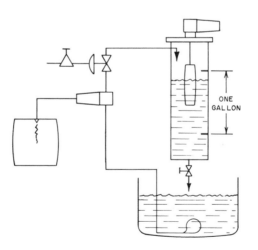

Fig. 20-15. Flow measurement—head-meter using d/p Cell transmitter. (A) Mark one gallon on large tank. (B) Calibrate d/p Cell 0 to ½ gpm. (C) With stopwatch, run for 2 minutes and compare reading with actual flow after adjusting control valve to allow full-scale ½-gpm flow. (D) Calculate error, if any. (E) Repeat at several flow rates.

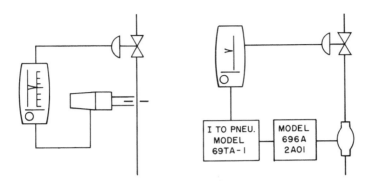

Fig. 20-16. Flow control using head measurement.

enters one end and is delayed for a number of seconds—depending on speed of rotation—before it exits at the other end. This unit makes it possible to simulate dead time and capacity as it occurs in many actual process situations. This device represents perhaps the ultimate in educational process units. It does represent a substantial investment.

From a practical point of view, building a unit similar to that shown in Figure 20-14 and making use of available instruments around the plant will provide an adequate teaching tool. A school engaged in teach-

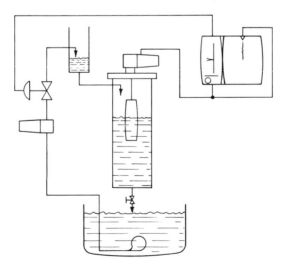

Fig. 20-17. Level control.

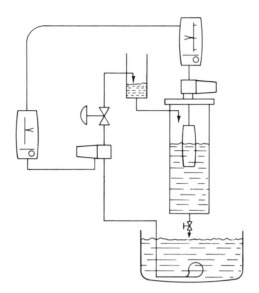

Fig. 20-18. Cascade control.

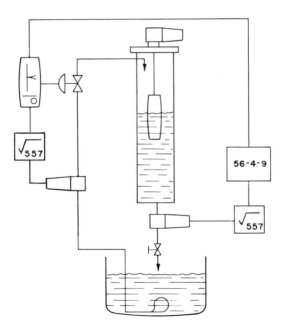

Fig. 20-19. Feedforward without trim.

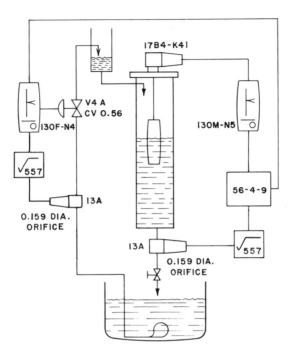

Fig. 20-20. Feedforward with trim.

ing process instrumentation and control will find a working loop a most valuable tool. The following problems are intended to stimulate possible uses of a process unit.

Problem 1 Figure 20-15. Flow measurement—head-meter using d/p Cell transmitter. (A) Mark 1 gallon on large tank. (B) Calibrate d/p Cell 0 to ½ gpm. (C) With stopwatch, run for 2 minutes and compare reading with actual flow after adjusting control valve to allow full-scale (½ gpm) flow. (D) Calculate error, if any. (E) Repeat at several flow rates.

Problem 2 Having calibrated the flowmeter, put the controller whose output positions the control valve on manual. With the system operating manually, position the valve at 10 percent increments from 0 to 100 percent and record the flow rate at each step. Now plot the characteristic curve for the valve and compare the results with those shown in Chapter 11.

Problem 3 Bypass the small tank as suggested in Question 20-7 and verify your answer.

Fig. 20-21. Educational process simulator unit.

Problem 4 Connect the system as shown in Figure 20-17 and carefully adjust the controller. Now use the cascade arrangement shown in Figure 20-18 and repeat the proper adjustment procedure. Explain any differences in results.

Problem 5 Create a feedforward system as shown in Figure 20-19. Once it is operating properly, adjust the zero on the flow transmitter to introduce an error. Observe the result. Repeat the same procedure using the

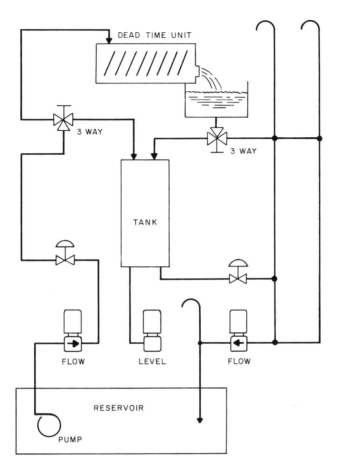

Fig. 20-22. Schematic—educational process unit.

system shown in Figure 20-20. Note the results and carefully explain any difference.

Questions

20-1. A big advantage in simulating a process is
   a. Pure dead time can be clearly demonstrated
   b. The noise factor is eliminated

    c. That it is relatively inexpensive and may be performed on a faster time base

    d. That one may witness every detail of cause and effect

20-2. The major difference between the pneumatic process simulator and the one that is electrically energized is that

    a. The pneumatic type outperforms the electrical type

    b. The electrical type outperforms the pneumatic type

    c. The pneumatic type operates much faster than the electrical

    d. The electrical type is physically smaller and easier to construct

20-3. The academic basis of all simulation techniques is

    a. All physical systems behave in essentially the same way with respect to time

    b. Electrical systems react faster than pneumatic systems

    c. Pneumatic systems are 3 to 15 times faster than real-time systems

    d. Electrical simulators react in a square root relationship to an equivalent pneumatic system

20-4. Proportional band has the following scaling factor for almost any process

    a. 1   **c.** 10

    b. 0   **d.** 100

20-5. The integral orifice d/p Cell flowmeter has the greatest readability

    a. At midscale

    b. Within 0 to 25 percent of scale

    c. Within 25 to 60 percent of scale

    d. On the upper 40 percent of scale

20-6. Assuming a properly adjusted controller is controlling a flow process, changes in load-flow rate may cause instability because

    a. The loop gain varies with flow rate

    b. The loop gain varies with valve position

    c. The location of the flow sensor is incorrect

    d. Not true. The loop remains stable.

20-7. If the small tank between the control valve and the main tank is bypassed, it becomes impossible to create oscillations or cycling. Why?

    a. It is similar to a flow process

    b. It does not have enough capacity and resistance elements to create sufficient phase shift

    c. The gain of the large tank is so small in value that the controller gain is inadequate for oscillation

    d. The statement is not true; the system will oscillate vigorously

20-8. The cascade control system exhibits several advantages, one of which is

    a. More efficiency
    b. Greater stability
    c. No possibility of cycling or oscillation
    d. Greater ease in adjustment

20-9. The disadvantage of feedforward without feedback trim control is
    a. Nonexistent
    b. Poor stability
    c. The errors accumulate and eventually the tank either goes dry or overflows
    d. Difficulty in making adjustments

20-10. The process employing feedforward control with feedback trim provides excellent results but
    a. Is very difficult to adjust
    b. Cannot correct for bad upsets
    c. Is very sluggish in its operation
    d. Must in practice be a process that can justify the expense required to provide this sophisticated control system

Table A-1. Unit Conversions

ATMOSPHERES—atm (Standard at sea-level pressure)
× 101.325	= Kilopascals (kPa) absolute
× 14.696	= Pounds-force per square inch absolute (psia)
× 76.00	= Centimetres of mercury (cmHg) at 0°C
× 29.92	= Inches of mercury (inHg) at 0°C
× 33.96	= Feet of water (ftH$_2$O) at 68°F
× 1.01325	= Bars (bar) absolute
× 1.0332	= Kilograms-force per square centimetre (kg/cm²) absolute
× 1.0581	= Tons-force per square foot (tonf/ft²) absolute
× 760	= Torr (torr) (= mmHg at 0°C)

BARRELS, LIQUID, U.S.—bbl
× 0.11924	= Cubic metres (m³)
× 31.5	= U.S. gallons (U.S. gal) liquid

BARRELS, PETROLEUM—bbl
× 0.15899	= Cubic metres (m³)
× 42	= U.S. gallons (U.S. gal) oil

BARS—bar
× 100	= Kilopascals (kPa)
× 14.504	= Pounds-force per square inch (psi)
× 33.52	= Feet of water (ftH$_2$O) at 68°F
× 29.53	= Inches of mercury (inHg) at 0°C

Table A-1. Unit Conversions (*continued*)

× 1.0197	= Kilograms-force per square centimetre (kg/cm²)
× 0.98692	= Atmospheres (atm) sea-level standard
× 1.0443	= Tons-force per square foot (tonf/ft²)
× 750.06	= Torr (torr) (= mmHg at 0°C)

BRITISH THERMAL UNITS—Btu (See note)

× 1055	= Joules (J)
× 778	= Foot-pounds-force (ft · lbf)
× 0.252	= Kilocalories (kcal)
× 107.6	= Kilogram-force-metres (kgf · m)
× 2.93 × 10⁻⁴	= Kilowatt-hours (kW · h)
× 3.93 × 10⁻⁴	= Horsepower-hours (hp · h)

BRITISH THERMAL UNITS PER MINUTE—Btu/min (See note)

× 17.58	= Watts (W)
× 12.97	= Foot-pounds-force per second (ft · lbf/s)
× 0.02358	= Horsepower (hp)

CENTARES

× 1	= Square metres (m²)

CENTIMETRES—cm

× 0.3937	= Inches (in)

CENTIMETRES OF MERCURY—cmHg, at 0°C

× 1.3332	= Kilopascals (kPa)
× 0.013332	= Bars (bar)
× 0.4468	= Feet of water (ftH₂O) at 68°F
× 5.362	= Inches of water (inH₂O) at 68°F
× 0.013595	= Kilograms-force per square centimetre (kg/cm²)
× 27.85	= Pounds-force per square foot (lbf/ft²)
× 0.19337	= Pounds-force per square inch (psi)
× 0.013158	= Atmospheres (atm) standard
× 10	= Torr (torr) (= mmHg at 0°C)

CENTIMETRES PER SECOND—cm/s

× 1.9685	= Feet per minute (ft/min)
× 0.03281	= Feet per second (ft/s)
× 0.03600	= Kilometres per hour (km/h)
× 0.6000	= Metres per minute (m/min)
× 0.02237	= Miles per hour (mph)

CUBIC CENTIMETRES—cm³

× 3.5315 × 10⁻⁵	= Cubic feet (ft³)
× 6.1024 × 10⁻²	= Cubic inches (in³)
× 1.308 × 10⁻⁶	= Cubic yards (yd³)
× 2.642 × 10⁻⁴	= U.S. gallons (U.S. gal)
× 2.200 × 10⁻⁴	= Imperial gallons (imp gal)
× 1.000 × 10⁻³	= Litres (l)

CUBIC FEET—ft³

× 0.02832	= Cubic metres (m³)
× 2.832 × 10⁴	= Cubic centimetres (cm³)

Table A-1. Unit Conversions (*continued*)

× 1728	= Cubic inches (in³)
× 0.03704	= Cubic yards (yd³)
× 7.481	= U.S. gallons (U.S. gal)
× 6.229	= Imperial gallons (imp gal)
× 28.32	= Litres (l)

CUBIC FEET PER MINUTE—cfm

× 472.0	= Cubic centimetres per second (cm³/s)
× 1.699	= Cubic metres per hour (m³/h)
× 0.4720	= Litres per second (l/s)
× 0.1247	= U.S. gallons per second (U.S. gps)
× 62.30	= Pounds of water per minute (lbH₂O/min) at 68°F

CUBIC FEET PER SECOND—cfs

× 0.02832	= Cubic metres per second (m³/s)
× 1.699	= Cubic metres per minute (m³/min)
× 448.8	= U.S. gallons per minute (U.S. gpm)
× 0.6463	= Million U.S. gallons per day (U.S. gpd)

CUBIC INCHES—in³

× 1.6387 × 10⁻⁵	= Cubic metres (m³)
× 16.387	= Cubic centimetres (cm³)
× 0.016387	= Litres (l)
× 5.787 × 10⁻⁴	= Cubic feet (ft³)
× 2.143 × 10⁻⁵	= Cubic yards (yd³)
× 4.329 × 10⁻³	= U.S. gallons (U.S. gal)
× 3.605 × 10⁻³	= Imperial gallons (imp gal)

CUBIC METRES—m³

× 1000	= Litres (l)
× 35.315	= Cubic feet (ft³)
× 61.024 × 10³	= Cubic inches (in³)
× 1.3080	= Cubic yards (yd³)
× 264.2	= U.S. gallons (U.S. gal)
× 220.0	= Imperial gallons (imp gal)

CUBIC METRES PER HOUR—m³/h

× 0.2778	= Litres per second (l/s)
× 2.778 × 10⁻⁴	= Cubic metres per second (m³/s)
× 4.403	= U.S. gallons per minute (U.S. gpm)

CUBIC METRES PER SECOND—m³/s

× 3600	= Cubic metres per hour (m³/h)
× 15.85 × 10³	= U.S. gallons per minute (U.S. gpm)

CUBIC YARDS—yd³

× 0.7646	= Cubic metres (m³)
× 764.6	= Litres (l)
× 7.646 × 10⁵	= Cubic centimetres (cm³)
× 27	= Cubic feet (ft³)
× 46,656	= Cubic inches (in³)
× 201.97	= U.S. gallons (U.S. gal)
× 168.17	= Imperial gallons (imp gal)

Table A-1. Unit Conversions (*continued*)

DEGREES, ANGULAR (°)
× 0.017453	= Radians (rad)
× 60	= Minutes (')
× 3600	= Seconds (")
× 1.111	= Grade (gon)

DEGREES PER SECOND, ANGULAR (°/s)
× 0.017453	= Radians per second (rad/s)
× 0.16667	= Revolutions per minute (r/min)
× 2.7778×10^{-3}	= Revolutions per second (r/s)

DRAMS (dr)
× 1.7718	= Grams (g)
× 27.344	= Grains (gr)
× 0.0625	= Ounces (oz)

FATHOMS
× 1.8288	= Metres (m)
× 6	= Feet (ft)

FEET—ft
× 0.3048	= Metres (m)
× 30.480	= Centimetres (cm)
× 12	= Inches (in)
× 0.3333	= Yards (yd)

FEET OF WATER—ftH$_2$O, at 68°F
× 2.984	= Kilopascals (kPa)
× 0.02984	= Bars (bar)
× 0.8811	= Inches of mercury (inHg) at 0°C
× 0.03042	= Kilograms-force per square centimetre (kg/cm²)
× 62.32	= Pounds-force per square foot (lbf/ft²)
× 0.4328	= Pounds-force per square inch (psi)
× 0.02945	= Standard atmospheres

FEET PER MINUTE—ft/min
× 0.5080	= Centimetres per second (cm/s)
× 0.01829	= Kilometres per hour (km/h)
× 0.3048	= Metres per minute (m/min)
× 0.016667	= Feet per second (ft/s)
× 0.01136	= Miles per hour (mph)

FEET PER SECOND PER SECOND—ft/s²
× 0.3048	= Metres per second per second (m/s²)
× 30.48	= Centimetres per second per second (cm/s²)

FOOT-POUNDS-FORCE—ft · lbf
× 1.356	= Joules (J)
× 1.285×10^{-3}	= British thermal units (Btu) (see note)
× 3.239×10^{-4}	= Kilocalories (kcal)
× 0.13825	= Kilogram-force-metres (kgf · m)
× 5.050×10^{-7}	= Horsepower-hours (hp · h)
× 3.766×10^{-7}	= Kilowatt-hours (kW · h)

Table A-1. Unit Conversions (*continued*)

GALLONS, U.S.—U.S. gal
× 3785.4	= Cubic centimetres (cm³)
× 3.7854	= Litres (l)
× 3.7854 × 10⁻³	= Cubic metres (m³)
× 231	= Cubic inches (in³)
× 0.13368	= Cubic feet (ft³)
× 4.951 × 10⁻³	= Cubic yards (yd³)
× 8	= Pints (pt) liquid
× 4	= Quarts (qt) liquid
× 0.8327	= Imperial gallons (imp gal)
× 8.328	= Pounds of water at 60°F in air
× 8.337	= Pounds of water at 60°F in vacuo

GALLONS, IMPERIAL—imp gal
× 4546	= Cubic centimetres (cm³)
× 4.546	= Litres (l)
× 4.546 × 10⁻³	= Cubic metres (m³)
× 0.16054	= Cubic feet (ft³)
× 5.946 × 10⁻³	= Cubic yards (yd³)
× 1.20094	= U.S. gallons (U.S. gal)
× 10.000	= Pounds of water at 62°F in air

GALLONS, PER MINUTE, U.S.—U.S. gpm
× 0.22715	= Cubic metres per hour (m³/h)
× 0.06309	= Litres per second (l/s)
× 8.021	= Cubic feet per hour (cfh)
× 2.228 × 10⁻³	= Cubic feet per second (cfs)

GRAINS—gr av. or troy
× 0.0648	= Grams (g)

GRAINS PER U.S. GALLON—gr/U.S. gal at 60°F
× 17.12	= Grams per cubic metre (g/m³)
× 17.15	= Parts per million by weight in water
× 142.9	= Pounds per million gallons

GRAINS PER IMPERIAL GALLON—gr/imp gal at 62°F
× 14.25	= Grams per cubic metre (g/m³)
× 14.29	= Parts per million by weight in water

GRAMS—g
× 15.432	= Grains (gr)
× 0.035274	= Ounces (oz) av.
× 0.032151	= Ounces (oz) troy
× 2.2046 × 10⁻³	= Pounds (lb)

GRAMS-FORCE—gf
× 9.807 × 10⁻³	= Newtons (N)

GRAMS-FORCE PER CENTIMETRE—gf/cm
× 98.07	= Newtons per metre (N/m)
× 5.600 × 10⁻³	= Pounds-force per inch (lbf/in)

Table A-1. Unit Conversions (*continued*)

GRAMS PER CUBIC CENTIMETRE—g/cm³
× 62.43	= Pounds per cubic foot (lb/ft³)
× 0.03613	= Pounds per cubic inch (lb/in³)

GRAMS PER LITRE—g/l
× 58.42	= Grains per U.S. gallon (gr/U.S. gal)
× 8.345	= Pounds per 1000 U.S. gallons
× 0.06243	= Pounds per cubic foot (lb/ft³)
× 1002	= Parts per million by mass (weight) in water at 60°F

HECTARES—ha
× 1.000 × 10⁴	= Square metres (m²)
× 1.0764 × 10⁵	= Square feet (ft²)

HORSEPOWER—hp
× 745.7	= Watts (W)
× 0.7457	= Kilowatts (kW)
× 33,000	= Foot-pounds-force per minute (ft · lbf/min)
× 550	= Foot-pounds-force per second (ft · lbf/s)
× 42.43	= British thermal units per minute (Btu/min) (see note)
× 10.69	= Kilocalories per minute (kcal/min)
× 1.0139	= Horsepower (metric)

HORSEPOWER—hp boiler
× 33,480	= British thermal units per hour (Btu/h) (see note)
× 9.809	= Kilowatts (kW)

HORSEPOWER-HOURS—hp · h
× 0.7457	= Kilowatt-hours (kW · h)
× 1.976 × 10⁶	= Foot-pounds-force (ft · lbf)
× 2545	= British thermal units (Btu) (see note)
× 641.5	= Kilocalories (kcal)
× 2.732 × 10⁵	= Kilogram-force-metres (kgf · m)

INCHES—in
× 2.540	= Centimetres (cm)

INCHES OF MERCURY—inHg at 0°C
× 3.3864	= Kilopascals (kPa)
× 0.03386	= Bars (bar)
× 1.135	= Feet of water (ftH₂O) at 68°F
× 13.62	= Inches of water (inH₂O) at 68°F
× 0.03453	= Kilograms-force per square centimetre (kg/cm²)
× 70.73	= Pounds-force per square foot (lbf/ft²)
× 0.4912	= Pounds-force per square inch (psi)
× 0.03342	= Standard atmospheres

INCHES OF WATER—inH₂O at 68°F
× 0.2487	= Kilopascals (kPa)
× 2.487 × 10⁻³	= Bars (bar)
× 0.07342	= Inches of mercury (inHg) at 0°C
× 2.535 × 10⁻³	= Kilograms-force per square centimetre (kg/cm²)
× 0.5770	= Ounces-force per square inch (ozf/in²)

Table A-1. Unit Conversions (*continued*)

× 5.193	= Pounds-force per square foot (lbf/ft²)
× 0.03606	= Pounds-force per square inch (psi)
× 2.454 × 10⁻³	= Standard atmospheres

JOULES—J

× 0.9484 × 10⁻³	= British thermal units (Btu) (see note)
× 0.2390	= Calories (cal) thermochemical
× 0.7376	= Foot-pounds-force (ft · lbf)
× 2.778 × 10⁻⁴	= Watt-hours (W · h)

KILOGRAMS—kg

× 2.2046	= Pounds (lb)
× 1.102 × 10⁻³	= Tons (ton) short

KILOGRAMS-FORCE—kgf

× 9.807	= Newtons (N)
× 2.205	= Pounds-force (lbf)

KILOGRAMS-FORCE PER METRE—kgf/m

× 9.807	= Newtons per metre (N/m)
× 0.6721	= Pounds-force per foot (lbf/ft)

KILOGRAMS-FORCE PER SQUARE CENTIMETRE—kg/cm²

× 98.07	= Kilopascals (kPa)
× 0.9807	= Bars (bar)
× 32.87	= Feet of water (ftH₂O) at 68°F
× 28.96	= Inches of mercury (inHg) at 0°C
× 2048	= Pounds-force per square foot (lbf/ft²)
× 14.223	= Pounds-force per square inch (psi)
× 0.9678	= Standard atmospheres

KILOGRAMS-FORCE PER SQUARE MILLIMETRE—kgf/mm²

× 9.807	= Megapascals (MPa)
× 1.000 × 10⁶	= Kilograms-force per square metre (kgf/m²)

KILOMETRES PER HOUR—km/h

× 27.78	= Centimetres per second (cm/s)
× 0.9113	= Feet per second (ft/s)
× 54.68	= Feet per minute (ft/min)
× 16.667	= Metres per minute (m/min)
× 0.53996	= International knots (kn)
× 0.6214	= Miles per hour (mph)

KILOMETRES PER HOUR PER SECOND—km · h⁻¹ · S⁻¹

× 0.2778	= Metres per second per second (m/s²)
× 27.78	= Centimetres per second per second (cm/s²)
× 0.9113	= Feet per second per second (ft/s²)

KILOMETRES PER SECOND—km/s

× 37.28	= Miles per minute (mi/min)

KILOPASCALS—kPa

× 10³	= Pascals (Pa) or newtons per square metre (N/m²)
× 0.1450	= Pounds-force per square inch (psi)
× 0.010197	= Kilograms-force per square centimetre (kg/cm²)

Table A-1. Unit Conversions (*continued*)

× 0.2953	= Inches of mercury (inHg) at 32°F
× 0.3351	= Feet of water (ftH$_2$O) at 68°F
× 4.021	= Inches of water (inH$_2$O) at 68°F

KILOWATTS—kW

× 4.425 × 10^4	= Foot-pounds-force per minute (ft · lbf/min)
× 737.6	= Foot-pounds-force per second (ft · lbf/s)
× 56.90	= British thermal units per minute (Btu/min) (see note)
× 14.33	= Kilocalories per minute (kcal/min)
× 1.3410	= Horsepower (hp)

KILOWATT-HOURS—kW · h

× 3.6 × 10^6	= Joules (J)
× 2.655 × 10^6	= Foot-pounds-force (ft · lbf)
× 3413	= British thermal units (Btu) (see note)
× 860	= Kilocalories (kcal)
× 3.671 × 10^5	= Kilogram-force metres (kgf · m)
× 1.3410	= Horsepower-hours (hp · h)

KNOTS—kn (International)

× 0.5144	= Metres per second (m/s)
× 1.151	= Miles per hour (mph)

LITRES—l

× 1000	= Cubic centimetres (cm^3)
× 0.035315	= Cubic feet (ft^3)
× 61.024	= Cubic inches (in^3)
× 1.308 × 10^{-3}	= Cubic yards (yd^3)
× 0.2642	= U.S. gallons (U.S. gal)
× 0.2200	= Imperial gallons (imp gal)

LITRES PER MINUTE—l/min

× 0.01667	= Litres per second (l/s)
× 5.885 × 10^{-4}	= Cubic feet per second (cfs)
× 4.403 × 10^{-3}	= U.S. gallons per second (U.S. gal/s)
× 3.666 × 10^{-3}	= Imperial gallons per second (imp gal/s)

LITRES PER SECOND—l/s

× 10^{-3}	= Cubic metres per second (m^3/s)
× 3.600	= Cubic metres per hour (m^3/h)
× 60	= Litres per minute (l/min)
× 15.85	= U.S. gallons per minute (U.S. gpm)
× 13.20	= Imperial gallons per minute (imp gpm)

MEGAPASCALS—MPa

× 10^6	= Pascals (Pa) or newtons per square metre (N/m^2)
× 10^3	= Kilopascals (kPa)
× 145.0	= Pounds-force per square inch (psi)
× 0.1020	= Kilograms-force per square millimetre (kgf/mm^2)

METRES—m

× 3.281	= Feet (ft)
× 39.37	= Inches (in)
× 1.0936	= Yards (yd)

Table A-1. Unit Conversions (*continued*)

METRES PER MINUTE—m/min
× 1.6667	= Centimetres per second (cm/s)
× 0.0600	= Kilometres per hour (km/h)
× 3.281	= Feet per minute (ft/min)
× 0.05468	= Feet per second (ft/s)
× 0.03728	= Miles per hour (mph)

METRES PER SECOND—m/s
× 3.600	= Kilometres per hour (km/h)
× 0.0600	= Kilometres per minute (km/min)
× 196.8	= Feet per minute (ft/min)
× 3.281	= Feet per second (ft/s)
× 2.237	= Miles per hour (mph)
× 0.03728	= Miles per minute (mi/min)

MICROMETRES—μm formerly micron
× 10^{-6}	= Metres (m)

MILES—mi
× 1.6093 × 10^{-3}	= Metres (m)
× 1.6093	= Kilometres (km)
× 5280	= Feet (ft)
× 1760	= Yards (yd)

MILES PER HOUR—mph
× 44.70	= Centimetres per second (cm/s)
× 1.6093	= Kilometres per hour (km/h)
× 26.82	= Metres per minute (m/min)
× 88	= Feet per minute (ft/min)
× 1.4667	= Feet per second (ft/s)
× 0.8690	= International knots (kn)

MILES PER MINUTE—mi/min
× 1.6093	= Kilometres per minute (km/min)
× 2682	= Centimetres per second (cm/s)
× 88	= Feet per second (ft/s)
× 60	= Miles per hour (mph)

MINUTES, ANGULAR—(')
× 2.909 × 10^{-4}	= Radians (rad)

NEWTONS—N
× 0.10197	= Kilograms-force (kgf)
× 0.2248	= Pounds-force (lbf)
× 7.233	= Poundals
× 10^5	= Dynes

OUNCES—oz av.
× 28.35	= Grams (g)
× 2.835 × 10^{-5}	= Tonnes (t) metric ton
× 16	= Drams (dr) av.
× 437.5	= Grains (gr)
× 0.06250	= Pounds (lb) av.

Table A-1. Unit Conversions (*continued*)

× 0.9115	= Ounces (oz) troy
× 2.790 × 10⁻⁵	= Tons (ton) long

OUNCES—oz troy

× 31.103	= Grams (g)
× 480	= Grains (gr)
× 20	= Pennyweights (dwt) troy
× 0.08333	= Pounds (lb) troy
× 0.06857	= Pounds (lb) av.
× 1.0971	= Ounces (oz) av.

OUNCES—oz U.S. fluid

× 0.02957	= Litres (l)
× 1.8046	= Cubic inches (in)

OUNCES-FORCE PER SQUARE INCH—ozf/in²

× 43.1	= Pascals (Pa)
× 0.06250	= Pounds-force per square inch (psi)
× 4.395	= Grams-force per square centimetre (gf/cm²)

PARTS PER MILLION BY MASS—mass (weight) in water

× 0.9991	= Grams per cubic metre (g/m³) at 15°C
× 0.0583	= Grains per U.S. gallon (gr/U.S. gal) at 60°F
× 0.0700	= Grains per imperial gallon (gr/imp gal) at 62°F
× 8.328	= Pounds per million U.S. gallons at 60°F

PASCALS—PA

× 1	= Newtons per square metre (N/m²)
× 1.450 × 10⁻⁴	= Pounds-force per square inch (psi)
× 1.0197 × 10⁻⁵	= Kilograms-force per square centimetre (kg/cm²)
× 10⁻³	= Kilopascals (kPa)

PENNYWEIGHTS—dwt troy

× 1.5552	= Grams (g)
× 24	= Grains (gr)

POISES—P

× 0.1000	= Newton-seconds per square metre (N · s/m²)
× 100	= Centipoises (cP)
× 2.0886 × 10⁻³	= Pound-force-seconds per square foot (lbf · s/ft²)
× 0.06721	= Pounds per foot second (lb/ft · s)

POUNDS-FORCE—lbf av.

× 4.448	= Newtons (N)
× 0.4536	= Kilograms-force (kgf)

POUNDS—lb av.

× 453.6	= Grams (g)
× 16	= Ounces (oz) av.
× 256	= Drams (dr) av.
× 7000	= Grains (gr)
× 5 × 10⁻⁴	= Tons (ton) short
× 1.2153	= Pounds (lb) troy

Table A-1. Unit Conversions (*continued*)

POUNDS—lb troy

× 373.2	= Grams (g)
× 12	= Ounces (oz) troy
× 240	= Pennyweights (dwt) troy
× 5760	= Grains (gr)
× 0.8229	= Pounds (lb) av.
× 13.166	= Ounces (oz) av.
× 3.6735 × 10^{-4}	= Tons (ton) long
× 4.1143 × 10^{-4}	= Tons (ton) short
× 3.7324 × 10^{-4}	= Tonnes (t) metric tons

POUNDS-MASS OF WATER AT 60°F

× 453.98	= Cubic centimetres (cm^3)
× 0.45398	= Litres (l)
× 0.01603	= Cubic feet (ft^3)
× 27.70	= Cubic inches (in^3)
× 0.1199	= U.S. gallons (U.S. gal)

POUNDS OF WATER PER MINUTE AT 60°F

× 7.576	= Cubic centimetres per second (cm^3/s)
× 2.675 × 10^{-4}	= Cubic feet per second (cfs)

POUNDS PER CUBIC FOOT—lb/ft³

× 16.018	= Kilograms per cubic metre (kg/m^3)
× 0.016018	= Grams per cubic centimetre (g/cm^3)
× 5.787 × 10^{-4}	= Pounds per cubic inch (lb/in^3)

POUNDS PER CUBIC INCH—lb/in³

× 2.768 × 10^4	= Kilograms per cubic metre (kg/m^3)
× 27.68	= Grams per cubic centimetre (g/cm^3)
× 1728	= Pounds per cubic foot (lb/ft^3)

POUNDS-FORCE PER FOOT—lbf/ft

× 14.59	= Newtons per metre (N/m)
× 1.488	= Kilograms-force per metre (kgf/m)
× 14.88	= Grams-force per centimetre (gf/cm)

POUNDS-FORCE PER SQUARE FOOT—lbf/ft²

× 47.88	= Pascals (Pa)
× 0.01605	= Feet of water (ftH_2O) at 68°F
× 4.882 × 10^{-4}	= Kilograms-force per square centimetre (kg/cm^2)
× 6.944 × 10^{-3}	= Pounds-force per square inch (psi)

POUNDS-FORCE PER SQUARE INCH—psi

× 6.895	= Kilopascals (kPa)
× 0.06805	= Standard atmospheres
× 2.311	= Feet of water (ftH_2O) at 68°F
× 27.73	= Inches of water (inH_2O) at 68°F
× 2.036	= Inches of mercury (inHg) at 0°C
× 0.07031	= Kilograms-force per square centimetre (kg/cm^2)

QUARTS—qt dry

× 1101	= Cubic centimetres (cm^3)
× 67.20	= Cubic inches (in^3)

Table A-1. Unit Conversions (*continued*)

QUARTS—qt liquid
× 946.4	= Cubic centimetres (cm³)
× 57.75	= Cubic inches (in³)

QUINTALS—obsolete metric mass term
× 100	= Kilograms (kg)
× 220.46	= Pounds (lb) U.S. av.
× 101.28	= Pounds (lb) Argentina
× 129.54	= Pounds (lb) Brazil
× 101.41	= Pounds (lb) Chile
× 101.47	= Pounds (lb) Mexico
× 101.43	= Pounds (lb) Peru

RADIANS—rad
× 57.30	= Degrees (°) angular

RADIANS PER SECOND—rad/s
× 57.30	= Degrees per second (°/s) angular

STANDARD CUBIC FEET PER MINUTE—scfm (at 14.696 psia and 60°F)
× 0.4474	= Litres per second (l/s) at standard conditions (760 mmHg and 0°C)
× 1.608	= Cubic metres per hour (m³/h) at standard conditions (760 mmHg and 0°C)

STOKES—St
× 10⁻⁴	= Square metres per second (m²/s)
× 1.076 × 10⁻³	= Square feet per second (ft²/s)

TONS-MASS—tonm long
× 1016	= Kilograms (kg)
× 2240	= Pounds (lb) av.
× 1.1200	= Tons (ton) short

TONNES—t metric ton, millier
× 1000	= Kilograms (kg)
× 2204.6	= Pounds (lb)

TONNES-FORCE—tf metric ton-force
× 980.7	= Newtons (N)

TONS—ton short
× 907.2	= Kilograms (kg)
× 0.9072	= Tonnes (t)
× 2000	= Pounds (lb) av.
× 32000	= Ounces (oz) av.
× 2430.6	= Pounds (lb) troy
× 0.8929	= Tons (ton) long

TONS OF WATER PER 24 HOURS AT 60°F
× 0.03789	= Cubic metres per hour (m³/h)
× 83.33	= Pounds of water per hour (lb/h H_2O) at 60°F
× 0.1668	= U.S. gallons per minute (U.S. gpm)
× 1.338	= Cubic feet per hour (cfh)

Table A-1. Unit Conversions (*continued*)

WATTS—W

× 0.05690	= British thermal units per minute (Btu/min) (see note)
× 44.25	= Foot-pounds-force per minute (ft · lbf/min)
× 0.7376	= Foot-pounds-force per second (ft · lbf/s)
× 1.341 × 10⁻³	= Horsepower (hp)
× 0.01433	= Kilocalories per minute (kcal/min)

WATT-HOURS—W · h

× 3600	= Joules (J)
× 3.413	= British thermal units (Btu) (see note)
× 2655	= Foot-pounds-force (ft · lbf)
× 1.341 × 10⁻³	= Horsepower-hours (hp · h)
× 0.860	= Kilocalories (kcal)
× 367.1	= Kilogram-force-metres (kgf · m)

NOTE: SIGNIFICANT FIGURES The precision to which a given conversion factor is known, and its application, determine the number of significant figures which should be used. While many handbooks and standards give factors contained in this table to six or more significant figures, the fact that different sources disagree, in many cases, in the fifth or further figure indicates that four or five significant figures represent the precision for these factors fairly. At present the accuracy of process instrumentation, analog or digital, is in the tenth percent region at best, thus needing only three significant figures. Hence this table is confined to four or five significant figures. The advent of the pocket calculator (and the use of digital computers in process instrumentation) tends to lead to use of as many figures as the calculator will handle. However, when this exceeds the precision of the data, or the accuracy of the application, such a practice is misleading and timewasting.

NOTE: BRITISH THERMAL UNIT When making calculations involving Btu it must be remembered that there are several definitions of the Btu. The first three significant figures of the conversion factors given in this table are common to most definitions of the Btu. However, if four or more significant figures are needed in the calculation, the appropriate handbooks and standards should be consulted to be sure the proper definition and factor are being used.

Table A-2. Temperature Conversions

C	*	F	C	*	F	C	*	F	C	*	F	C	*	F
−273.15	−459.67		−17.2	1	33.8	10.6	51	123.8	43	110	230	266	510	950
−268	−450		−16.7	2	35.6	11.1	52	125.6	49	120	248	271	520	968
−262	−440		−16.1	3	37.4	11.7	53	127.4	54	130	266	277	530	986
−257	−430		−15.6	4	39.2	12.2	54	129.2	60	140	284	282	540	1004
−251	−420		−15.0	5	41.0	12.8	55	131.0	66	150	302	288	550	1022
−246	−410		−14.4	6	42.8	13.3	56	132.8	71	160	320	293	560	1040
−240	−400		−13.9	7	44.6	13.9	57	134.6	77	170	338	299	570	1058
−234	−390		−13.3	8	46.4	14.4	58	136.4	82	180	356	304	580	1076
−229	−380		−12.8	9	48.2	15.0	59	138.2	88	190	374	310	590	1094
−223	−370		−12.2	10	50.0	15.6	60	140.0	93	200	392	316	600	1112
−218	−360		−11.7	11	51.8	16.1	61	141.8	99	210	410	321	610	1130
−212	−350		−11.1	12	53.6	16.7	62	143.6				327	620	1148
−207	−340		−10.6	13	55.4	17.2	63	145.4				332	630	1166
−201	−330		−10.0	14	57.2	17.8	64	147.2				338	640	1184
−196	−320		−9.4	15	59.0	18.3	65	149.0				343	650	1202
−190	−310		−8.9	16	60.8	18.9	66	150.8	100	212	413	349	660	1220
−184	−300		−8.3	17	62.6	19.4	67	152.6				354	670	1238
−179	−290		−7.8	18	64.4	20.0	68	154.4				360	680	1256
−173	−280		−7.2	19	66.2	20.6	69	156.2				366	690	1274
−169	−273	−459.4	−6.7	20	68.0	21.1	70	158.0				371	700	1292
−168	−270	−454	−6.1	21	69.8	21.7	71	159.8				377	710	1310
−162	−260	−436	−5.6	22	71.6	22.2	72	161.6	104	220	428	382	720	1328
−157	−250	−418	−5.0	23	73.4	22.8	73	163.4	110	230	446	388	730	1346
−151	−240	−400	−4.4	24	75.2	23.3	74	165.2	116	240	464	393	740	1364
−146	−230	−382	−3.9	25	77.0	23.9	75	167.0	121	250	482	399	750	1382
−140	−220	−364	−3.3	26	78.8	24.4	76	168.8	127	260	500	404	760	1400
−134	−210	−346	−2.8	27	80.6	25.0	77	170.6	132	270	518	410	770	1418
−129	−200	−328	−2.2	28	82.4	25.6	78	172.4	138	280	536	416	780	1436
−123	−190	−310	−1.7	29	84.2	26.1	79	174.2	143	290	554	421	790	1454
−118	−180	−292	−1.1	30	86.0	26.7	80	176.0	149	300	572	427	800	1472
−112	−170	−274	−0.6	31	87.8	27.2	81	177.8	154	310	590	432	810	1490
−107	−160	−256	0	32	89.6	27.8	82	179.6	160	320	608	438	820	1508
−101	−150	−238	0.6	33	91.4	28.3	83	181.4	166	330	626	443	830	1526
−95.6	−140	−220	1.1	34	93.2	28.9	84	183.2	171	340	644	449	840	1544
−90.0	−130	−202	1.7	35	95.0	29.4	85	185.0	177	350	662	454	850	1562
−84.4	−120	−184	2.2	36	96.8	30.0	86	186.8	182	360	680	460	860	1580
−78.9	−110	−166	2.8	37	98.6	30.6	87	188.6	188	370	698	466	870	1598
−73.3	−100	−148	3.3	38	100.4	31.1	88	190.4	193	380	716	471	880	1616
−67.8	−90	−130	3.9	39	102.2	31.7	89	192.2	199	390	734	477	890	1634
−62.2	−80	−112	4.4	40	104.0	32.2	90	194.0	204	400	752	482	900	1652
−56.7	−70	−94	5.0	41	105.8	32.8	91	195.8	210	410	770	488	910	1670
−51.1	−60	−76	5.6	42	107.6	33.3	92	197.6	216	420	788	493	920	1688
−45.6	−50	−58	6.1	43	109.4	33.9	93	199.4	221	430	806	499	930	1706
−40.0	−40	−40	6.7	44	111.2	34.4	94	201.2	227	440	824	504	940	1724
−34.4	−30	−22	7.2	45	113.0	35.0	95	203.0	232	450	842	510	950	1742
−28.9	−20	−4	7.8	46	114.8	35.6	96	204.8	238	460	860	516	960	1760
−23.3	−10	14	8.3	47	116.6	36.1	97	206.6	243	470	878	521	970	1778
−17.8	0	32	8.9	48	118.4	36.7	98	208.4	249	480	896	527	980	1796
			9.4	49	120.2	37.2	99	210.2	254	490	914	532	990	1814
			10.0	50	122.0	37.8	100	212.0	260	500	932	538	1000	1832

* In the center column, find the temperature to be converted. The equivalent temperature is in the left column, if converting to Celsius, and in the right column, if converting to Fahrenheit.

Interpolation Values

C	*	F	C	*	F
0.56	1	1.8	3.33	6	10.8
1.11	2	3.6	3.89	7	12.6
1.67	3	5.4	4.44	8	14.4
2.22	4	7.2	5.00	9	16.2
2.78	5	9.0	5.56	10	18.0

Table A-2. Temperature Conversions (*continued*)

C	*	F	C	*	F	C	*	F	C	*	F
543	1010	1850	821	1510	2750	1099	2010	3650	1377	2510	4550
549	1020	1868	827	1520	2768	1104	2020	3668	1382	2520	4568
554	1030	1886	832	1530	2786	1110	2030	3686	1388	2530	4586
560	1040	1904	838	1540	2804	1116	2040	3704	1393	2540	4604
566	1050	1922	843	1550	2822	1121	2050	3722	1399	2550	4622
571	1060	1940	849	1560	2840	1127	2060	3740	1404	2560	4640
577	1070	1958	854	1570	2858	1132	2070	3758	1410	2570	4658
582	1080	1976	860	1580	2876	1138	2080	3776	1416	2580	4676
588	1090	1994	866	1590	2894	1143	2090	3794	1421	2590	4694
593	1100	2012	871	1600	2912	1149	2100	3812	1427	2600	4712
599	1110	2030	877	1610	2930	1154	2110	3830	1432	2610	4730
604	1120	2048	882	1620	2948	1160	2120	3848	1438	2620	4748
610	1130	2066	888	1630	2966	1166	2130	3866	1443	2630	4766
616	1140	2084	893	1640	2984	1171	2140	3884	1449	2640	4784
621	1150	2102	899	1650	3002	1177	2150	3902	1454	2650	4802
627	1160	2120	904	1660	3020	1182	2160	3920	1460	2660	4820
632	1170	2138	910	1670	3038	1188	2170	3938	1466	2670	4838
638	1180	2156	916	1680	3056	1193	2180	3956	1471	2680	4856
643	1190	2174	921	1690	3074	1199	2190	3974	1477	2690	4874
649	1200	2192	927	1700	3092	1204	2200	3992	1482	2700	4892
654	1210	2210	932	1710	3110	1210	2210	4010	1488	2710	4910
660	1220	2228	938	1720	3128	1216	2220	4028	1493	2720	4928
666	1230	2246	943	1730	3146	1221	2230	4046	1499	2730	4946
671	1240	2264	949	1740	3164	1227	2240	4064	1504	2740	4964
677	1250	2282	954	1750	3182	1232	2250	4082	1510	2750	4982
682	1260	2300	960	1760	3200	1238	2260	4100	1516	2760	5000
688	1270	2318	966	1770	3218	1243	2270	4118	1521	2770	5018
693	1280	2336	971	1780	3236	1249	2280	4136	1527	2780	5036
699	1290	2354	977	1790	3254	1254	2290	4154	1532	2790	5054
704	1300	2372	982	1800	3272	1260	2300	4172	1538	2800	5072
710	1310	2390	988	1810	3290	1266	2310	4190	1543	2810	5090
716	1320	2408	993	1820	3308	1271	2320	4208	1549	2820	5108
721	1330	2426	999	1830	3326	1277	2330	4226	1554	2830	5126
727	1340	2444	1004	1840	3344	1282	2340	4244	1560	2840	5144
732	1350	2462	1010	1850	3362	1288	2350	4262	1566	2850	5162
738	1360	2480	1016	1860	3380	1293	2360	4280	1571	2860	5180
743	1370	2498	1021	1870	3398	1299	2370	4298	1577	2870	5198
749	1380	2516	1027	1880	3416	1304	2380	4316	1582	2880	5216
754	1390	2534	1032	1890	3434	1310	2390	4334	1588	2890	5234
760	1400	2552	1038	1900	3452	1316	2400	4352	1593	2900	5252
766	1410	2570	1043	1910	3470	1321	2410	4370	1599	2910	5270
771	1420	2588	1049	1920	3488	1327	2420	4388	1604	2920	5288
777	1430	2606	1054	1930	3506	1332	2430	4406	1610	2930	5306
782	1440	2624	1060	1940	3524	1338	2440	4424	1616	2940	5324
788	1450	2642	1066	1950	3542	1343	2450	4442	1621	2950	5342
793	1460	2660	1071	1960	3560	1349	2460	4460	1627	2960	5360
799	1470	2678	1077	1970	3578	1354	2470	4478	1632	2970	5378
804	1480	2696	1082	1980	3596	1360	2480	4496	1638	2980	5396
810	1490	2714	1088	1990	3614	1366	2490	4514	1643	2990	5414
816	1500	2732	1093	2000	3632	1371	2500	4532	1649	3000	5432

TEMPERATURE CONVERSION FORMULAS

DEGREES CELSIUS (FORMERLY CENTIGRADE) C	DEGREES FAHRENHEIT—F	DEGREES RÉAUMUR—R
$C + 273.15 = K$ Kelvin	$F + 459.67 = $ Rankine	$R \times {}^5/_4 = C$ Celsius
$(C \times {}^9/_5) + 32 = F$ Fahrenheit	$(F - 32) \times {}^5/_9 = C$ Celsius	$(R \times {}^9/_4) + 32 = F$ Fahrenheit
$C \times {}^4/_5 = R$ Réaumur	$(F - 32) \times {}^4/_9 = R$ Réaumur	

Table A-3. Approximate Specific
Gravities of Some Common Liquids
under Normal Conditions of
Pressure and Temperature

Liquid	Specific Gravity
Acid	
Hydrochloric, 31.5%	1.05
Muriatic, 40%	1.20
Nitric, 91%	1.50
Sulphuric, 87%	1.80
Sulphuric, 100%	1.83
Alcohol	
Ethyl, 100%	0.79
Methyl, 100%	0.80
Benzine	0.73–0.75
Chloroform	1.50
Corn Syrup	1.40–1.47
Crude Oil	0.78–0.92
Ether	0.74
Ethylene Glycol	1.125
Fish and Animal Oils	0.88–0.96
Freon	1.37–1.49
Fuel Oils	0.82–0.95
Gasoline	0.68–0.75
Glycerine, 100%	1.26
Kerosene	0.78–0.82
Lubricating Oils	0.88–0.94
Milk	1.02–1.05
Mercury	13.546
Molasses	1.40–1.49
Tar	1.07–1.30
Varnish	0.9
Vegetable Oils	0.90–0.98
Water	1.0
Water at 100°C	0.96
Water, Ice	0.88–0.92
Water, Sea	1.02–1.03

Table A-4. Standard Dimensions for Welded or Seamless Steel Pipe

Nominal Pipe Size Inches	OD Inches mm	INTERNAL DIAMETER			Threads Per Inch	Nominal Weight lb/ft
		Inches mm Sch 10	Inches mm Sch 40	Inches mm Sch 80		
⅛	.405	.307	.269	.215	27	0.244
	10.29	7.80	6.83	5.46		
¼	.540	.410	.364	.302	18	0.424
	13.72	10.41	9.25	7.67		
⅜	.675	.545	.493	.423	18	0.567
	17.15	13.84	12.52	10.74		
½	.840	.674	.622	.546	14	0.850
	21.34	17.12	15.80	13.87		
¾	1.050	.884	.824	.742	14	1.130
	26.67	22.45	20.93	18.85		
1	1.315	1.097	1.049	.957	11½	1.678
	33.40	27.86	26.64	24.31		
1¼	1.660	1.442	1.380	1.278	11½	2.272
	42.16	36.63	35.05	32.46		
1½	1.900	1.682	1.610	1.500	11½	2.717
	48.26	42.72	40.89	38.10		
2	2.375	2.157	2.067	1.939	11½	3.652
	60.33	54.79	52.50	49.25		
2½	2.875	2.635	2.469	2.323	8	5.793
	73.03	66.93	62.71	59.00		
3	3.500	3.260	3.068	2.900	8	7.575
	88.90	82.80	77.93	73.66		
4	4.500	4.260	4.026	3.826	8	0.790
	114.3	108.2	102.3	97.18		
6	6.625	6.357	6.065	5.761	8	18.974
	168.3	161.5	154.1	146.3		
8	8.625	8.329	7.981	7.625	8	28.554
	219.1	211.6	202.7	193.7		
10	10.750	10.420	10.020	9.564	8	40.483
	273.1	264.7	254.5	242.9		
12	12.750	12.390	11.938	11.376	8	43.773
	323.9	314.7	303.2	289.0		

Table A-5. Properties of Saturated Steam and Saturated Water

Press.	Temp.	Volume, ft³/lbm			Enthalpy, Btu/lbm			Entropy, Btu/lbm x F			Energy, Btu/lbm	
psia	F	Water	Evap.	Steam	Water	Evap.	Steam	Water	Evap.	Steam	Water	Steam
		v_f	v_{fg}	v_g	h_f	h_{fg}	h_g	s_f	s_{fg}	s_g	u_f	u_g
3208.2	705.47	0.05078	0.00000	0.05078	906.0	0.0	906.0	1.0612	0.0000	1.0612	875.9	875.9
3094.3	700.0	0.03662	0.03857	0.07519	822.4	172.7	995.2	0.9901	0.1490	1.1390	801.5	952.2
3000.0	695.33	0.03428	0.05073	0.08500	801.8	218.4	1020.3	0.9728	0.1891	1.1619	782.8	973.1
2708.6	680.0	0.03037	0.08080	0.11117	758.5	310.1	1068.5	0.9365	0.2720	1.2086	743.2	1012.8
2500.0	668.11	0.02859	0.10209	0.13068	731.7	361.6	1093.3	0.9139	0.3206	1.2345	718.5	1032.9
2365.7	660.0	0.02768	0.11663	0.14431	714.9	392.1	1107.0	0.8995	0.3502	1.2498	702.8	1043.9
2059.9	640.0	0.02595	0.15427	0.18021	679.1	454.6	1133.7	0.8686	0.4134	1.2821	669.2	1065.0
2000.0	635.80	0.02565	0.16266	0.18831	672.1	466.2	1138.3	0.8625	0.4256	1.2881	662.6	1068.6
1786.9	620.0	0.02466	0.19615	0.22081	646.9	506.3	1153.2	0.8403	0.4689	1.3092	638.8	1080.2
1543.2	600.0	0.02364	0.24384	0.26747	617.1	550.6	1167.7	0.8134	0.5196	1.3330	610.4	1091.4
1500.0	596.20	0.02346	0.25372	0.27719	611.7	558.4	1170.1	0.8085	0.5288	1.3373	605.2	1093.1
1326.17	580.0	0.02279	0.29937	0.32216	589.1	589.9	1179.0	0.7876	0.5673	1.3550	583.5	1099.9
1200.0	567.19	0.02232	0.34013	0.36245	571.9	613.0	1184.8	0.7714	0.5969	1.3683	566.9	1104.3
1133.38	560.0	0.02207	0.36507	0.38714	562.4	625.3	1187.7	0.7625	0.6132	1.3757	557.8	1106.5
1000.0	544.58	0.02159	0.42436	0.44596	542.6	650.4	1192.9	0.7434	0.6476	1.3910	538.6	1110.4
962.79	540.0	0.02146	0.44367	0.46513	536.8	657.5	1194.3	0.7378	0.6577	1.3954	532.9	1111.4
812.53	520.0	0.02091	0.53864	0.55956	512.0	687.0	1199.0	0.7133	0.7013	1.4146	508.8	1115.0
800.0	518.21	0.02087	0.54809	0.56896	509.8	689.6	1199.4	0.7111	0.7051	1.4163	506.7	1115.2
680.86	500.0	0.02043	0.65448	0.67492	487.9	714.3	1202.2	0.6890	0.7443	1.4333	485.4	1117.2
600.0	486.20	0.02013	0.74962	0.76975	471.7	732.0	1203.7	0.6723	0.7738	1.4461	469.5	1118.2
566.15	480.0	0.02000	0.79716	0.81717	464.5	739.6	1204.1	0.6648	0.7871	1.4518	462.4	1118.5
500.0	467.01	0.01975	0.90787	0.92762	449.5	755.1	1204.7	0.6490	0.8148	1.4639	447.7	1118.8
466.87	460.0	0.01961	0.97463	0.99424	441.5	763.2	1204.8	0.6405	0.8299	1.4704	439.8	1118.9
400.0	444.60	0.01934	1.1416	1.610	424.2	780.4	1204.6	0.6217	0.8630	1.4847	422.7	1118.7
381.54	440.0	0.01926	1.1976	1.2169	419.0	785.4	1204.4	0.6161	0.8729	1.4890	417.6	1118.5
308.780	420.0	0.01894	1.4808	1.4997	396.9	806.2	1203.1	0.5915	0.9165	1.5080	395.8	1117.5
300.0	417.35	0.01889	1.5238	1.5427	394.0	808.9	1202.9	0.5882	0.9223	1.5105	392.9	1117.2
250.0	400.97	0.01865	1.8245	1.8432	376.1	825.0	1201.1	0.5679	0.9585	1.5264	375.3	1115.8
247.259	400.0	0.01864	1.8444	1.8630	375.1	825.9	1201.0	0.5667	0.9607	1.5274	374.3	1115.7
200.0	381.80	0.01839	2.2689	2.2873	355.5	842.8	1198.3	0.5438	1.0016	1.5454	354.8	1113.7
195.729	380.0	0.01836	2.3170	2.3353	353.6	844.5	1198.0	0.5416	1.0057	1.5473	352.9	1113.5
153.010	360.0	0.01811	2.9392	2.9573	332.3	862.1	1194.4	0.5161	1.0517	1.5678	331.8	1110.6
150.0	358.43	0.01809	2.9958	3.0139	330.6	863.4	1194.1	0.5141	1.0554	1.5695	330.1	1110.4
120.0	341.27	0.01789	3.7097	3.7275	312.6	877.8	1190.4	0.4919	1.0960	1.5879	312.2	1107.6
117.992	340.0	0.01787	3.7699	3.7878	311.3	878.8	1190.1	0.4902	1.0990	1.5892	310.9	1107.4
100.0	327.82	0.01774	4.4133	4.4310	298.5	888.6	1187.2	0.4743	1.1284	1.6027	298.2	1105.2
89.643	320.0	0.01766	4.8961	4.9138	290.4	894.8	1185.2	0.4640	1.1477	1.6116	290.1	1103.7
80.0	312.04	0.01757	5.4536	5.4711	282.1	900.9	1183.1	0.4534	1.1675	1.6208	281.9	1102.1
70.0	302.93	0.01748	6.1875	6.2050	272.7	907.8	1180.6	0.4411	1.1905	1.6316	272.5	1100.2
67.005	300.0	0.01745	6.4483	6.4658	269.7	910.0	1179.7	0.4372	1.1979	1.6351	269.5	1099.6
60.0	292.71	0.017383	7.1562	7.1736	262.2	915.4	1177.6	0.4273	1.2167	1.6440	262.0	1098.0
50.0	281.02	0.017274	8.4967	8.5140	250.2	923.9	1174.1	0.4112	1.2474	1.6586	250.1	1095.3
49.200	280.0	0.017264	8.6267	8.6439	249.2	924.6	1173.8	0.4098	1.2501	1.6599	249.1	1095.1
40.0	267.25	0.017151	10.479	10.496	236.1	933.6	1169.8	0.3921	1.2844	1.6765	236.0	1092.3
35.427	260.0	0.017089	11.745	11.762	228.8	938.6	1167.4	0.3819	1.3043	1.6862	228.6	1090.3
30.0	250.34	0.017009	13.727	13.744	218.9	945.2	1164.1	0.3682	1.3313	1.6995	218.8	1087.9
25.0	240.07	0.016927	16.284	16.301	208.52	952.1	1160.6	0.3535	1.3607	1.7141	208.4	1085.2
24.968	240.0	0.016926	16.304	16.321	208.45	952.1	1160.6	0.3533	1.3609	1.7142	208.3	1085.2
20.0	227.96	0.016834	20.070	20.087	196.27	960.1	1156.3	0.3358	1.3962	1.7320	196.21	1082.0
17.186	220.0	0.016772	23.131	23.148	188.23	965.2	1153.4	0.3241	1.4201	1.7442	188.18	1079.8
15.0	213.03	0.016726	26.274	26.290	181.21	969.7	1150.9	0.3137	1.4415	1.7552	181.16	1077.8
14.696	212.00	0.016719	26.782	26.799	180.17	970.3	1150.5	0.3121	1.4447	1.7568	180.12	1077.6
11.526	200.0	0.016637	33.622	33.639	168.09	977.9	1146.0	0.2940	1.4824	1.7764	168.05	1074.2
10.0	193.21	0.016592	38.404	38.420	161.26	982.1	1143.3	0.2836	1.5043	1.7879	161.23	1072.3
8.0	182.86	0.016527	47.328	47.345	150.87	988.5	1139.3	0.2676	1.5384	1.8060	150.84	1069.2
7.5110	180.0	0.016510	50.208	50.225	148.00	990.2	1138.2	0.2631	1.5480	1.8111	147.98	1068.4
6.0	170.05	0.016451	61.967	61.984	138.03	996.2	1134.2	0.2474	1.5820	1.8294	138.01	1065.4
5.0	162.24	0.016407	73.515	73.532	130.20	1000.9	1131.1	0.2349	1.6094	1.8443	130.18	1063.1
4.7414	160.0	0.016395	77.27	77.29	127.96	1002.2	1130.2	0.2313	1.6174	1.8487	127.94	1062.4
4.0	152.96	0.016358	90.63	90.64	120.92	1006.4	1127.3	0.2199	1.6428	1.8626	120.90	1060.2
3.0	141.47	0.016300	118.71	118.73	109.42	1013.2	1122.6	0.2009	1.6854	1.8864	109.41	1056.7
2.8892	140.0	0.016293	122.98	123.00	107.95	1014.0	1122.0	0.1985	1.6910	1.8895	107.94	1056.2
2.0	126.07	0.016230	173.74	173.76	94.03	1022.1	1116.2	0.1750	1.7450	1.9200	94.03	1051.8
1.6927	120.0	0.016204	203.25	203.26	87.97	1025.6	1113.6	0.1646	1.7693	1.9339	87.96	1049.9
1.0	101.74	0.016136	333.59	333.60	69.732	1036.1	1105.8	0.1326	1.8455	1.9781	69.73	1044.1
0.94924	100.0	0.016130	350.4	350.4	67.999	1037.1	1105.1	0.1295	1.8530	1.9825	68.00	1043.5
0.50683	80.0	0.016072	633.3	633.3	48.037	1048.4	1096.4	0.0932	1.9426	2.0359	48.036	1037.0
0.25611	60.0	0.016033	1207.6	1207.6	28.060	1059.7	1087.7	0.0555	2.0391	2.0946	28.060	1030.5
0.12163	40.0	0.016019	2445.8	2445.8	8.027	1071.0	1079.0	0.0162	2.1432	2.1594	8.027	1024.0
0.08865	32.018	0.016022	3302.4	3302.4	0.0003	1075.5	1075.5	0.0000	2.1872	2.1872	0.000	1021.3

Derived and Abridged from the 1967 ASME Steam Tables. Copyright 1967 by the American Society of Mechanical Engineers.

Table A-6. Resistance Values of Copper Wire and Thermocouple Extension Wire

RESISTANCE VALUES FOR SOLID ROUND ANNEALED COPPER WIRES AT 20 DEGREES CELSIUS

Number American Wire Gauge	Resistance Ohms per 1000 Feet
30	103.2
29	81.8
28	64.9
27	51.5
26	40.8
25	32.4
24	25.7
23	20.4
22	16.1
21	12.8
20	10.2
19	8.05
18	6.39
17	5.06
16	4.02
15	3.18
14	2.53

RESISTANCE CONVERSION

$$R_t = R_o[1 + \alpha (t - t_o)]$$

R_t = Resistance at operating temperature, t

R_o = Resistance at a known temperature, t_o

t = Operating temperature

t_o = Temperature for a known resistance and α

α = Temperature coefficient of resistance at t_o (0.00393 per degree C)

Table A-6. (*continued*) Resistance Values of Thermocouple and Duplex Extension Wire (Ohms/ft at 70°F)

Gauge	I/C (J)	C/A (K)	CR/C	CU/C (T)
2	0.005	0.009	0.011	—
6	0.014	0.023	0.027	—
8	0.022	0.037	0.044	—
11	0.043	0.074	0.087	—
12	0.054	0.095	0.112	0.047
14	0.089	0.147	0.175	0.074
16	0.137	0.234	0.279	0.117
18	0.221	0.380	0.448	0.190
20	0.350	0.586	0.706	0.297
22	0.550	0.940	1.120	0.457
24	0.87	1.49	1.77	0.754
28	2.261	3.81	4.55	1.92
30	3.568	6.05	7.23	3.04

Diameter	PLT/PLT + 10% RH. (S)	PLT/PLT + 13% RH. (R)
0.020	0.46	0.48
0.022	0.38	0.39
0.024	0.32	0.33

Table A-7. Velocity and Pressure Drop through Pipe (Air)

FLOW RATE (cfm)	ΔP (PER 100 ft)	FLOW RATE (cfm)	ΔP (PER 100 ft)
¾" Pipe		1" Pipe	
0.769	0.037	1.282	0.029
1.282	0.094	3.204	0.156
2.563	0.345	4.486	0.293
3.204	0.526	6.408	0.578
4.486	1.0	8.971	1.10
5.767	1.62	12.82	2.21
7.690	2.85	19.22	4.87
10.25	4.96	28.84	10.8
12.82	7.69	35.24	16.0
19.22	17.0	44.87	25.8

Table A-7. Velocity and Pressure Drop through Pipe (Air) (*continued*)

FLOW RATE (cfm)	ΔP (PER 100 ft)	FLOW RATE (cfm)	ΔP (PER 100 ft)
	1½" Pipe		2" Pipe
4.486	0.035	8.971	0.036
7.690	0.094	16.02	0.107
11.53	0.203	25.63	0.264
22.43	0.727	38.45	0.573
28.84	1.19	48.06	0.887
41.65	2.42	60.88	1.40
54.47	4.09	76.90	2.21
64.08	5.61	102.5	3.90
89.71	10.9	121.8	5.47
115.3	18.0	166.6	10.1
	3" Pipe		4" Pipe
28.84	0.045	57.67	0.042
41.65	0.090	76.9	0.073
54.47	0.151	102.5	0.127
70.49	0.248	128.2	0.197
96.12	0.451	179.4	0.377
121.8	0.715	256.3	0.757
166.6	1.32	512.6	2.94
230.7	2.50	897.1	8.94
448.6	9.23	1410	22.2
769	27.1	1794	36.0
	6" Pipe		8" Pipe
121.8	0.023	256.3	0.023
166.6	0.041	448.6	0.068
205.1	0.061	576.7	0.111
384.5	0.204	897.1	0.262
576.7	0.450	1153	0.427
897.1	1.07	1538	0.753
1410	2.62	1794	1.02
1794	4.21	2307	1.68
2307	6.96	2820	2.50
3332	14.5	3588	4.04

Table A-7. Velocity and Pressure Drop through Pipe (Air) (*continued*)

FLOW RATE (cfm)	ΔP (PER 100 ft)	FLOW RATE (cfm)	ΔP (PER 100 ft)
10" Pipe		*12" Pipe*	
448.6	0.022	640.8	0.018
640.8	0.043	769.0	0.025
897.1	0.082	1025	0.044
1282	0.164	1282	0.067
1538	0.234	1538	0.096
1922	0.364	1794	0.129
2307	0.520	2051	0.167
3076	0.918	2563	0.260
3588	1.25	3076	0.371
3845	1.42	3588	0.505

STEAM OR VAPOR FLOW

Determination of pressure loss through schedule-40 pipe for a steam or vapor can be accomplished by calculating the equivalent liquid velocity.

$$\text{Velocity (feet per second)} = \frac{(4.245)(W)(v)}{D^2 \times 1000}$$

W = Flow rate (pounds per hour)
D = Nominal pipe diameter (inches)
v = Specific volume (cubic feet per pound)

By determining the velocity, pressure loss can be found by referring to the tables for liquid losses. E.g., if a 3-inch pipe is used, and the equivalent liquid velocity is 8.68 feet per second, the pressure loss is 3.7 psi/100 feet.

AIR FLOW

Following is a tabulation of pressure drop per 100 feet of schedule-40 pipe for air at 100 psi and 60°F.

Note: If air at temperatures other than 60°F or pressure other than 100 psi is used, the pressure drop can be determined by multiplying the values given in the table by

$$\frac{100 + 14.7}{P + 14.7} \quad \text{or} \quad \frac{460 + t}{520}$$

where P is the actual pressure in psi and t is the actual temperature in degrees Fahrenheit. Likewise, if the fluid is not air and has a specific gravity other than 1.0, multiply the pressure drop by $\frac{G}{1.0}$, where G is the actual specific gravity.

Table A-8. Thermocouple Temperature/mV Data For Platinum vs. Platinum-10% Rhodium Thermocouples, Type S Based on International Practical Temperature Scale of 1968 (IPTS-68) and DIN 43710, Table 5, Sept. 1977

Reference Junction: At 32°F, mV = 0

Temperature in Degrees Fahrenheit vs. Absolute mV

°F	mV	°F	mV	°F	mV	°F	mV	°F	mV	°F	mV
-50	-0.218	500	1.962	1050	4.888	1600	8.126	2150	11.667	2700	15.362
-40	-0.194	510	2.011	1060	4.944	1610	8.188	2160	11.734	2710	15.429
-30	-0.170	520	2.061	1070	5.000	1620	8.250	2170	11.800	2720	15.496
-20	-0.145	530	2.111	1080	5.057	1630	8.312	2180	11.867	2730	15.563
-10	-0.119	540	2.161	1090	5.113	1640	8.374	2190	11.934	2740	15.630
0	-0.092	550	2.211	1100	5.169	1650	8.436	2200	12.001	2750	15.697
10	-0.064	560	2.262	1110	5.226	1660	8.498	2210	12.067	2760	15.763
20	-0.035	570	2.313	1120	5.283	1670	8.560	2220	12.134	2770	15.830
30	-0.006	580	2.363	1130	5.339	1680	8.623	2230	12.201	2780	15.897
40	0.024	590	2.414	1140	5.396	1690	8.685	2240	12.268	2790	15.963
50	0.055	600	2.465	1150	5.453	1700	8.748	2250	12.335	2800	16.030
60	0.087	610	2.517	1160	5.510	1710	8.811	2260	12.402	2810	16.096
70	0.119	620	2.568	1170	5.567	1720	8.874	2270	12.469	2820	16.163
80	0.152	630	2.620	1180	5.625	1730	8.937	2280	12.536	2830	16.229
90	0.186	640	2.672	1190	5.682	1740	9.000	2290	12.604	2840	16.296
100	0.221	650	2.723	1200	5.740	1750	9.063	2300	12.671	2850	16.362
110	0.256	660	2.775	1210	5.797	1760	9.126	2310	12.738	2860	16.428
120	0.291	670	2.828	1220	5.855	1770	9.189	2320	12.805	2870	16.494
130	0.328	680	2.880	1230	5.913	1780	9.253	2330	12.872	2880	16.560
140	0.365	690	2.932	1240	5.971	1790	9.317	2340	12.940	2890	16.626
150	0.402	700	2.985	1250	6.029	1800	9.380	2350	13.007	2900	16.692
160	0.440	710	3.037	1260	6.087	1810	9.444	2360	13.074	2910	16.758
170	0.478	720	3.090	1270	6.146	1820	9.508	2370	13.142	2920	16.824
180	0.517	730	3.143	1280	6.204	1830	9.572	2380	13.209	2930	16.890
190	0.557	740	3.196	1290	6.263	1840	9.636	2390	13.276	2940	16.955
200	0.597	750	3.249	1300	6.321	1850	9.700	2400	13.344	2950	17.021
210	0.637	760	3.302	1310	6.380	1860	9.764	2410	13.411	2960	17.086
220	0.678	770	3.356	1320	6.439	1870	9.829	2420	13.478	2970	17.152
230	0.719	780	3.409	1330	6.498	1880	9.893	2430	13.546	2980	17.217
240	0.761	790	3.463	1340	6.557	1890	9.958	2440	13.613	2990	17.282
250	0.803	800	3.516	1350	6.616	1900	10.023	2450	13.681	3000	17.347
260	0.846	810	3.570	1360	6.675	1910	10.087	2460	13.748	3010	17.412
270	0.889	820	3.624	1370	6.734	1920	10.152	2470	13.815	3020	17.477
280	0.932	830	3.678	1380	6.794	1930	10.217	2480	13.883	3030	17.542
290	0.976	840	3.732	1390	6.853	1940	10.282	2490	13.950	3040	17.607
300	1.020	850	3.786	1400	6.913	1950	10.348	2500	14.018	3050	17.672
310	1.064	860	3.840	1410	6.972	1960	10.413	2510	14.085	3060	17.736
320	1.109	870	3.895	1420	7.032	1970	10.478	2520	14.152	3070	17.801
330	1.154	880	3.949	1430	7.092	1980	10.544	2530	14.220	3080	17.865
340	1.199	890	4.004	1440	7.152	1990	10.609	2540	14.287	3090	17.929
350	1.245	900	4.058	1450	7.212	2000	10.675	2550	14.354	3100	17.993
360	1.291	910	4.113	1460	7.272	2010	10.740	2560	14.422	3110	18.056
370	1.337	920	4.168	1470	7.333	2020	10.806	2570	14.489	3120	18.119
380	1.384	930	4.223	1480	7.393	2030	10.872	2580	14.556	3130	18.182
390	1.431	940	4.278	1490	7.454	2040	10.938	2590	14.624	3140	18.245
400	1.478	950	4.333	1500	7.514	2050	11.004	2600	14.691	3150	18.307
410	1.525	960	4.388	1510	7.575	2060	11.070	2610	14.758	3160	18.369
420	1.573	970	4.443	1520	7.636	2070	11.136	2620	14.826	3170	18.431
430	1.620	980	4.498	1530	7.697	2080	11.202	2630	14.893	3180	18.492
440	1.669	990	4.554	1540	7.758	2090	11.268	2640	14.960	3190	18.552
450	1.717	1000	4.609	1550	7.819	2100	11.335	2650	15.027	3200	18.612
460	1.765	1010	4.665	1560	7.880	2110	11.401	2660	15.094		
470	1.814	1020	4.721	1570	7.942	2120	11.467	2670	15.161		
480	1.863	1030	4.776	1580	8.003	2130	11.534	2680	15.228		
490	1.912	1040	4.832	1590	8.065	2140	11.600	2690	15.295		

Table A-9. Thermocouple Temperature/mV Data: Type S
Reference Junction: At 0°C, mV = 0
Temperature in Degrees Celsius vs. Absolute mV

°C	mV	°C	mV	°C	mV	°C	mV	°C	mV	°C	mV	°C	mV
-50	-0.236	225	1.654	500	4.234	775	7.074	1050	10.165	1325	13.458	1600	16.771
-45	-0.215	230	1.698	505	4.283	780	7.128	1055	10.224	1330	13.519	1605	16.830
-40	-0.194	235	1.741	510	4.333	785	7.182	1060	10.282	1335	13.579	1610	16.890
-35	-0.173	240	1.785	515	4.382	790	7.236	1065	10.341	1340	13.640	1615	16.949
-30	-0.150	245	1.829	520	4.432	795	7.291	1070	10.400	1345	13.701	1620	17.008
-25	-0.127	250	1.873	525	4.482	800	7.345	1075	10.459	1350	13.761	1625	17.067
-20	-0.103	255	1.917	530	4.532	805	7.399	1080	10.517	1355	13.822	1630	17.125
-15	-0.078	260	1.962	535	4.582	810	7.454	1085	10.576	1360	13.883	1635	17.184
-10	-0.053	265	2.006	540	4.632	815	7.508	1090	10.635	1365	13.943	1640	17.243
-5	-0.027	270	2.051	545	4.682	820	7.563	1095	10.694	1370	14.004	1645	17.302
0	0.0	275	2.096	550	4.732	825	7.618	1100	10.754	1375	14.065	1650	17.360
5	0.027	280	2.141	555	4.782	830	7.672	1105	10.813	1380	14.125	1655	17.419
10	0.055	285	2.186	560	4.832	835	7.727	1110	10.872	1385	14.186	1660	17.477
15	0.084	290	2.232	565	4.883	840	7.782	1115	10.931	1390	14.247	1665	17.536
20	0.113	295	2.277	570	4.933	845	7.837	1120	10.991	1395	14.307	1670	17.594
25	0.142	300	2.323	575	4.984	850	7.892	1125	11.050	1400	14.368	1675	17.652
30	0.173	305	2.368	580	5.034	855	7.948	1130	11.110	1405	14.429	1680	17.711
35	0.203	310	2.414	585	5.085	860	8.003	1135	11.169	1410	14.489	1685	17.769
40	0.235	315	2.460	590	5.136	865	8.058	1140	11.229	1415	14.550	1690	17.826
45	0.266	320	2.506	595	5.186	870	8.114	1145	11.288	1420	14.610	1695	17.884
50	0.299	325	2.553	600	5.237	875	8.169	1150	11.348	1425	14.671	1700	17.942
55	0.331	330	2.599	605	5.288	880	8.225	1155	11.408	1430	14.731	1705	17.999
60	0.365	335	2.646	610	5.339	885	8.281	1160	11.467	1435	14.792	1710	18.056
65	0.398	340	2.692	615	5.391	890	8.336	1165	11.527	1440	14.852	1715	18.113
70	0.432	345	2.739	620	5.442	895	8.392	1170	11.587	1445	14.913	1720	18.170
75	0.467	350	2.786	625	5.493	900	8.448	1175	11.647	1450	14.973	1725	18.226
80	0.502	355	2.833	630	5.544	905	8.504	1180	11.707	1455	15.034	1730	18.282
85	0.537	360	2.880	635	5.596	910	8.560	1185	11.767	1460	15.094	1735	18.338
90	0.573	365	2.927	640	5.648	915	8.617	1190	11.827	1465	15.155	1740	18.394
95	0.609	370	2.974	645	5.700	920	8.673	1195	11.887	1470	15.215	1745	18.449
100	0.645	375	3.022	650	5.751	925	8.729	1200	11.947	1475	15.275	1750	18.504
105	0.682	380	3.069	655	5.803	930	8.786	1205	12.007	1480	15.336	1755	18.558
110	0.719	385	3.117	660	5.855	935	8.842	1210	12.067	1485	15.396	1760	18.612
115	0.757	390	3.164	665	5.907	940	8.899	1215	12.128	1490	15.456	1765	18.666
120	0.795	395	3.212	670	5.960	945	8.956	1220	12.188	1495	15.516		
125	0.833	400	3.260	675	6.012	950	9.012	1225	12.248	1500	15.576		
130	0.872	405	3.308	680	6.064	955	9.069	1230	12.308	1505	15.637		
135	0.910	410	3.356	685	6.117	960	9.126	1235	12.369	1510	15.697		
140	0.950	415	3.404	690	6.169	965	9.183	1240	12.429	1515	15.757		
145	0.989	420	3.452	695	6.222	970	9.240	1245	12.489	1520	15.817		
150	1.029	425	3.500	700	6.274	975	9.298	1250	12.550	1525	15.877		
155	1.069	430	3.549	705	6.327	980	9.355	1255	12.610	1530	15.937		
160	1.109	435	3.597	710	6.380	985	9.412	1260	12.671	1535	15.997		
165	1.149	440	3.645	715	6.433	990	9.470	1265	12.731	1540	16.057		
170	1.190	445	3.694	720	6.486	995	9.527	1270	12.792	1545	16.116		
175	1.231	450	3.743	725	6.539	1000	9.585	1275	12.852	1550	16.176		
180	1.273	455	3.791	730	6.592	1005	9.642	1280	12.913	1555	16.236		
185	1.314	460	3.840	735	6.645	1010	9.700	1285	12.973	1560	16.296		
190	1.356	465	3.889	740	6.699	1015	9.758	1290	13.034	1565	16.355		
195	1.398	470	3.938	745	6.752	1020	9.816	1295	13.094	1570	16.415		
200	1.440	475	3.987	750	6.805	1025	9.874	1300	13.155	1575	16.474		
205	1.482	480	4.036	755	6.859	1030	9.932	1305	13.216	1580	16.534		
210	1.525	485	4.086	760	6.913	1035	9.990	1310	13.276	1585	16.593		
215	1.568	490	4.135	765	6.966	1040	10.048	1315	13.337	1590	16.653		
220	1.611	495	4.184	770	7.020	1045	10.107	1320	13.397	1595	16.712		

Table A-10. Thermocouple Temperature/mV Data For Platinum–6% Rhodium vs. Platinum–30% Rhodium Thermocouples, Type B Based on International Practical Temperature Scale of 1968 (IPTS-68)

Reference Junction: At 32°F, mV = 0

Temperature in Degrees Fahrenheit vs. Absolute mV

°F	mV	°F	mV	°F	mV	°F	mV	°F	mV	°F	mV
0	0.006	600	0.479	1200	2.093	1800	4.672	2400	8.014	3000	11.829
10	0.004	610	0.497	1210	2.128	1810	4.722	2410	8.075	3010	11.894
20	0.002	620	0.515	1220	2.164	1820	4.772	2420	8.136	3020	11.959
30	0.000	630	0.534	1230	2.201	1830	4.823	2430	8.197	3030	12.024
40	-0.001	640	0.553	1240	2.237	1840	4.873	2440	8.258	3040	12.089
50	-0.002	650	0.572	1250	2.274	1850	4.924	2450	8.319	3050	12.154
60	-0.002	660	0.592	1260	2.311	1860	4.975	2460	8.381	3060	12.219
70	-0.003	670	0.612	1270	2.348	1870	5.027	2470	8.442	3070	12.284
80	-0.002	680	0.632	1280	2.385	1880	5.078	2480	8.504	3080	12.349
90	-0.002	690	0.652	1290	2.423	1890	5.130	2490	8.566	3090	12.413
100	-0.001	700	0.673	1300	2.461	1900	5.182	2500	8.628	3100	12.478
110	0.000	710	0.694	1310	2.499	1910	5.234	2510	8.690	3110	12.543
120	0.002	720	0.716	1320	2.538	1920	5.286	2520	8.752	3120	12.608
130	0.004	730	0.737	1330	2.576	1930	5.339	2530	8.814	3130	12.672
140	0.006	740	0.759	1340	2.615	1940	5.391	2540	8.877	3140	12.737
150	0.009	750	0.782	1350	2.655	1950	5.444	2550	8.939	3150	12.801
160	0.012	760	0.804	1360	2.694	1960	5.497	2560	9.002	3160	12.866
170	0.015	770	0.827	1370	2.734	1970	5.551	2570	9.065	3170	12.930
180	0.019	780	0.851	1380	2.774	1980	5.604	2580	9.128	3180	12.995
190	0.023	790	0.874	1390	2.814	1990	5.658	2590	9.191	3190	13.059
200	0.027	800	0.898	1400	2.855	2000	5.712	2600	9.254	3200	13.124
210	0.032	810	0.922	1410	2.896	2010	5.766	2610	9.317	3210	13.188
220	0.037	820	0.947	1420	2.937	2020	5.820	2620	9.380	3220	13.252
230	0.043	830	0.972	1430	2.978	2030	5.875	2630	9.443	3230	13.316
240	0.049	840	0.997	1440	3.019	2040	5.930	2640	9.507	3240	13.380
250	0.055	850	1.022	1450	3.061	2050	5.984	2650	9.570	3250	13.444
260	0.061	860	1.048	1460	3.103	2060	6.039	2660	9.634	3260	13.508
270	0.068	870	1.074	1470	3.146	2070	6.095	2670	9.697	3270	13.572
280	0.075	880	1.100	1480	3.188	2080	6.150	2680	9.761	3280	13.635
290	0.083	890	1.127	1490	3.231	2090	6.206	2690	9.825	3290	13.699
300	0.090	900	1.153	1500	3.274	2100	6.262	2700	9.889	3300	13.763
310	0.099	910	1.181	1510	3.317	2110	6.318	2710	9.953		
320	0.107	920	1.208	1520	3.361	2120	6.374	2720	10.017		
330	0.116	930	1.236	1530	3.404	2130	6.430	2730	10.081		
340	0.125	940	1.264	1540	3.448	2140	6.487	2740	10.145		
350	0.135	950	1.292	1550	3.492	2150	6.543	2750	10.210		
360	0.144	960	1.321	1560	3.537	2160	6.600	2760	10.274		
370	0.155	970	1.350	1570	3.581	2170	6.657	2770	10.338		
380	0.165	980	1.379	1580	3.626	2180	6.714	2780	10.403		
390	0.176	990	1.409	1590	3.672	2190	6.772	2790	10.467		
400	0.187	1000	1.438	1600	3.717	2200	6.829	2800	10.532		
410	0.199	1010	1.468	1610	3.762	2210	6.887	2810	10.596		
420	0.210	1020	1.499	1620	3.808	2220	6.945	2820	10.661		
430	0.223	1030	1.529	1630	3.854	2230	7.003	2830	10.726		
440	0.235	1040	1.560	1640	3.901	2240	7.061	2840	10.790		
450	0.248	1050	1.591	1650	3.947	2250	7.120	2850	10.855		
460	0.261	1060	1.623	1660	3.994	2260	7.178	2860	10.920		
470	0.275	1070	1.655	1670	4.041	2270	7.237	2870	10.985		
480	0.288	1080	1.687	1680	4.088	2280	7.296	2880	11.050		
490	0.303	1090	1.719	1690	4.136	2290	7.355	2890	11.115		
500	0.317	1100	1.752	1700	4.183	2300	7.414	2900	11.179		
510	0.332	1110	1.785	1710	4.231	2310	7.473	2910	11.244		
520	0.347	1120	1.818	1720	4.279	2320	7.533	2920	11.309		
530	0.362	1130	1.851	1730	4.327	2330	7.592	2930	11.374		
540	0.378	1140	1.885	1740	4.376	2340	7.652	2940	11.439		
550	0.394	1150	1.919	1750	4.425	2350	7.712	2950	11.504		
560	0.410	1160	1.953	1760	4.474	2360	7.772	2960	11.569		
570	0.427	1170	1.988	1770	4.523	2370	7.833	2970	11.634		
580	0.444	1180	2.022	1780	4.572	2380	7.893	2980	11.699		
590	0.462	1190	2.058	1790	4.622	2390	7.953	2990	11.764		

Table A-11. Thermocouple Temperature/mV Data: Type B
Reference Junction: At 0°C, mV = 0
Temperature in Degrees Celsius vs. Absolute mV

°C	mV	°C	mV	°C	mV	°C	mV	°C	mV
0	0.0	400	0.786	800	3.154	1200	6.783	1600	11.257
10	-0.002	410	0.827	810	3.231	1210	6.887	1610	11.374
20	-0.003	420	0.870	820	3.308	1220	6.991	1620	11.491
30	-0.002	430	0.913	830	3.387	1230	7.096	1630	11.608
40	-0.000	440	0.957	840	3.466	1240	7.202	1640	11.725
50	0.002	450	1.002	850	3.546	1250	7.308	1650	11.842
60	0.006	460	1.048	860	3.626	1260	7.414	1660	11.959
70	0.011	470	1.095	870	3.708	1270	7.521	1670	12.076
80	0.017	480	1.143	880	3.790	1280	7.628	1680	12.193
90	0.025	490	1.192	890	3.873	1290	7.736	1690	12.310
100	0.033	500	1.241	900	3.957	1300	7.845	1700	12.426
110	0.043	510	1.292	910	4.041	1310	7.953	1710	12.543
120	0.053	520	1.344	920	4.126	1320	8.063	1720	12.659
130	0.065	530	1.397	930	4.212	1330	8.172	1730	12.776
140	0.078	540	1.450	940	4.298	1340	8.283	1740	12.892
150	0.092	550	1.505	950	4.386	1350	8.393	1750	13.008
160	0.107	560	1.560	960	4.474	1360	8.504	1760	13.124
170	0.123	570	1.617	970	4.562	1370	8.616	1770	13.239
180	0.140	580	1.674	980	4.652	1380	8.727	1780	13.354
190	0.159	590	1.732	990	4.742	1390	8.839	1790	13.470
200	0.178	600	1.791	1000	4.833	1400	8.952	1800	13.585
210	0.199	610	1.851	1010	4.924	1410	9.065	1810	13.699
220	0.220	620	1.912	1020	5.016	1420	9.178	1820	13.814
230	0.243	630	1.974	1030	5.109	1430	9.291		
240	0.266	640	2.036	1040	5.202	1440	9.405		
250	0.291	650	2.100	1050	5.297	1450	9.519		
260	0.317	660	2.164	1060	5.391	1460	9.634		
270	0.344	670	2.230	1070	5.487	1470	9.748		
280	0.372	680	2.296	1080	5.583	1480	9.863		
290	0.401	690	2.363	1090	5.680	1490	9.979		
300	0.431	700	2.430	1100	5.777	1500	10.094		
310	0.462	710	2.499	1110	5.875	1510	10.210		
320	0.494	720	2.569	1120	5.973	1520	10.325		
330	0.527	730	2.639	1130	6.073	1530	10.441		
340	0.561	740	2.710	1140	6.172	1540	10.558		
350	0.596	750	2.782	1150	6.273	1550	10.674		
360	0.632	760	2.855	1160	6.374	1560	10.790		
370	0.669	770	2.928	1170	6.475	1570	10.907		
380	0.707	780	3.003	1180	6.577	1580	11.024		
390	0.746	790	3.078	1190	6.680	1590	11.141		

Table A-12. Thermocouple Temperature/mV Data
Reference Junction: At 0°C, mV = 0
Temperature in Degrees Celsius vs. Absolute mV

°C	mV	°C	mV	°C	mV	°C	mV	°C	mV	°C	mV	°C	mV
-50	-0.226	225	1.692	500	4.471	775	7.642	1050	11.170	1325	14.976	1600	18.842
-45	-0.207	230	1.738	505	4.525	780	7.703	1055	11.237	1330	15.047	1605	18.912
-40	-0.188	235	1.784	510	4.580	785	7.765	1060	11.304	1335	15.117	1610	18.981
-35	-0.167	240	1.830	515	4.634	790	7.826	1065	11.372	1340	15.188	1615	19.050
-30	-0.145	245	1.876	520	4.689	795	7.887	1070	11.439	1345	15.258	1620	19.119
-25	-0.123	250	1.923	525	4.744	800	7.949	1075	11.507	1350	15.329	1625	19.188
-20	-0.100	255	1.970	530	4.799	805	8.010	1080	11.574	1355	15.399	1630	19.257
-15	-0.076	260	2.017	535	4.854	810	8.072	1085	11.642	1360	15.470	1635	19.326
-10	-0.051	265	2.064	540	4.910	815	8.134	1090	11.710	1365	15.540	1640	19.395
-5	-0.026	270	2.111	545	4.965	820	8.196	1095	11.778	1370	15.611	1645	19.464
0	0.0	275	2.159	550	5.021	825	8.258	1100	11.846	1375	15.682	1650	19.533
5	0.027	280	2.207	555	5.076	830	8.320	1105	11.914	1380	15.752	1655	19.602
10	0.054	285	2.255	560	5.132	835	8.383	1110	11.983	1385	15.823	1660	19.670
15	0.082	290	2.303	565	5.188	840	8.445	1115	12.051	1390	15.893	1665	19.739
20	0.111	295	2.351	570	5.244	845	8.508	1120	12.119	1395	15.964	1670	19.807
25	0.141	300	2.400	575	5.300	850	8.570	1125	12.188	1400	16.035	1675	19.875
30	0.171	305	2.449	580	5.356	855	8.633	1130	12.257	1405	16.105	1680	19.944
35	0.201	310	2.498	585	5.413	860	8.696	1135	12.325	1410	16.176	1685	20.012
40	0.232	315	2.547	590	5.469	865	8.759	1140	12.394	1415	16.247	1690	20.080
45	0.264	320	2.596	595	5.526	870	8.822	1145	12.463	1420	16.317	1695	20.148
50	0.296	325	2.646	600	5.582	875	8.885	1150	12.532	1425	16.388	1700	20.215
55	0.329	330	2.695	605	5.639	880	8.949	1155	12.600	1430	16.458	1705	20.283
60	0.363	335	2.745	610	5.696	885	9.012	1160	12.669	1435	16.529	1710	20.350
65	0.397	340	2.795	615	5.753	890	9.076	1165	12.739	1440	16.599	1715	20.417
70	0.431	345	2.845	620	5.810	895	9.140	1170	12.808	1445	16.670	1720	20.483
75	0.466	350	2.896	625	5.867	900	9.203	1175	12.877	1450	16.741	1725	20.550
80	0.501	355	2.946	630	5.925	905	9.267	1180	12.946	1455	16.811	1730	20.616
85	0.537	360	2.997	635	5.982	910	9.331	1185	13.016	1460	16.882	1735	20.682
90	0.573	365	3.048	640	6.040	915	9.395	1190	13.085	1465	16.952	1740	20.748
95	0.610	370	3.099	645	6.098	920	9.460	1195	13.154	1470	17.022	1745	20.813
100	0.647	375	3.150	650	6.155	925	9.524	1200	13.224	1475	17.093	1750	20.878
105	0.685	380	3.201	655	6.213	930	9.589	1205	13.293	1480	17.163	1755	20.942
110	0.723	385	3.252	660	6.272	935	9.653	1210	13.363	1485	17.234	1760	21.006
115	0.761	390	3.304	665	6.330	940	9.718	1215	13.433	1490	17.304	1765	21.070
120	0.800	395	3.355	670	6.388	945	9.783	1220	13.502	1495	17.374		
125	0.839	400	3.407	675	6.447	950	9.848	1225	13.572	1500	17.445		
130	0.879	405	3.459	680	6.505	955	9.913	1230	13.642	1505	17.515		
135	0.919	410	3.511	685	6.564	960	9.978	1235	13.712	1510	17.585		
140	0.959	415	3.563	690	6.623	965	10.043	1240	13.782	1515	17.655		
145	1.000	420	3.616	695	6.682	970	10.109	1245	13.852	1520	17.726		
150	1.041	425	3.668	700	6.741	975	10.174	1250	13.922	1525	17.796		
155	1.082	430	3.721	705	6.800	980	10.240	1255	13.992	1530	17.866		
160	1.124	435	3.774	710	6.860	985	10.305	1260	14.062	1535	17.936		
165	1.166	440	3.826	715	6.919	990	10.371	1265	14.132	1540	18.006		
170	1.208	445	3.879	720	6.979	995	10.437	1270	14.202	1545	18.076		
175	1.251	450	3.933	725	7.039	1000	10.503	1275	14.272	1550	18.146		
180	1.294	455	3.986	730	7.098	1005	10.569	1280	14.343	1555	18.216		
185	1.337	460	4.039	735	7.158	1010	10.636	1285	14.413	1560	18.286		
190	1.380	465	4.093	740	7.218	1015	10.702	1290	14.483	1565	18.355		
195	1.424	470	4.146	745	7.279	1020	10.768	1295	14.554	1570	18.425		
200	1.468	475	4.200	750	7.339	1025	10.835	1300	14.624	1575	18.495		
205	1.512	480	4.254	755	7.399	1030	10.902	1305	14.694	1580	18.564		
210	1.557	485	4.308	760	7.460	1035	10.968	1310	14.765	1585	18.634		
215	1.602	490	4.362	765	7.521	1040	11.035	1315	14.835	1590	18.703		
220	1.647	495	4.416	770	7.582	1045	11.102	1320	14.906	1595	18.773		

Table A-13. Thermocouple Temperature/mV Data: Type R
Reference Junction: At 0°C, mV = 0
Temperature in Degrees Celsius vs. Absolute mV

°C	mV	°C	mV	°C	mV	°C	mV	°C	mV	°C	mV	°C	mV
-50	-0.226	225	1.692	500	4.471	775	7.642	1050	11.170	1325	14.976	1600	18.842
-45	-0.207	230	1.738	505	4.525	780	7.703	1055	11.237	1330	15.047	1605	18.912
-40	-0.188	235	1.784	510	4.580	785	7.765	1060	11.304	1335	15.117	1610	18.981
-35	-0.167	240	1.830	515	4.634	790	7.826	1065	11.372	1340	15.188	1615	19.050
-30	-0.145	245	1.876	520	4.689	795	7.887	1070	11.439	1345	15.258	1620	19.119
-25	-0.123	250	1.923	525	4.744	800	7.949	1075	11.507	1350	15.329	1625	19.188
-20	-0.100	255	1.970	530	4.799	805	8.010	1080	11.574	1355	15.399	1630	19.257
-15	-0.076	260	2.017	535	4.854	810	8.072	1085	11.642	1360	15.470	1635	19.326
-10	-0.051	265	2.064	540	4.910	815	8.134	1090	11.710	1365	15.540	1640	19.395
-5	-0.026	270	2.111	545	4.965	820	8.196	1095	11.778	1370	15.611	1645	19.464
0	0.0	275	2.159	550	5.021	825	8.258	1100	11.846	1375	15.682	1650	19.533
5	0.027	280	2.207	555	5.076	830	8.320	1105	11.914	1380	15.752	1655	19.602
10	0.054	285	2.255	560	5.132	835	8.383	1110	11.983	1385	15.823	1660	19.670
15	0.082	290	2.303	565	5.188	840	8.445	1115	12.051	1390	15.893	1665	19.739
20	0.111	295	2.351	570	5.244	845	8.508	1120	12.119	1395	15.964	1670	19.807
25	0.141	300	2.400	575	5.300	850	8.570	1125	12.188	1400	16.035	1675	19.875
30	0.171	305	2.449	580	5.356	855	8.633	1130	12.257	1405	16.105	1680	19.944
35	0.201	310	2.498	585	5.413	860	8.696	1135	12.325	1410	16.176	1685	20.012
40	0.232	315	2.547	590	5.469	865	8.759	1140	12.394	1415	16.247	1690	20.080
45	0.264	320	2.596	595	5.526	870	8.822	1145	12.463	1420	16.317	1695	20.148
50	0.296	325	2.646	600	5.582	875	8.885	1150	12.532	1425	16.388	1700	20.215
55	0.329	330	2.695	605	5.639	880	8.949	1155	12.600	1430	16.458	1705	20.283
60	0.363	335	2.745	610	5.696	885	9.012	1160	12.669	1435	16.529	1710	20.350
65	0.397	340	2.795	615	5.753	890	9.076	1165	12.739	1440	16.599	1715	20.417
70	0.431	345	2.845	620	5.810	895	9.140	1170	12.808	1445	16.670	1720	20.483
75	0.466	350	2.896	625	5.867	900	9.203	1175	12.877	1450	16.741	1725	20.550
80	0.501	355	2.946	630	5.925	905	9.267	1180	12.946	1455	16.811	1730	20.616
85	0.537	360	2.997	635	5.982	910	9.331	1185	13.016	1460	16.882	1735	20.682
90	0.573	365	3.048	640	6.040	915	9.395	1190	13.085	1465	16.952	1740	20.748
95	0.610	370	3.099	645	6.098	920	9.460	1195	13.154	1470	17.022	1745	20.813
100	0.647	375	3.150	650	6.155	925	9.524	1200	13.224	1475	17.093	1750	20.878
105	0.685	380	3.201	655	6.213	930	9.589	1205	13.293	1480	17.163	1755	20.942
110	0.723	385	3.252	660	6.272	935	9.653	1210	13.363	1485	17.234	1760	21.006
115	0.761	390	3.304	665	6.330	940	9.718	1215	13.433	1490	17.304	1765	21.070
120	0.800	395	3.355	670	6.388	945	9.783	1220	13.502	1495	17.374		
125	0.839	400	3.407	675	6.447	950	9.848	1225	13.572	1500	17.445		
130	0.879	405	3.459	680	6.505	955	9.913	1230	13.642	1505	17.515		
135	0.919	410	3.511	685	6.564	960	9.978	1235	13.712	1510	17.585		
140	0.959	415	3.563	690	6.623	965	10.043	1240	13.782	1515	17.655		
145	1.000	420	3.616	695	6.682	970	10.109	1245	13.852	1520	17.726		
150	1.041	425	3.668	700	6.741	975	10.174	1250	13.922	1525	17.796		
155	1.082	430	3.721	705	6.800	980	10.240	1255	13.992	1530	17.866		
160	1.124	435	3.774	710	6.860	985	10.305	1260	14.062	1535	17.936		
165	1.166	440	3.826	715	6.919	990	10.371	1265	14.132	1540	18.006		
170	1.208	445	3.879	720	6.979	995	10.437	1270	14.202	1545	18.076		
175	1.251	450	3.933	725	7.039	1000	10.503	1275	14.272	1550	18.146		
180	1.294	455	3.986	730	7.098	1005	10.569	1280	14.343	1555	18.216		
185	1.337	460	4.039	735	7.158	1010	10.636	1285	14.413	1560	18.286		
190	1.380	465	4.093	740	7.218	1015	10.702	1290	14.483	1565	18.355		
195	1.424	470	4.146	745	7.279	1020	10.768	1295	14.554	1570	18.425		
200	1.468	475	4.200	750	7.339	1025	10.835	1300	14.624	1575	18.495		
205	1.512	480	4.254	755	7.399	1030	10.902	1305	14.694	1580	18.564		
210	1.557	485	4.308	760	7.460	1035	10.968	1310	14.765	1585	18.634		
215	1.602	490	4.362	765	7.521	1040	11.035	1315	14.835	1590	18.703		
220	1.647	495	4.416	770	7.582	1045	11.102	1320	14.906	1595	18.773		

Table A-14. Thermocouple Temperature/mV Data For Nickel-Chromium vs. Nickel-Aluminum (Chromel-Alumel) Thermocouples, Type K Based on International Practical Temperature Scale of 1968 (IPTS-68) and DIN 43710, Table 4, Sept. 1977

Reference Junction: At 32°F, mV = 0

Temperature in Degrees Fahrenheit vs. Absolute mV

°F	mV	°F	mV	°F	mV	°F	mV	°F	mV
-450	-6.456	150	2.666	750	16.349	1350	30.475	1950	43.798
-440	-6.447	160	2.896	760	16.583	1360	30.706	1960	44.010
-430	-6.431	170	3.127	770	16.818	1370	30.937	1970	44.222
-420	-6.409	180	3.358	780	17.053	1380	31.168	1980	44.434
-410	-6.380	190	3.589	790	17.288	1390	31.399	1990	44.645
-400	-6.344	200	3.819	800	17.523	1400	31.629	2000	44.856
-390	-6.301	210	4.049	810	17.759	1410	31.859	2010	45.066
-380	-6.251	220	4.279	820	17.994	1420	32.088	2020	45.276
-370	-6.195	230	4.508	830	18.230	1430	32.317	2030	45.486
-360	-6.133	240	4.737	840	18.466	1440	32.546	2040	45.695
-350	-6.064	250	4.964	850	18.702	1450	32.775	2050	45.904
-340	-5.989	260	5.192	860	18.938	1460	33.003	2060	46.113
-330	-5.908	270	5.418	870	19.174	1470	33.231	2070	46.321
-320	-5.822	280	5.643	880	19.410	1480	33.459	2080	46.529
-310	-5.730	290	5.868	890	19.646	1490	33.686	2090	46.737
-300	-5.632	300	6.092	900	19.883	1500	33.913	2100	46.944
-290	-5.529	310	6.316	910	20.120	1510	34.140	2110	47.150
-280	-5.421	320	6.539	920	20.356	1520	34.366	2120	47.356
-270	-5.308	330	6.761	930	20.593	1530	34.593	2130	47.562
-260	-5.190	340	6.984	940	20.830	1540	34.818	2140	47.767
-250	-5.067	350	7.205	950	21.066	1550	35.044	2150	47.972
-240	-4.939	360	7.427	960	21.303	1560	35.269	2160	48.177
-230	-4.806	370	7.649	970	21.540	1570	35.494	2170	48.381
-220	-4.669	380	7.870	980	21.777	1580	35.718	2180	48.584
-210	-4.527	390	8.092	990	22.014	1590	35.942	2190	48.787
-200	-4.381	400	8.314	1000	22.251	1600	36.166	2200	48.990
-190	-4.230	410	8.537	1010	22.488	1610	36.390	2210	49.192
-180	-4.075	420	8.759	1020	22.725	1620	36.613	2220	49.394
-170	-3.917	430	8.983	1030	22.961	1630	36.836	2230	49.595
-160	-3.754	440	9.206	1040	23.198	1640	37.058	2240	49.796
-150	-3.587	450	9.430	1050	23.435	1650	37.280	2250	49.996
-140	-3.417	460	9.655	1060	23.672	1660	37.502	2260	50.196
-130	-3.242	470	9.880	1070	23.908	1670	37.724	2270	50.395
-120	-3.065	480	10.106	1080	24.145	1680	37.945	2280	50.594
-110	-2.883	490	10.333	1090	24.382	1690	38.166	2290	50.792
-100	-2.699	500	10.560	1100	24.618	1700	38.387	2300	50.990
-90	-2.511	510	10.787	1110	24.854	1710	38.607	2310	51.187
-80	-2.320	520	11.015	1120	25.091	1720	38.827	2320	51.384
-70	-2.126	530	11.243	1130	25.327	1730	39.046	2330	51.580
-60	-1.929	540	11.472	1140	25.563	1740	39.266	2340	51.776
-50	-1.729	550	11.702	1150	25.799	1750	39.485	2350	51.971
-40	-1.527	560	11.931	1160	26.034	1760	39.703	2360	52.165
-30	-1.322	570	12.161	1170	26.270	1770	39.922	2370	52.360
-20	-1.114	580	12.392	1180	26.505	1780	40.140	2380	52.553
-10	-0.904	590	12.623	1190	26.740	1790	40.358	2390	52.747
0	-0.692	600	12.854	1200	26.975	1800	40.575	2400	52.939
10	-0.478	610	13.085	1210	27.210	1810	40.792	2410	53.132
20	-0.262	620	13.317	1220	27.445	1820	41.009	2420	53.324
30	-0.044	630	13.549	1230	27.679	1830	41.225	2430	53.515
40	0.176	640	13.781	1240	27.914	1840	41.442	2440	53.706
50	0.397	650	14.013	1250	28.148	1850	41.657	2450	53.897
60	0.619	660	14.246	1260	28.382	1860	41.873	2460	54.087
70	0.843	670	14.479	1270	28.615	1870	42.088	2470	54.277
80	1.068	680	14.712	1280	28.849	1880	42.303	2480	54.466
90	1.294	690	14.945	1290	29.082	1890	42.518	2490	54.656
100	1.520	700	15.178	1300	29.315	1900	42.732	2500	54.845
110	1.748	710	15.412	1310	29.547	1910	42.946		
120	1.977	720	15.646	1320	29.780	1920	43.159		
130	2.206	730	15.880	1330	30.012	1930	43.373		
140	2.436	740	16.114	1340	30.244	1940	43.585		

Table A-15. Thermocouple Temperature/mV Data: Type K
Reference Junction: At 0°C, mV = 0
Temperature in Degrees Celsius vs. Absolute mV

°C	mV	°C	mV	°C	mV	°C	mV	°C	mV	°C	mV
-270	-6.458	5	0.198	280	11.381	555	22.985	830	34.502	1105	45.297
-265	-6.452	10	0.397	285	11.587	560	23.198	835	34.705	1110	45.486
-260	-6.441	15	0.597	290	11.793	565	23.411	840	34.909	1115	45.675
-255	-6.425	20	0.798	295	12.000	570	23.624	845	35.111	1120	45.863
-250	-6.404	25	1.000	300	12.207	575	23.837	850	35.314	1125	46.051
-245	-6.377	30	1.203	305	12.415	580	24.050	855	35.516	1130	46.238
-240	-6.344	35	1.407	310	12.623	585	24.263	860	35.718	1135	46.425
-235	-6.306	40	1.611	315	12.831	590	24.476	865	35.920	1140	46.612
-230	-6.262	45	1.817	320	13.039	595	24.689	870	36.121	1145	46.799
-225	-6.213	50	2.022	325	13.247	600	24.902	875	36.323	1150	46.985
-220	-6.158	55	2.229	330	13.456	605	25.114	880	36.524	1155	47.171
-215	-6.099	60	2.436	335	13.665	610	25.327	885	36.724	1160	47.356
-210	-6.035	65	2.643	340	13.874	615	25.539	890	36.925	1165	47.542
-205	-5.965	70	2.850	345	14.083	620	25.751	895	37.125	1170	47.726
-200	-5.891	75	3.058	350	14.292	625	25.964	900	37.325	1175	47.911
-195	-5.813	80	3.266	355	14.502	630	26.176	905	37.524	1180	48.095
-190	-5.730	85	3.473	360	14.712	635	26.387	910	37.724	1185	48.279
-185	-5.642	90	3.681	365	14.922	640	26.599	915	37.923	1190	48.462
-180	-5.550	95	3.888	370	15.132	645	26.811	920	38.122	1195	48.645
-175	-5.454	100	4.095	375	15.342	650	27.022	925	38.320	1200	48.828
-170	-5.354	105	4.302	380	15.552	655	27.234	930	38.519	1205	49.010
-165	-5.249	110	4.508	385	15.763	660	27.445	935	38.717	1210	49.192
-160	-5.141	115	4.714	390	15.974	665	27.656	940	38.915	1215	49.374
-155	-5.029	120	4.919	395	16.184	670	27.867	945	39.112	1220	49.555
-150	-4.912	125	5.124	400	16.395	675	28.078	950	39.310	1225	49.736
-145	-4.792	130	5.327	405	16.607	680	28.288	955	39.507	1230	49.916
-140	-4.669	135	5.533	410	16.818	685	28.498	960	39.703	1235	50.096
-135	-4.541	140	5.733	415	17.029	690	28.709	965	39.900	1240	50.276
-130	-4.410	145	5.936	420	17.241	695	28.919	970	40.096	1245	50.455
-125	-4.276	150	6.137	425	17.453	700	29.128	975	40.292	1250	50.633
-120	-4.138	155	6.338	430	17.664	705	29.338	980	40.488	1255	50.812
-115	-3.997	160	6.539	435	17.876	710	29.547	985	40.684	1260	50.990
-110	-3.852	165	6.739	440	18.088	715	29.756	990	40.879	1265	51.167
-105	-3.704	170	6.939	445	18.301	720	29.965	995	41.074	1270	51.344
-100	-3.553	175	7.139	450	18.513	725	30.174	1000	41.269	1275	51.521
-95	-3.399	180	7.338	455	18.725	730	30.383	1005	41.463	1280	51.697
-90	-3.242	185	7.538	460	18.938	735	30.591	1010	41.657	1285	51.873
-85	-3.082	190	7.737	465	19.150	740	30.799	1015	41.851	1290	52.049
-80	-2.920	195	7.937	470	19.363	745	31.007	1020	42.045	1295	52.224
-75	-2.754	200	8.137	475	19.576	750	31.214	1025	42.239	1300	52.398
-70	-2.586	205	8.336	480	19.788	755	31.422	1030	42.432	1305	52.573
-65	-2.416	210	8.537	485	20.001	760	31.629	1035	42.625	1310	52.747
-60	-2.243	215	8.737	490	20.214	765	31.836	1040	42.817	1315	52.920
-55	-2.067	220	8.938	495	20.427	770	32.042	1045	43.010	1320	53.093
-50	-1.889	225	9.139	500	20.640	775	32.249	1050	43.202	1325	53.266
-45	-1.709	230	9.341	505	20.853	780	32.455	1055	43.394	1330	53.439
-40	-1.527	235	9.543	510	21.066	785	32.661	1060	43.585	1335	53.611
-35	-1.342	240	9.745	515	21.280	790	32.866	1065	43.777	1340	53.782
-30	-1.156	245	9.948	520	21.493	795	33.072	1070	43.968	1345	53.954
-25	-0.968	250	10.151	525	21.706	800	33.277	1075	44.159	1350	54.125
-20	-0.777	255	10.355	530	21.919	805	33.482	1080	44.349	1355	54.296
-15	-0.585	260	10.560	535	22.132	810	33.686	1085	44.539	1360	54.466
-10	-0.392	265	10.764	540	22.346	815	33.891	1090	44.729	1365	54.637
-5	-0.197	270	10.969	545	22.559	820	34.095	1095	44.919	1370	54.807
0	0.0	275	11.175	550	22.772	825	34.299	1100	45.108		

Table A-16. Thermocouple Temperature/mV Data For Iron vs. Copper-Nickel (Iron-Constantan) Thermocouples, Type J Based on International Practical Temperature Scale of 1968 (IPTS-68)

Reference Junction: At 32°F, mV = 0

Temperature in Degrees Fahrenheit vs. Absolute mV

°F	mV	°F	mV	°F	mV	°F	mV	°F	mV	°F	mV	°F	mV	°F	mV
-350	-8.137	-25	-1.562	300	7.947	625	17.953	950	27.949	1275	38.544	1600	50.059	1900	60.163
-345	-8.085	-20	-1.428	305	8.100	630	18.107	955	28.105	1280	38.716	1605	50.235	1905	60.326
-340	-8.030	-15	-1.293	310	8.253	635	18.260	960	28.261	1285	38.888	1610	50.411	1910	60.488
-335	-7.973	-10	-1.158	315	8.407	640	18.414	965	28.417	1290	39.061	1615	50.586	1915	60.650
-330	-7.915	-5	-1.022	320	8.560	645	18.567	970	28.573	1295	39.234	1620	50.761	1920	60.812
-325	-7.854	0	-0.885	325	8.714	650	18.721	975	28.730	1300	39.407	1625	50.936	1925	60.974
-320	-7.791	5	-0.748	330	8.867	655	18.874	980	28.887	1305	39.581	1630	51.110	1930	61.135
-315	-7.726	10	-0.611	335	9.021	660	19.027	985	29.044	1310	39.754	1635	51.284	1935	61.297
-310	-7.659	15	-0.473	340	9.175	665	19.180	990	29.201	1315	39.928	1640	51.458	1940	61.459
-305	-7.590	20	-0.334	345	9.329	670	19.334	995	29.358	1320	40.103	1645	51.632	1945	61.620
-300	-7.519	25	-0.195	350	9.483	675	19.487	1000	29.515	1325	40.277	1650	51.805	1950	61.781
-295	-7.447	30	-0.056	355	9.636	680	19.640	1005	29.673	1330	40.452	1655	51.978	1955	61.943
-290	-7.372	35	0.084	360	9.790	685	19.793	1010	29.831	1335	40.627	1660	52.151	1960	62.104
-285	-7.296	40	0.224	365	9.944	690	19.947	1015	29.989	1340	40.802	1665	52.324	1965	62.265
-280	-7.218	45	0.365	370	10.098	695	20.100	1020	30.147	1345	40.978	1670	52.496	1970	62.426
-275	-7.139	50	0.507	375	10.252	700	20.253	1025	30.305	1350	41.154	1675	52.668	1975	62.587
-270	-7.057	55	0.648	380	10.407	705	20.406	1030	30.464	1355	41.329	1680	52.840	1980	62.748
-265	-6.974	60	0.791	385	10.561	710	20.559	1035	30.623	1360	41.506	1685	53.012	1985	62.909
-260	-6.890	65	0.933	390	10.715	715	20.713	1040	30.782	1365	41.682	1690	53.183	1990	63.070
-255	-6.804	70	1.076	395	10.869	720	20.866	1045	30.941	1370	41.859	1695	53.354	1995	63.231
-250	-6.716	75	1.220	400	11.023	725	21.019	1050	31.100	1375	42.035	1700	53.525	2000	63.392
-245	-6.627	80	1.363	405	11.177	730	21.172	1055	31.260	1380	42.212	1705	53.695	2005	63.552
-240	-6.536	85	1.507	410	11.332	735	21.325	1060	31.420	1385	42.390	1710	53.865	2010	63.713
-235	-6.444	90	1.652	415	11.486	740	21.478	1065	31.580	1390	42.567	1715	54.035	2015	63.874
-230	-6.350	95	1.797	420	11.640	745	21.631	1070	31.740	1395	42.744	1720	54.205	2020	64.034
-225	-6.255	100	1.942	425	11.794	750	21.785	1075	31.901	1400	42.922	1725	54.374	2025	64.195
-220	-6.159	105	2.088	430	11.949	755	21.938	1080	32.061	1405	43.100	1730	54.544	2030	64.355
-215	-6.061	110	2.233	435	12.103	760	22.091	1085	32.222	1410	43.278	1735	54.712	2035	64.516
-210	-5.962	115	2.380	440	12.257	765	22.244	1090	32.384	1415	43.456	1740	54.881	2040	64.676
-205	-5.861	120	2.526	445	12.411	770	22.397	1095	32.545	1420	43.635	1745	55.049	2045	64.837
-200	-5.760	125	2.673	450	12.566	775	22.551	1100	32.707	1425	43.813	1750	55.218	2050	64.997
-195	-5.657	130	2.820	455	12.720	780	22.704	1105	32.869	1430	43.992	1755	55.385	2055	65.158
-190	-5.553	135	2.967	460	12.874	785	22.857	1110	33.031	1435	44.171	1760	55.553	2060	65.318
-185	-5.447	140	3.115	465	13.029	790	23.010	1115	33.194	1440	44.350	1765	55.720	2065	65.478
-180	-5.341	145	3.263	470	13.183	795	23.164	1120	33.356	1445	44.529	1770	55.888	2070	65.638
-175	-5.233	150	3.411	475	13.337	800	23.317	1125	33.519	1450	44.709	1775	56.055	2075	65.799
-170	-5.124	155	3.560	480	13.491	805	23.471	1130	33.683	1455	44.888	1780	56.221	2080	65.959
-165	-5.014	160	3.708	485	13.645	810	23.624	1135	33.846	1460	45.067	1785	56.388	2085	66.119
-160	-4.903	165	3.857	490	13.800	815	23.777	1140	34.010	1465	45.247	1790	56.554	2090	66.279
-155	-4.791	170	4.006	495	13.954	820	23.931	1145	34.174	1470	45.426	1795	56.720	2095	66.440
-150	-4.678	175	4.156	500	14.108	825	24.085	1150	34.339	1475	45.606	1800	56.886	2100	66.600
-145	-4.563	180	4.305	505	14.262	830	24.238	1155	34.503	1480	45.785	1805	57.051	2105	66.760
-140	-4.448	185	4.455	510	14.416	835	24.392	1160	34.668	1485	45.965	1810	57.217	2110	66.920
-135	-4.332	190	4.605	515	14.570	840	24.546	1165	34.834	1490	46.144	1815	57.382	2115	67.080
-130	-4.215	195	4.755	520	14.724	845	24.699	1170	34.999	1495	46.324	1820	57.547	2120	67.240
-125	-4.097	200	4.906	525	14.878	850	24.853	1175	35.165	1500	46.503	1825	57.712	2125	67.400
-120	-3.978	205	5.057	530	15.032	855	25.007	1180	35.331	1505	46.682	1830	57.876	2130	67.559
-115	-3.858	210	5.207	535	15.186	860	25.161	1185	35.498	1510	46.861	1835	58.041	2135	67.719
-110	-3.737	215	5.358	540	15.340	865	25.315	1190	35.664	1515	47.040	1840	58.205	2140	67.879
-105	-3.615	220	5.509	545	15.494	870	25.469	1195	35.831	1520	47.219	1845	58.369	2145	68.039
-100	-3.492	225	5.661	550	15.648	875	25.623	1200	35.999	1525	47.398	1850	58.533	2150	68.198
-95	-3.369	230	5.812	555	15.802	880	25.778	1205	36.166	1530	47.577	1855	58.697	2155	68.358
-90	-3.245	235	5.964	560	15.956	885	25.932	1210	36.334	1535	47.755	1860	58.860	2160	68.517
-85	-3.120	240	6.116	565	16.110	890	26.087	1215	36.502	1540	47.934	1865	59.024	2165	68.677
-80	-2.994	245	6.268	570	16.264	895	26.241	1220	36.671	1545	48.112	1870	59.187	2170	68.836
-75	-2.867	250	6.420	575	16.417	900	26.396	1225	36.840	1550	48.290	1875	59.350	2175	68.995
-70	-2.740	255	6.572	580	16.571	905	26.551	1230	37.009	1555	48.468	1880	59.513	2180	69.155
-65	-2.612	260	6.724	585	16.725	910	26.705	1235	37.178	1560	48.645	1885	59.676	2185	69.314
-60	-2.483	265	6.877	590	16.879	915	26.860	1240	37.348	1565	48.823	1890	59.838	2190	69.472
-55	-2.353	270	7.029	595	17.032	920	27.016	1245	37.518	1570	49.000	1895	60.001		
-50	-2.223	275	7.182	600	17.186	925	27.171	1250	37.688	1575	49.177				
-45	-2.092	280	7.335	605	17.339	930	27.326	1255	37.859	1580	49.354				
-40	-1.960	285	7.488	610	17.493	935	27.482	1260	38.030	1585	49.531				
-35	-1.828	290	7.641	615	17.646	940	27.637	1265	38.201	1590	49.707				
-30	-1.695	295	7.794	620	17.800	945	27.793	1270	38.372	1595	49.883				

Table A-17. Thermocouple Temperature/mV Data: Type J
Reference Junction: At 0°C, mV = 0
Temperature in Degrees Celsius vs. Absolute mV

°C	mV	°C	mV	°C	mV	°C	mV	°C	mV
-210	-8.096	90	4.725	390	21.295	690	38.510	990	57.349
-205	-7.996	95	4.996	395	21.570	695	38.819	995	57.646
-200	-7.890	100	5.268	400	21.846	700	39.130	1000	57.942
-195	-7.778	105	5.540	405	22.122	705	39.442	1005	58.238
-190	-7.659	110	5.812	410	22.397	710	39.754	1010	58.533
-185	-7.533	115	6.085	415	22.673	715	40.068	1015	58.827
-180	-7.402	120	6.359	420	22.949	720	40.382	1020	59.121
-175	-7.265	125	6.633	425	23.225	725	40.697	1025	59.415
-170	-7.122	130	6.907	430	23.501	730	41.013	1030	59.708
-165	-6.974	135	7.182	435	23.777	735	41.329	1035	60.001
-160	-6.821	140	7.457	440	24.054	740	41.647	1040	60.293
-155	-6.663	145	7.732	445	24.330	745	41.965	1045	60.585
-150	-6.499	150	8.008	450	24.607	750	42.283	1050	60.876
-145	-6.331	155	8.284	455	24.884	755	42.602	1055	61.168
-140	-6.159	160	8.560	460	25.161	760	42.922	1060	61.459
-135	-5.982	165	8.837	465	25.438	765	43.242	1065	61.749
-130	-5.801	170	9.113	470	25.716	770	43.563	1070	62.039
-125	-5.615	175	9.390	475	25.994	775	43.885	1075	62.330
-120	-5.426	180	9.667	480	26.272	780	44.207	1080	62.619
-115	-5.233	185	9.944	485	26.551	785	44.529	1085	62.909
-110	-5.036	190	10.222	490	26.829	790	44.852	1090	63.199
-105	-4.836	195	10.499	495	27.109	795	45.175	1095	63.488
-100	-4.632	200	10.777	500	27.388	800	45.498	1100	63.777
-95	-4.425	205	11.054	505	27.668	805	45.821	1105	64.066
-90	-4.215	210	11.332	510	27.949	810	46.144	1110	64.355
-85	-4.001	215	11.609	515	28.230	815	46.467	1115	64.644
-80	-3.785	220	11.887	520	28.511	820	46.790	1120	64.933
-75	-3.566	225	12.165	525	28.793	825	47.112	1125	65.222
-70	-3.344	230	12.442	530	29.075	830	47.434	1130	65.510
-65	-3.120	235	12.720	535	29.358	835	47.755	1135	65.799
-60	-2.892	240	12.998	540	29.642	840	48.076	1140	66.087
-55	-2.663	245	13.275	545	29.926	845	48.397	1145	66.375
-50	-2.431	250	13.553	550	30.210	850	48.716	1150	66.664
-45	-2.197	255	13.830	555	30.496	855	49.036	1155	66.952
-40	-1.960	260	14.108	560	30.782	860	49.354	1160	67.240
-35	-1.722	265	14.385	565	31.068	865	49.672	1165	67.527
-30	-1.481	270	14.663	570	31.356	870	49.989	1170	67.815
-25	-1.239	275	14.940	575	31.644	875	50.305	1175	68.103
-20	-0.995	280	15.217	580	31.933	880	50.621	1180	68.390
-15	-0.748	285	15.494	585	32.222	885	50.936	1185	68.677
-10	-0.501	290	15.771	590	32.513	890	51.249	1190	68.964
-5	-0.251	295	16.048	595	32.804	895	51.562	1195	69.250
0	0.0	300	16.325	600	33.096	900	51.875	1200	69.536
5	0.253	305	16.602	605	33.389	905	52.186		
10	0.507	310	16.879	610	33.683	910	52.496		
15	0.762	315	17.155	615	33.977	915	52.806		
20	1.019	320	17.432	620	34.273	920	53.115		
25	1.277	325	17.708	625	34.569	925	53.422		
30	1.536	330	17.984	630	34.867	930	53.729		
35	1.797	335	18.260	635	35.155	935	54.035		
40	2.058	340	18.537	640	35.464	940	54.341		
45	2.321	345	18.813	645	35.764	945	54.645		
50	2.585	350	19.089	650	36.066	950	54.948		
55	2.849	355	19.364	655	36.368	955	55.251		
60	3.115	360	19.640	660	36.671	960	55.553		
65	3.381	365	19.916	665	36.975	965	55.854		
70	3.649	370	20.192	670	37.280	970	56.155		
75	3.917	375	20.467	675	37.586	975	56.454		
80	4.186	380	20.743	680	37.893	980	56.753		
85	4.455	385	21.019	685	38.201	985	57.051		

Table A-18. Thermocouple Temperature/mV Data For Copper vs. Copper-Nickel (Copper-Constantan) Thermocouples, Type T Based on International Practical Temperature Scale of 1968 (IPTS-68)

Reference Junction: At 32°F, mV = 0

Temperature in Degrees Fahrenheit vs. Absolute mV

°F	mV	°F	mV	°F	mV	°F	mV	°F	mV
-450	-6.254	-200	-4.149	50	0.391	300	6.647	550	14.153
-440	-6.240	-190	-4.009	60	0.611	310	6.926	560	14.474
-430	-6.217	-180	-3.864	70	0.834	320	7.207	570	14.795
-420	-6.187	-170	-3.717	80	1.060	330	7.490	580	15.118
-410	-6.150	-160	-3.565	90	1.288	340	7.775	590	15.443
-400	-6.105	-150	-3.410	100	1.518	350	8.062	600	15.769
-390	-6.053	-140	-3.251	110	1.752	360	8.350	610	16.096
-380	-5.995	-130	-3.089	120	1.988	370	8.641	620	16.424
-370	-5.930	-120	-2.923	130	2.226	380	8.933	630	16.753
-360	-5.860	-110	-2.753	140	2.467	390	9.227	640	17.084
-350	-5.785	-100	-2.581	150	2.711	400	9.523	650	17.416
-340	-5.705	-90	-2.405	160	2.958	410	9.820	660	17.750
-330	-5.620	-80	-2.225	170	3.206	420	10.120	670	18.084
-320	-5.532	-70	-2.042	180	3.458	430	10.420	680	18.420
-310	-5.439	-60	-1.856	190	3.711	440	10.723	690	18.757
-300	-5.341	-50	-1.667	200	3.967	450	11.027	700	19.095
-290	-5.240	-40	-1.475	210	4.225	460	11.333	710	19.434
-280	-5.135	-30	-1.279	220	4.486	470	11.640	720	19.774
-270	-5.025	-20	-1.081	230	4.749	480	11.949	730	20.116
-260	-4.912	-10	-0.879	240	5.014	490	12.260	740	20.458
-250	-4.794	0	-0.674	250	5.281	500	12.572	750	20.801
-240	-4.673	10	-0.467	260	5.550	510	12.885		
-230	-4.548	20	-0.256	270	5.821	520	13.200		
-220	-4.419	30	-0.043	280	6.094	530	13.516		
-210	-4.286	40	0.173	290	6.369	540	13.834		

Reference Junction: At 0°C, mV = 0

Temperature in Degrees Celsius vs. Absolute mV

°C	mV	°C	mV	°C	mV	°C	mV	°C	mV
-270	-6.258	-120	-3.923	30	1.196	180	8.235	330	16.621
-265	-6.248	-115	-3.791	35	1.403	185	8.495	335	16.919
-260	-6.232	-110	-3.656	40	1.611	190	8.757	340	17.217
-255	-6.209	-105	-3.519	45	1.822	195	9.021	345	17.516
-250	-6.181	-100	-3.378	50	2.035	200	9.286	350	17.816
-245	-6.146	-95	-3.235	55	2.250	205	9.553	355	18.118
-240	-6.105	-90	-3.089	60	2.467	210	9.820	360	18.420
-235	-6.059	-85	-2.939	65	2.687	215	10.090	365	18.723
-230	-6.007	-80	-2.788	70	2.908	220	10.360	370	19.027
-225	-5.950	-75	-2.633	75	3.131	225	10.632	375	19.332
-220	-5.889	-70	-2.475	80	3.357	230	10.905	380	19.638
-215	-5.823	-65	-2.315	85	3.584	235	11.180	385	19.945
-210	-5.753	-60	-2.152	90	3.813	240	11.456	390	20.252
-205	-5.680	-55	-1.987	95	4.044	245	11.733	395	20.560
-200	-5.603	-50	-1.819	100	4.277	250	12.011	400	20.869
-195	-5.522	-45	-1.648	105	4.512	255	12.291		
-190	-5.439	-40	-1.475	110	4.749	260	12.572		
-185	-5.351	-35	-1.299	115	4.987	265	12.854		
-180	-5.261	-30	-1.121	120	5.227	270	13.137		
-175	-5.167	-25	-0.940	125	5.469	275	13.421		
-170	-5.069	-20	-0.757	130	5.712	280	13.707		
-165	-4.969	-15	-0.571	135	5.957	285	13.993		
-160	-4.865	-10	-0.383	140	6.204	290	14.281		
-155	-4.758	-5	-0.193	145	6.452	295	14.570		
-150	-4.648	0	0.0	150	6.702	300	14.860		
-145	-4.535	5	0.195	155	6.954	305	15.151		
-140	-4.419	10	0.391	160	7.207	310	15.443		
-135	-4.299	15	0.589	165	7.462	315	15.736		
-130	-4.177	20	0.789	170	7.718	320	16.030		
-125	-4.051	25	0.992	175	7.975	325	16.325		

Table A-19. Thermocouple Temperature/mV Data For Nickel-Chromium vs. Copper-Nickel (Chromel-Constantan) Thermocouples, Type E Based on International Practical Temperature Scale of 1968 (IPTS-68)

Reference Junction: At 0°C, mV = 0

Temperature in Degrees Celsius vs. Absolute mV

°C	mV	°C	mV	°C	mV	°C	mV
-270	-9.835	80	4.983	430	31.350	780	59.451
-260	-9.797	90	5.646	440	32.155	790	60.237
-250	-9.719	100	6.317	450	32.960	800	61.022
-240	-9.604	110	6.996	460	33.767	810	61.806
-230	-9.455	120	7.683	470	34.574	820	62.588
-220	-9.274	130	8.377	480	35.382	830	63.368
-210	-9.063	140	9.078	490	36.190	840	64.147
-200	-8.824	150	9.787	500	36.999	850	64.924
-190	-8.561	160	10.501	510	37.808	860	65.700
-180	-8.273	170	11.222	520	38.617	870	66.473
-170	-7.963	180	11.949	530	39.426	880	67.245
-160	-7.631	190	12.681	540	40.236	890	68.015
-150	-7.279	200	13.419	550	41.045	900	68.783
-140	-6.907	210	14.161	560	41.853	910	69.549
-130	-6.516	220	14.909	570	42.662	920	70.313
-120	-6.107	230	15.661	580	43.470	930	71.075
-110	-5.680	240	16.417	590	44.278	940	71.835
-100	-5.237	250	17.178	600	45.085	950	72.593
-90	-4.777	260	17.942	610	45.891	960	73.350
-80	-4.301	270	18.710	620	46.697	970	74.104
-70	-3.811	280	19.481	630	47.502	980	74.857
-60	-3.306	290	20.256	640	48.306	990	75.608
-50	-2.787	300	21.033	650	49.109	1000	76.358
-40	-2.254	310	21.814	660	49.911		
-30	-1.709	320	22.597	670	50.713		
-20	-1.151	330	23.383	680	51.513		
-10	-0.581	340	24.171	690	52.312		
0	0.0	350	24.961	700	53.110		
10	0.591	360	25.754	710	53.907		
20	1.192	370	26.549	720	54.703		
30	1.801	380	27.345	730	55.498		
40	2.419	390	28.143	740	56.291		
50	3.047	400	28.943	750	57.083		
60	3.683	410	29.744	760	57.873		
70	4.329	420	30.546	770	58.663		

Table A-20. Thermocouple Temperature/mV Data For Nickel-Chromium vs. Copper-Nickel (Chromel-Constantan) Thermocouples, Type E Based on International Practical Temperature Scale of 1968 (IPTS-68)

Reference Junction: At 32°F, mV = 0

Temperature in Degrees Fahrenheit vs. Absolute mV

°F	mV	°F	mV	°F	mV	°F	mV	°F	mV
-450	-9.830	50	0.591	550	20.083	1050	42.303	1550	64.406
-440	-9.809	60	0.924	560	20.514	1060	42.752	1560	64.838
-430	-9.775	70	1.259	570	20.947	1070	43.201	1570	65.269
-420	-9.729	80	1.597	580	21.380	1080	43.650	1580	65.700
-410	-9.672	90	1.937	590	21.814	1090	44.098	1590	66.130
-400	-9.604	100	2.281	600	22.248	1100	44.547	1600	66.559
-390	-9.526	110	2.627	610	22.684	1110	44.995	1610	66.988
-380	-9.437	120	2.977	620	23.120	1120	45.443	1620	67.416
-370	-9.338	130	3.329	630	23.558	1130	45.891	1630	67.844
-360	-9.229	140	3.683	640	23.996	1140	46.339	1640	68.271
-350	-9.112	150	4.041	650	24.434	1150	46.786	1650	68.698
-340	-8.986	160	4.401	660	24.873	1160	47.234	1660	69.124
-330	-8.852	170	4.764	670	25.313	1170	47.681	1670	69.549
-320	-8.710	180	5.130	680	25.754	1180	48.127	1680	69.974
-310	-8.561	190	5.498	690	26.195	1190	48.574	1690	70.398
-300	-8.404	200	5.869	700	26.637	1200	49.020	1700	70.821
-290	-8.240	210	6.242	710	27.079	1210	49.466	1710	71.244
-280	-8.069	220	6.618	720	27.522	1220	49.911	1720	71.667
-270	-7.891	230	6.996	730	27.966	1230	50.357	1730	72.088
-260	-7.707	240	7.377	740	28.409	1240	50.802	1740	72.509
-250	-7.516	250	7.760	750	28.854	1250	51.246	1750	72.930
-240	-7.319	260	8.145	760	29.299	1260	51.691	1760	73.350
-230	-7.116	270	8.532	770	29.744	1270	52.135	1770	73.769
-220	-6.907	280	8.922	780	30.190	1280	52.578	1780	74.188
-210	-6.692	290	9.314	790	30.636	1290	53.022	1790	74.606
-200	-6.471	300	9.708	800	31.082	1300	53.465	1800	75.024
-190	-6.245	310	10.103	810	31.529	1310	53.907	1810	75.441
-180	-6.013	320	10.501	820	31.976	1320	54.349	1820	75.858
-170	-5.776	330	10.901	830	32.423	1330	54.791	1830	76.274
-160	-5.534	340	11.302	840	32.871	1340	55.233		
-150	-5.287	350	11.706	850	33.319	1350	55.674		
-140	-5.034	360	12.111	860	33.767	1360	56.115		
-130	-4.777	370	12.518	870	34.215	1370	56.555		
-120	-4.515	380	12.926	880	34.664	1380	56.995		
-110	-4.248	390	13.336	890	35.113	1390	57.434		
-100	-3.976	400	13.748	900	35.562	1400	57.873		
-90	-3.700	410	14.161	910	36.011	1410	58.312		
-80	-3.419	420	14.576	920	36.460	1420	58.750		
-70	-3.134	430	14.992	930	36.909	1430	59.188		
-60	-2.845	440	15.410	940	37.358	1440	59.626		
-50	-2.552	450	15.829	950	37.808	1450	60.063		
-40	-2.254	460	16.249	960	38.257	1460	60.499		
-30	-1.953	470	16.670	970	38.707	1470	60.935		
-20	-1.648	480	17.093	980	39.157	1480	61.371		
-10	-1.339	490	17.517	990	39.606	1490	61.806		
0	-1.026	500	17.942	1000	40.056	1500	62.240		
10	-0.709	510	18.368	1010	40.505	1510	62.675		
20	-0.389	520	18.795	1020	40.955	1520	63.108		
30	-0.065	530	19.223	1030	41.404	1530	63.542		
40	0.262	540	19.653	1040	41.853	1540	63.974		

Table A-21. Platinum RTD Temperature/Resistance Data Based on DIN Curve 43760, 9-68, (Foxboro PR 238) Degrees Celsius vs. Absolute Ohms
These tables show the value of resistance (in Absolute Ohms) between the black and white terminals or leads of RTD type sensors at different temperatures (in degrees Celsius).

°C	Ohms	°C	Ohms	°C	Ohms	°C	Ohms	°C	Ohms
		0	100.00	225	185.00	450	264.14	675	337.43
-220	10.41	5	101.95	230	186.82	455	265.83	680	338.99
-215	12.35	10	103.90	235	188.64	460	267.52	685	340.55
-210	14.36	15	105.85	240	190.46	465	269.21	690	342.10
-205	16.43	20	107.79	245	192.27	470	270.89	695	343.66
-200	18.53	25	109.73	250	194.08	475	272.57	700	345.21
-195	20.65	30	111.67	255	195.89	480	274.25	705	346.76
-190	22.78	35	113.61	260	197.70	485	275.92	710	348.30
-185	24.92	40	115.54	265	199.50	490	277.60	715	349.84
-180	27.05	45	117.47	270	201.30	495	279.27	720	351.38
-175	29.17	50	119.40	275	203.09	500	280.93	725	352.92
-170	31.28	55	121.32	280	204.88	505	282.60	730	354.45
-165	33.38	60	123.24	285	206.68	510	284.26	735	355.98
-160	35.48	65	125.16	290	208.46	515	285.91	740	357.51
-155	37.57	70	127.07	295	210.25	520	287.57	745	359.03
-150	39.65	75	128.98	300	212.03	525	289.22	750	360.55
-145	41.73	80	130.89	305	213.81	530	290.87	755	362.07
-140	43.80	85	132.80	310	215.58	535	292.51	760	363.59
-135	45.87	90	134.70	315	217.36	540	294.16	765	365.10
-130	47.93	95	136.60	320	219.13	545	295.80	770	366.61
-125	49.99	100	138.50	325	220.90	550	297.43	775	368.12
-120	52.04	105	140.39	330	222.66	555	299.07	780	369.62
-115	54.09	110	142.28	335	224.42	560	300.70	785	371.12
-110	56.13	115	144.18	340	226.18	565	302.33	790	372.62
-105	58.17	120	146.06	345	227.94	570	303.95	795	374.12
-100	60.20	125	147.94	350	229.69	575	305.58	800	375.61
-95	62.23	130	149.82	355	231.44	580	307.20	805	377.10
-90	64.25	135	151.70	360	233.19	585	308.81	810	378.59
-85	66.27	140	153.57	365	234.93	590	310.43	815	380.07
-80	68.28	145	155.45	370	236.67	595	312.04	820	381.55
-75	70.29	150	157.32	375	238.41	600	313.65	825	383.03
-70	72.29	155	159.18	380	240.15	605	315.25	830	384.50
-65	74.29	160	161.04	385	241.88	610	316.86	835	385.98
-60	76.28	165	162.90	390	243.61	615	318.46	840	387.45
-55	78.27	170	164.76	395	245.34	620	320.05	845	388.91
-50	80.25	175	166.62	400	247.06	625	321.65	850	390.38
-45	82.23	180	168.47	405	248.78	630	323.24		
-40	84.21	185	170.32	410	250.50	635	324.83		
-35	86.19	190	172.16	415	252.21	640	326.41		
-30	88.17	195	174.00	420	253.93	645	327.99		
-25	90.15	200	175.84	425	255.64	650	329.57		
-20	92.13	205	177.68	430	257.34	655	331.15		
-15	94.10	210	179.51	435	259.05	660	332.72		
-10	96.07	215	181.34	440	260.75	665	334.29		
-5	98.04	220	183.17	445	262.45	670	335.86		

Table A-22. Platinum RTD Temperature/Resistance Data Calculated from DIN 43760 Platinum RTD Curve Data (Foxboro PR 239) Degrees Fahrenheit vs. Absolute Ohms

These tables show the value of resistance (in Absolute Ohms) between the black and white terminals or leads of RTD type sensors at different temperatures (in degrees Fahrenheit).

°F	Ohm	°F	Ohm	°F	Ohm	°F	Ohm	°F	Ohm
-360	11.26								
-350	13.47	50	103.90	450	187.63	850	265.64	1250	337.95
-340	15.74	60	106.07	460	189.65	860	267.52	1260	339.68
-330	18.06	70	108.22	470	191.67	870	269.40	1270	341.41
-320	20.41	80	110.38	480	193.68	880	271.26	1280	343.14
-310	22.78	90	112.53	490	195.69	890	273.13	1290	344.87
-300	25.16	100	114.68	500	197.70	900	274.99	1300	346.59
-290	27.52	110	116.83	510	199.70	910	276.86	1310	348.30
-280	29.87	120	118.97	520	201.70	920	278.71	1320	350.01
-270	32.22	130	121.11	530	203.69	930	280.56	1330	351.72
-260	34.55	140	123.24	540	205.68	940	282.42	1340	353.43
-250	36.87	150	125.37	550	207.67	950	284.26	1350	355.13
-240	39.19	160	127.49	560	209.65	960	286.09	1360	356.83
-230	41.50	170	129.62	570	211.63	970	287.94	1370	358.52
-220	43.80	180	131.74	580	213.61	980	289.77	1380	360.21
-210	46.10	190	133.86	590	215.58	990	291.60	1390	361.90
-200	48.39	200	135.97	600	217.56	1000	293.43	1400	363.59
-190	50.67	210	138.08	610	219.52	1010	295.25	1410	365.27
-180	52.95	220	140.18	620	221.49	1020	297.07	1420	366.95
-170	55.22	230	142.28	630	223.44	1030	298.89	1430	368.62
-160	57.49	240	144.39	640	225.40	1040	300.70	1440	370.29
-150	59.75	250	146.48	650	227.35	1050	302.51	1450	371.95
-140	62.01	260	148.57	660	229.30	1060	304.31	1460	373.62
-130	64.25	270	150.66	670	231.25	1070	306.12	1470	375.28
-120	66.49	280	152.74	680	233.19	1080	307.92	1480	376.94
-110	68.73	290	154.82	690	235.12	1090	309.71	1490	378.59
-100	70.96	300	156.91	700	237.06	1100	311.50	1500	380.24
-90	73.18	310	158.97	710	238.99	1110	313.29	1510	381.88
-80	75.40	320	161.04	720	240.92	1120	315.07	1520	383.52
-70	77.61	330	163.11	730	242.84	1130	316.86	1530	385.16
-60	79.81	340	165.17	740	244.76	1140	318.64	1540	386.80
-50	82.01	350	167.24	750	246.68	1150	320.41	1550	388.42
-40	84.21	360	169.29	760	248.59	1160	322.18	1560	390.05
-30	86.41	370	171.34	770	250.50	1170	323.95		
-20	88.61	380	173.39	780	252.40	1180	325.71		
-10	90.81	390	175.43	790	254.31	1190	327.46		
0	93.01	400	177.48	800	256.21	1200	329.22		
10	95.19	410	179.51	810	258.10	1210	330.98		
20	97.38	420	181.54	820	259.99	1220	332.72		
30	99.57	430	183.58	830	261.88	1230	334.46		
40	101.73	440	185.61	840	263.76	1240	336.21		

Table A-23

CURVE CR-228 FOR COPPER RTDs, FAHRENHEIT TEMPERATURE
VS ABSOLUTE OHMS

Deg F	Ohms	Deg F	Ohms	Deg F	Ohms
0	8.359				
10	8.573	110	10.712	210	12.852
20	8.787	120	10.926	220	13.066
30	9.000	130	11.140	230	13.280
40	9.215	140	11.354	240	13.494
50	9.428	150	11.568	250	13.708
60	9.642	160	11.782	260	13.922
70	9.856	170	11.996	270	14.136
77	10.005				
80	10.070	180	12.210	280	14.350
90	10.284	190	12.424	290	14.564
100	10.498	200	12.638	300	14.778

CURVE CR-229 FOR COPPER RTDs, CELSIUS TEMPERATURE
VS ABSOLUTE OHMS

Deg C	Ohms	Deg C	Ohms	Deg C	Ohms
0	9.042				
5	9.235	55	11.160	105	13.086
10	9.428	60	11.353	110	13.279
15	9.620	65	11.546	115	13.472
20	9.813	70	11.739	120	13.665
25	10.005	75	11.931	125	13.857
30	10.198	80	12.124	130	14.050
35	10.390	85	12.316	135	14.242
40	10.583	90	12.509	140	14.435
45	10.775	95	12.701	145	14.627
50	10.968	100	12.894	150	14.820

One absolute Ohm equals 0.999505 international ohm.

Table A-24

CURVE NR-227 FOR FOXBORO RTDs, CELSIUS TEMPERATURE VS ABSOLUTE OHMS
(SEE NOTES REGARDING 217.657-OHM RESISTOR.)

Deg C	Ohms	Deg C	Ohms	Deg C	Ohms	Deg C	Ohms	Deg C	Ohms
		−112.5	189.249			102.5	286.520	215.0	355.962
		−110.0	190.164			105.0	287.905	217.5	357.683
		−107.5	191.083	0	235.116	107.5	289.296	220.0	359.412
		−105.0	192.007			110.0	290.695	222.5	361.149
−215.0	155.297	−102.5	192.935			112.5	292.100	225.0	362.895
−212.5	156.047	−100.0	193.868	+2.5	236.253	115.0	293.512	227.5	364.649
−210.0	156.802	−97.5	194.805	5.0	237.395	117.5	294.931	230.0	366.412
−207.5	157.560	−95.0	195.747	7.5	238.543	120.0	296.357	232.5	368.183
−205.0	158.322	−92.5	196.693	10.0	239.696	122.5	297.790	235.0	369.963
−202.5	159.087	−90.0	197.644	12.5	240.855	125.0	299.229	237.5	371.752
−200.0	159.856	−87.5	198.600	15.0	242.019	127.5	300.676	240.0	373.549
−197.5	160.629	−85.0	199.560	17.5	243.189	130.0	302.129	242.5	375.355
−195.0	161.405	−82.5	200.524	20.0	244.364	132.5	303.590	245.0	377.170
−192.5	162.186	−80.0	201.494	22.5	245.546	135.0	305.057	247.5	378.993
−190.0	162.970	−77.5	202.468	25.0	246.733	137.5	306.532	250.0	380.825
−187.5	163.757	−75.0	203.447	27.5	247.926	140.0	308.014	252.5	382.666
−185.0	164.549	−72.5	204.430	30.0	249.124	142.5	309.503	255.0	384.516
−182.5	165.344	−70.0	205.418	32.5	250.328	145.0	310.999	257.5	386.375
−180.0	166.144	−67.5	206.411	35.0	251.539	147.5	312.503	260.0	388.243
−177.5	166.947	−65.0	207.409	37.5	252.755	150.0	314.013	262.5	390.119
−175.0	167.754	−62.5	208.412	40.0	253.977	152.5	315.531	265.0	392.005
−172.5	168.565	−60.0	209.419	42.5	255.204	155.0	317.057	267.5	393.900
−170.0	169.380	−57.5	210.432	45.0	256.438	157.5	318.589	270.0	395.805
−167.5	170.199	−55.0	211.449	47.5	257.678	160.0	320.130	272.5	397.718
−165.0	171.022	−52.5	212.471	50.0	258.924	162.5	321.678	275.0	399.640
−162.5	171.848	−50.0	213.498	52.5	260.175	165.0	323.233	277.5	401.572
−160.0	172.679	−47.5	214.530	55.0	261.433	167.5	324.795	280.0	403.514
−157.5	173.514	−45.0	215.567	57.5	262.697	170.0	326.365	282.5	405.465
−155.0	174.353	−42.5	216.610	60.0	263.967	172.5	327.943	285.0	407.425
−152.5	175.196	−40.0	217.657	62.5	265.243	175.0	329.528	287.5	409.394
−150.0	176.042	−37.5	218.709	65.0	266.525	177.5	331.121	290.0	411.373
−147.5	176.893	−35.0	219.766	67.5	267.813	180.0	332.722	292.5	413.362
−145.0	177.749	−32.5	220.829	70.0	269.108	182.5	334.330	295.0	415.360
−142.5	178.608	−30.0	221.896	72.5	270.409	185.0	335.947	297.5	417.368
−140.0	179.471	−27.5	222.969	75.0	271.716	187.5	337.571	300.0	419.386
−137.5	180.339	−25.0	224.047	77.5	273.030	190.0	339.203	302.5	421.413
−135.0	181.211	−22.5	225.130	80.0	274.350	192.5	340.842	305.0	423.451
−132.5	182.087	−20.0	226.218	82.5	275.676	195.0	342.490	307.5	425.498
−130.0	182.967	−17.5	227.312	85.0	277.009	197.5	344.146	310.0	427.555
−127.5	183.851	−15.0	228.411	87.5	278.348	200.0	345.809	312.5	429.622
−125.0	184.740	−12.5	229.515	90.0	279.693	202.5	347.481	315.0	431.699
−122.5	185.633	−10.0	230.624	92.5	281.045	205.0	349.161	317.5	433.785
−120.0	186.531	−7.5	231.739	95.0	282.404	207.5	350.849	320.0	435.882
−117.5	187.433	−5.0	232.859	97.5	283.769	210.0	352.545		
−115.0	188.339	−2.5	233.985	100.0	285.141	212.5	354.249		

Table A-25

CURVE NR-226 FOR FOXBORO RTDs, FAHRENHEIT TEMPERATURE VS ABSOLUTE OHMS
(ONE ABSOLUTE OHM = 0.999505 INTERNATIONAL OHM)

Deg F	Ohms	Deg F	Ohms	Deg F	Ohms	Deg F	Ohms	Deg F	Ohms
		−125	198.706					405	350.661
−320	161.233	−120	199.774					410	352.545
−315	162.099	−115	200.847	35	235.874			415	354.439
−310	162.970	−110	201.926	40	237.141	215	286.059	420	356.343
−305	163.845	−105	203.011	45	238.415	220	287.596	425	358.258
−300	164.725	−100	204.102	50	239.696	225	289.142	430	360.183
−295	165.610	− 95	205.198	55	240.984	230	290.695	435	362.118
−290	166.500	− 90	206.301	60	242.279	235	292.257	440	364.064
−285	167.395	− 85	207.409	65	243.580	240	293.827	445	366.020
−280	168.294	− 80	208.524	70	244.889	245	295.406	450	367.986
−275	169.199	− 75	209.644	75	246.205	250	296.993	455	369.963
−270	170.108	− 70	210.771	80	247.527	255	298.589	460	371.951
−265	171.022	− 65	211.903	85	248.857	260	300.193	465	373.949
−260	171.940	− 60	213.042	90	250.194	265	301.806	470	375.959
−255	172.864	− 55	214.187	95	251.539	270	303.428	475	377.979
−250	173.793	− 50	215.337	100	252.890	275	305.057	480	380.009
−245	174.727	− 45	216.494	105	254.249	280	306.696	485	382.051
−240	175.666	− 40	217.657	110	255.615	285	308.344	490	384.104
−235	176.609	− 35	218.826	115	256.988	290	310.001	495	386.168
−230	177.558	− 30	220.002	120	258.369	295	311.667	500	388.243
−225	178.512	− 25	221.184	125	259.757	300	313.341	505	390.329
−220	179.471	− 20	222.372	130	261.153	305	315.025	510	392.426
−215	180.436	− 15	223.567	135	262.556	310	316.717	515	394.534
−210	181.405	− 10	224.768	140	263.967	315	318.419	520	396.654
−205	182.380	− 5	225.976	145	265.385	320	320.130	525	398.785
−200	183.360			150	266.811	325	321.850	530	400.928
−195	184.345			155	268.244	330	323.579	535	403.082
−190	185.335	0	227.190	160	269.685	335	325.318	540	405.247
−185	186.331			165	271.134	340	327.066	545	407.425
−180	187.332			170	272.591	345	328.823	550	409.614
−175	188.339	+ 5	228.411	175	274.056	350	330.590	555	411.814
−170	189.351	10	229.638	180	275.528	355	332.366	560	414.027
−165	190.368	15	230.872	185	277.009	360	334.152	565	416.252
−160	191.391	20	232.112	190	278.497	365	335.947	570	418.488
−155	192.419	25	233.359	195	279.993	370	337.752	575	420.736
−150	193.453	30	234.613	200	281.498	375	339.566	580	422.997
−145	194.492			205	283.010	380	341.391	585	425.270
−140	195.537			210	284.531	385	343.225	590	427.555
−135	196.588	32	235.117	212	285.141	390	345.069	595	429.852
−130	197.644					395	346.923	600	432.162
						400	348.787		

NOTE: For narrow span instruments or instruments with midscale values below −40°F or −40°C a special value resistor is used in place of the 217.657 ohm resistor. Padder panels with these special value resistors are identified with a RED dot and the actual resistance value is marked on the end of the coil. This special value should be used instead of 217.657 ohms.

Table A-26. Fahrenheit Table of Relative Humidity or Per Cent of Saturation

Dry Bulb Deg. F.	1	2	3	4	5	6	7	8	9	10	11	12	13	14	15	16	17	18	19	20	22	24	26	28	30	32	34	36	38	40	45	50	55	60	70	Dry Bulb Deg. F.
30	89	78	67	56	46	36	26	16	6	0	0																									30
35	91	81	72	63	54	45	36	27	19	10	2	0																								35
40	92	83	75	68	60	52	45	37	29	22	15	7	0	0	0																					40
45	93	86	78	71	64	57	51	44	38	31	25	18	12	6	0	0	0																			45
50	93	87	80	74	67	61	55	49	43	38	32	27	21	16	10	5	0	0																		50
55	94	88	82	76	70	65	59	54	49	43	38	33	28	23	19	14	9	5	0	0																55
60	94	89	83	78	73	68	63	58	53	48	43	39	34	30	26	21	17	13	9	5	0															60
65	95	90	85	80	75	70	66	61	56	52	48	44	39	35	31	27	24	20	16	12	5	0														65
70	95	90	86	81	77	72	68	64	59	55	51	48	44	40	36	33	29	25	22	19	12	6	0	0												70
75	96	91	86	82	78	74	70	66	62	58	54	51	47	44	40	37	34	30	27	24	18	12	7	1	0											75
80	96	91	87	83	79	75	72	68	64	61	57	54	50	47	44	41	38	35	32	29	23	18	12	7	3	0										80
85	96	92	88	84	80	76	73	70	66	63	59	56	53	50	47	44	41	38	35	32	27	22	17	13	8	4	0	0								85
90	96	92	89	85	81	78	74	71	68	65	61	58	55	52	49	47	44	41	39	36	31	26	22	17	13	9	5	1	0							90
95	96	93	89	85	82	79	75	72	69	66	63	60	57	54	52	49	46	43	42	38	34	30	25	21	17	13	9	6	2	0						95
100	96	93	89	86	83	80	77	73	70	68	65	62	59	56	54	51	49	46	44	41	37	33	28	24	21	17	13	10	7	4						100
102	96	93	89	86	83	80	77	74	71	69	65	63	60	58	55	52	50	48	46	43	38	34	30	26												102
104	96	93	90	86	83	80	77	74	71	69	65	63	60	58	55	52	50	48	46	43	39	35	31	27	23											104
106	96	93	90	87	83	80	77	74	72	69	66	63	60	58	56	53	51	48	46	44	40	36	32	28	24	21										106
108	96	93	90	87	84	81	78	75	72	70	66	64	61	59	56	54	51	49	47	45	41	37	33	29	26	22	19									108
110	96	93	90	87	84	81	78	75	72	70	67	64	62	60	57	55	52	50	48	46	41	37	34	30	27	23	20	17								110
112	96	93	90	87	84	81	78	75	73	70	67	65	62	60	57	55	53	51	49	47	42	38	35	31	28	24	21	18	15							112
114	97	93	90	87	84	81	78	75	73	71	68	65	63	61	58	56	53	51	49	47	43	39	35	32	28	25	22	19	16	13						114
116	97	93	90	88	84	82	79	76	74	71	68	66	63	61	59	56	54	52	50	48	44	40	36	33	29	26	24	20	17	14						116
118	97	93	91	88	85	82	79	76	74	71	68	66	64	62	59	57	54	53	51	49	44	41	37	34	30	27	24	21	19	15						118
120	97	94	91	88	85	82	79	77	74	72	69	66	64	62	60	57	55	53	51	49	45	41	38	34	31	28	25	22	19	16	10					120
122	97	94	91	88	85	82	79	77	75	72	69	67	65	63	60	58	56	54	52	50	46	42	38	35	32	29	26	23	20	17	12					122
124	97	94	91	88	85	83	80	77	75	72	70	67	65	63	61	58	56	54	52	51	46	43	39	36	33	29	27	24	21	18	13					124
126	97	94	91	88	86	83	80	78	75	73	70	68	65	64	61	59	57	55	53	51	47	43	40	37	33	30	28	25	22	18	13					126
128	97	94	91	89	86	83	80	78	76	73	71	68	66	64	61	59	57	55	53	52	47	44	40	37	34	31	28	25	23	20	15					128
130	97	94	91	89	86	83	80	78	76	73	71	68	66	64	62	60	58	56	54	52	48	44	41	38	35	32	29	26	24	21	15	10				130
132	97	94	92	89	86	83	81	78	76	74	71	69	67	65	62	60	58	56	54	53	49	45	42	39	35	32	30	27	24	22	16	11				132
134	97	94	92	89	86	84	81	79	76	74	71	69	67	65	63	61	59	57	55	53	49	46	42	39	36	33	31	28	25	23	17	12				134
136	97	94	92	89	86	84	81	79	77	74	72	69	67	65	63	61	59	57	55	53	50	46	43	40	37	34	31	28	26	24	18	13				136
138	97	94	92	89	86	84	81	79	77	74	72	70	68	66	63	62	60	58	56	54	50	47	43	40	37	35	32	29	27	24	19	14				138
140	97	94	92	89	87	84	81	79	77	75	72	70	68	66	64	62	60	58	56	54	51	47	44	41	38	35	33	30	28	25	19	14	10			140
142	97	94	92	89	87	84	82	80	77	75	73	70	68	66	64	62	60	58	57	55	51	48	44	42	39	36	33	30	28	26	20	15	11			142
144	97	95	92	89	87	84	82	80	78	75	73	71	69	67	65	63	61	59	57	55	52	48	45	42	39	36	34	31	29	26	21	16	11			144
146	97	95	92	90	87	85	82	80	78	76	73	71	69	67	65	63	61	59	57	56	52	49	45	43	40	37	35	32	29	27	21	17	12			146
148	97	95	92	90	87	85	82	80	78	76	73	71	69	67	65	63	61	60	58	56	53	49	46	43	40	38	35	32	30	28	22	17	13			148
150	98	95	92	90	87	85	82	80	78	76	74	72	70	68	66	64	62	60	58	57	53	50	47	44	41	38	36	33	31	28	23	18	13			150
152	98	95	93	90	88	85	83	81	79	76	74	72	70	68	66	65	63	61	59	57	53	50	47	44	41	39	36	33	31	29	24	19	14	10		152
154	98	95	93	90	88	85	83	81	79	77	74	72	70	68	66	65	63	61	59	57	54	50	47	44	42	39	37	34	32	29	24	19	15	11		154
156	98	95	93	90	88	85	83	81	79	77	75	73	71	69	67	65	63	61	60	58	54	51	48	45	42	40	37	34	32	30	25	20	15	11		156
158	98	95	93	90	88	86	83	81	79	77	75	73	71	69	67	65	63	61	60	58	55	51	48	45	43	40	38	35	33	30	25	21	16	12		158
160	98	95	93	90	88	86	83	81	79	77	75	73	71	69	67	65	64	62	60	58	55	52	49	46	43	41	38	35	33	30	26	21	17	13		160
162	98	95	93	90	88	86	84	82	80	77	75	73	71	69	68	66	64	62	61	59	55	52	49	46	44	41	39	36	34	31	26	22	17	13		162
164	98	95	93	91	88	86	84	82	80	78	75	73	72	70	68	66	64	62	61	59	56	52	49	47	44	41	39	36	34	32	26	22	18	14		164
166	98	95	93	91	88	86	84	82	80	78	76	74	72	70	68	66	65	63	61	59	56	53	50	47	44	42	40	37	35	32	27	23	19	14		166
168	98	95	93	91	88	86	84	82	80	78	76	74	72	70	68	67	65	63	61	60	56	53	50	47	45	42	40	37	35	33	28	23	19	15		168
170	98	95	93	91	89	86	84	82	80	78	76	74	72	70	69	67	65	63	61	60	57	53	51	48	45	43	40	38	35	33	28	24	19	15		170
172	98	95	93	91	89	86	84	82	81	78	76	74	73	71	69	67	66	64	62	60	57	54	51	48	46	43	41	38	36	34	28	24	20	16		172
174	98	95	93	91	89	87	84	83	81	78	76	75	73	71	69	67	66	64	62	61	57	54	51	49	46	43	41	39	36	34	29	24	20	16		174
176	98	96	94	91	89	87	85	83	81	79	77	75	73	71	70	68	66	64	63	61	58	55	52	49	47	44	42	39	37	35	30	25	21	17	10	176
178	98	96	94	91	89	87	85	83	81	79	77	75	73	72	70	68	66	64	63	61	58	55	52	49	47	44	42	39	37	35	30	25	21	17	11	178
180	98	96	94	91	89	87	85	83	81	79	77	75	74	72	70	68	67	65	63	62	58	55	52	50	47	45	42	40	38	35	30	26	22	18	11	180
182	98	96	94	91	89	87	85	83	81	79	77	75	74	72	70	68	67	65	63	62	59	56	53	50	48	45	43	40	38	36	31	26	22	18	12	182
184	98	96	94	92	89	87	85	83	82	79	77	76	74	72	70	69	67	65	64	62	59	56	53	50	48	45	43	41	38	36	31	27	22	19	12	184
186	98	96	94	92	90	87	85	83	82	80	78	76	74	73	71	69	67	66	64	62	59	56	53	51	49	46	44	41	39	37	32	27	23	19	13	186
188	98	96	94	92	90	87	85	84	82	80	78	76	74	73	71	69	68	66	64	63	59	57	54	51	49	46	44	41	39	37	32	27	23	20	13	188
190	98	96	94	92	90	88	85	84	82	80	78	76	75	73	71	70	68	66	65	62	60	57	54	52	50	47	44	42	40	38	33	28	24	20	14	190
200	98	96	94	92	90	88	86	84	82	80	79	77	75	74	72	70	69	67	66	64	61	58	55	53	51	48	46	43	41	39	34	30	26	22	16	200
205	98	96	94	92	90	88	86	84	83	81	79	77	76	74	72	71	69	68	66	65	62	59	56	54	51	49	46	44	42	40	35	31	27	23	17	205
210	98	96	94	92	90	88	87	85	83	81	80	78	76	75	73	71	70	68	67	65	62	60	57	54	52	49	47	45	43	41	36	32	28	24	18	210

Table A-27. Celsius Table of Relative Humidity or Per Cent of Saturation

Dry Bulb Deg. C.	0.5	1.0	1.5	2.0	2.5	3.0	3.5	4.0	4.5	5	6	7	8	9	10	11	12	13	14	15	16	18	20	22	24	26	28	30	32	34	36	38	40	Dry Bulb Deg. C.
2	92	83	75	67	59	52	43	36	27	20																								2
4	93	85	77	70	63	56	48	41	34	28	15																							4
6	94	87	80	73	66	60	54	47	41	35	23	11																						6
8	94	87	81	74	68	62	56	50	45	39	28	17																						8
10	94	88	82	76	71	65	60	54	49	44	34	23	14																					10
12	94	89	84	78	73	68	63	58	53	48	38	30	21	12	4																			12
14	95	90	84	79	74	69	65	60	55	51	41	33	24	16	10																			14
16	95	90	85	81	76	71	67	62	58	54	45	37	29	21	14	7																		16
18	95	90	86	82	78	73	69	65	61	57	49	42	35	27	20	13	6																	18
20	96	91	87	82	78	74	70	66	62	58	51	44	36	30	23	17	11																	20
22	96	92	87	83	79	75	72	68	64	60	53	46	40	34	27	21	16	11																22
24	96	92	88	85	81	77	74	70	66	63	56	49	43	37	31	26	21	14	10															24
26	96	92	89	85	81	77	74	71	67	64	57	51	45	39	34	28	23	18	13															26
28	96	92	89	85	82	78	75	72	68	65	59	53	47	42	37	31	26	21	17	13														28
30	96	93	89	86	82	79	76	73	70	67	61	55	50	44	39	35	30	24	20	16	12													30
32	96	93	90	86	83	80	77	74	71	68	62	56	51	46	41	36	32	27	23	19	15													32
34	97	93	90	87	84	81	77	74	71	69	63	58	53	48	43	38	34	30	26	22	18	10												34
36	97	93	90	87	84	81	78	75	72	70	64	59	54	50	45	41	36	32	28	24	21	13												36
38	97	94	90	87	84	81	79	76	73	70	65	60	56	51	46	42	38	34	30	26	23	16	10											38
40	97	94	91	88	85	82	79	76	74	71	66	61	57	52	48	44	40	36	32	29	25	19	13											40
42	97	94	91	88	85	82	80	77	74	72	67	62	58	53	49	45	41	38	34	31	27	21	15											42
44	97	94	91	88	86	83	80	77	75	73	68	63	59	54	50	47	43	39	36	32	29	23	17	12										44
46	97	94	91	89	86	83	81	78	76	73	68	64	60	55	52	48	44	41	37	34	31	25	19	14										46
48	97	94	92	89	86	84	81	78	76	74	69	65	61	56	53	49	45	42	39	35	33	27	21	16	12									48
50	97	94	92	89	87	84	82	79	77	75	70	65	62	57	52	48	44	40	37	34	30	26	21	18	14									50
52	97	94	92	89	87	84	82	79	77	75	70	66	62	58	55	51	48	44	41	38	35	30	25	20	16	11								52
54	97	95	92	90	87	85	82	80	78	76	71	67	63	59	56	52	49	45	42	39	36	31	26	21	17	13								54
56	97	95	92	90	87	85	83	80	78	76	72	68	64	60	57	53	50	46	43	40	38	32	27	23	19	15	11							56
58	97	95	93	90	88	85	83	80	79	77	72	68	64	61	57	54	51	47	44	42	39	33	29	24	20	16	12							58
60	98	95	93	90	88	86	83	81	79	77	73	69	65	62	58	55	52	48	45	43	40	35	30	26	21	18	14	11						60
62	98	95	93	91	88	86	84	81	79	78	73	69	66	62	59	56	53	49	46	43	41	36	31	27	23	19	15	12						62
64	98	95	93	91	88	86	84	82	80	78	74	70	66	63	59	56	53	50	47	44	42	37	32	28	24	20	17	13						64
66	98	95	93	91	89	86	84	82	80	78	74	70	67	64	60	57	54	51	48	45	43	38	33	29	25	21	18	15	12					66
68	98	95	93	91	89	87	85	82	81	79	75	71	67	64	61	58	55	52	49	46	44	39	34	30	26	22	19	16	13					68
70	98	96	93	91	89	87	85	83	81	79	75	71	68	65	61	58	55	52	50	47	44	40	35	31	27	23	20	17	14	11				70
72	98	96	94	92	89	87	85	83	81	80	76	72	69	65	62	59	56	53	50	48	45	40	36	32	28	24	21	18	15	12				72
74	98	96	94	92	90	87	85	83	82	80	76	72	69	66	63	60	57	54	51	48	46	41	37	33	29	25	22	19	16	13	11			74
76	98	96	94	92	90	88	86	84	82	80	76	73	70	66	63	60	57	55	52	50	47	43	38	34	30	27	24	21	18	15	13	10		76
78	98	96	94	92	90	88	86	84	82	81	77	73	70	67	64	61	58	55	52	50	47	43	39	34	30	27	24	21	18	15	13	10		78
80	98	96	94	92	90	88	86	84	83	81	77	74	71	67	64	61	58	56	53	50	48	43	39	35	31	28	24	22	19	16	14	11		80
82	98	96	94	92	90	88	86	84	83	81	77	74	71	68	65	62	59	56	54	51	49	44	40	36	32	29	25	22	20	17	15	12	10	82
84	98	96	94	92	90	88	86	85	83	81	78	74	71	68	65	62	59	57	54	52	49	45	40	37	33	29	26	23	20	18	16	13	11	84
86	98	96	94	92	91	89	87	85	83	82	78	75	72	69	66	63	60	57	55	53	50	46	41	37	34	30	27	24	21	19	17	15	13	86
88	98	96	95	93	91	89	87	85	83	82	78	75	72	69	66	63	60	58	55	53	51	46	42	38	34	31	28	25	22	19	17	15	13	88
90	98	96	95	93	91	89	87	85	84	82	79	76	73	69	67	64	61	58	56	53	51	47	42	39	35	32	28	26	23	20	18	16	14	90
92	98	97	95	93	91	89	87	86	84	82	79	76	73	70	67	64	61	59	56	54	52	47	43	39	36	32	29	26	24	21	19	16	14	92
94	99	97	95	93	91	89	88	86	84	83	79	76	73	70	67	65	62	59	57	54	52	48	44	40	36	33	30	27	24	22	19	17	15	94
96	99	97	95	93	91	90	88	86	84	83	80	76	74	70	68	65	62	60	57	55	53	49	45	41	37	34	31	28	25	22	20	18	16	96
98	99	97	95	93	92	90	88	86	85	83	80	77	74	71	68	65	63	60	58	55	53	49	45	41	38	34	31	28	26	23	21	19	16	98
100	99	97	95	92	92	90	88	86	85	83	80	77	74	71	68	66	63	60	58	56	54	49	45	42	38	35	32	29	26	24	22	19	17	100

Table A-28. Seal Fluids

Material	Sp Gr	Pour Point °F or Freezing Point	APPROXIMATE VAPOR PRESSURE MILLIMETERS OF MERCURY				Manufacturer or Source
			100°F	160°F	300°F	500°F	
Aquaseal[1]	1.05	Approx. −31 FP	.002	.020	1.2	100	Widely available
Nujol or other highly refined neutral mineral oil	.87–.90		.0002	.004			Any drug store
Fluorolube[2] FS-5	1.868	−60 P.P.	0.1	0.6	15	30	Hooker Industrial
S-30	1.927	12 P.P.	0.003	0.02	1.5	20	Chemicals
HO-125	1.953	60 P.P.	0.002	0.01	0.6	15	
Mercury	13.6	−38.8 C	.005	.05	2.6	96	Widely available
Halocarbon[3] 4–11	1.85	−110 P.P.	0.18	1.7	60	HIGH	Halocarbon Products Corp.
11–21	1.90	0 P.P.	<0.01	0.12	10	HIGH	
10–25	1.95	35 P.P.	<0.01	<0.01	1.2	190	

[1] Dibutylphthalate
[2] Trifluorovinyl Chloride Polymers
[3] Chlorotrifluoroethylene

In the measurement of pressure and flow, it is sometimes desirable to protect the instruments against dangerous and corrosive fluids—such as sulphuric, hydrochloric, and nitric acids; strong sodium or potassium hydroxide; the halogens, and many of the halogenated hydrocarbons. In these applications it is necessary to fill the connecting leads between the instrument and the source of measurement with a seal fluid which will remain stable under a wide range of operating conditions. Seal fluids should have the following characteristics:

1. Higher specific gravity than the measured fluid.

2. Inert to attack by, immiscible with, and not a solvent for, the measured fluid.

3. Stability at high ambient temperatures.

4. High boiling point, low vapor pressure.

5. Low pour or freeze point, low viscosity and flat viscosity curve.

6. Nontoxic on contact or inhalation, freedom from obnoxious odors.

No one fluid will meet the wide variety of conditions imposed by many installations. The seal fluids listed above have been reported by users or field service engineers as materials which have given satisfactory service. The data pertaining to these materials was abstracted from manufacturers' publications or technical data, and is assumed to be reliable.

Unless a user has had practical experience with any of these fluids, they should be considered on a quasi-experimental basis until the validity of their use is confirmed.

Table A-29. Psychrometric Chart

BAROMETRIC PRESSURE 29.92 in. Hg

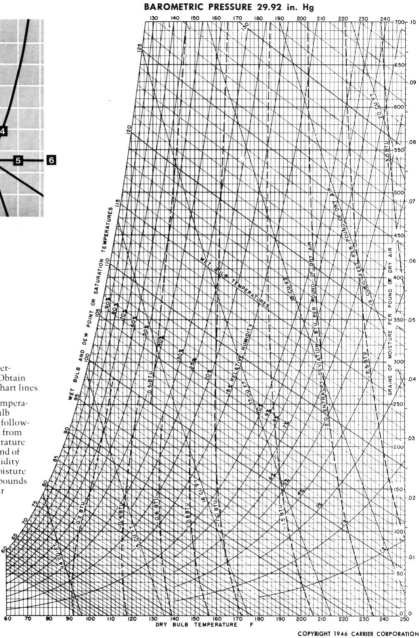

COPYRIGHT 1946 CARRIER CORPORATION
REPRINTED WITH PERMISSION

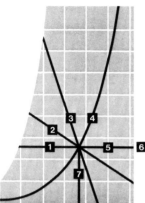

Legend

1 Dew Point Temperature
2 Wet Bulb Temperature
3 Cubic Feet Per Pound
4 Relative Humidity
5 Grains of Moisture
6 Pounds of Water
7 Dry Bulb Temperature

How to use chart

Locate point on chart at intersection of any two values. Obtain
other values by following chart lines.

Example: at a dry bulb temperature 7 of 172 F and a wet bulb
temperature 2 of 102 F, the following values may be obtained from
the chart: dew point temperature
1 87.5 F; cubic feet per pound of
dry air 3 16.7; relative humidity
4 10.3 percent; grains of moisture
per pound of dry air 5 201; pounds
of water per pound of dry air
6 .0287.

479

Accuracy. Conformity to an indicated, standard, or true value, usually expressed as a percentage (of span, reading, or upper-range value) deviation from the indicated, standard, or true value.

Amplification. The dimensionless ratio of output/input in a device intended by design to have this ratio greater than unity.

Amplifier. A device whose output is, by design, an enlarged reproduction of the input signal and which is energized from a source other than the signal. See *Gain.*

Amplitude ratio. The ratio of the magnitude of a steady state sinusoidal output relative to the input.

Analog computer. A computer operating on continuous variables. Compare *Digital computer.*

Analog simulator. An electronic, pneumatic, or mechanical device that solves problems by simulation of the physical system under study using electrical or physical variables to represent process variables.

Attenuation. A decrease in signal magnitude—the reciprocal of gain.

Automatic controller. A device, or combination of devices, which measures the value of a variable, quantity, or condition and operates to correct or limit deviation of this measured value from a selected (set point) reference.

Automatic control system. An operable arrangement of one or more automatic controllers along with their associated equipment connected in closed loops with one or more processes.

Automation. The act or method of making a processing or manufacturing system perform without the necessity of operator intervention or supervision. The common word designating the state of being automatic.

Block. A set of things, such as words, characters, digits, or parameters handled as a unit.

Bode diagram. A plot of log-gain and phase-angle values on a log-frequency base, for an element, loop, or output transfer function. It also comprises similar functional plots of the complex variable.

Breakpoint. The junction of two confluent straight line segments of a plotted curve.

Bulk storage. An auxiliary memory device with storage capacity many orders of magnitude greater than working memory; for example, disk file, drum, magnetic tape drives.

Bus. One or more conductors used to transfer signals or power.

Capacitance. The property that may be expressed as the time integral of flow rate (heat, electric current, and so on) to or from a storage divided by the associated potential change.

Capacity. Measure of capability to store liquid volume, mass, heat, information, or any form of energy or matter.

Cascade control system. A control system in which the output of one controller is the set point of another.

Chip. An integrated circuit.

Closed loop (feedback loop). Several automatic control units and the process connected so as to provide a signal path that includes a forward path, a feedback path, and a summing point. The controlled variable is consistently measured, and if it deviates from that which has been prescribed, corrective action is applied to the final element in such direction as to return the controlled variable to the desired value.

Compiler. A program that translates a higher level language like "BASIC" or FORTRAN into assembly language or machine language, which the CPU can execute.

Computer. A device that performs mathematical calculations. It may range from a simple device (such as a slide rule) to a very complicated one (such as a digital computer). In process control, the computer is either an analog mechanism set up to perform a continuous calculation on one or more input signals and to provide an output as a function of time without relying on external assistance (human prompting), or a digital device used in direct digital control (DDC).

Control accuracy. The degree of correspondence between the controlled variable and the desired value after stability has been achieved.

Control loop. Starts at the process in the form of a measurement or variable, is monitored, and returns to the process in the form of a manipulated variable or "valve position" being controlled by some means.

Control point. The value at which the controlled system or process settles out or stabilizes. It may or may not agree with the set point (instruction) applied to the controller.

Control system. A system in which deliberate guidance or manipulation is used to achieve a prescribed value of a variable.

Controlled system. That part of the system under control—the process.

Controlling means. The elements in a control system that contribute to the required corrective action.

CPU. Central processer unit—the portion of a computer that decodes the instructions, performs the actual computations, and keeps order in the execution of programs.

Cycling. A periodic change in the factor under control, usually resulting in equal excursions above and below the control point of sinusoidal wave shape—oscillation.

Damping. Progressive reduction in the amplitude of cycling of a system. Critically damped describes a system that is damped just enough to prevent overshoot following an abrupt stimulus.

Data. A general term to denote any information that can be processed.

Dead band. The change through which the input to an instrument can be varied without initiating instrument response.

Dead time, instrument. The time that elapses while the input to an instrument varies sufficiently to pass through the dead band zone and causes the instrument to respond.

Derivative action. Control action in which the rate of change of the error signal determines the amplitude of the corrective action applied. It is calibrated in time units. When subjected to a ramp change, the derivative output precedes the straight proportional action by this time.

Deviation. The departure from a desired value. The system deviation that exists after transients have expired is synonymous with offset.

Digital computer. A computer operating on data in the form of digits—discrete quantities of variable rather than continuous.

Dynamic behavior. Behavior as a function of time.

Equilibrium. The condition of a system when all inputs and outputs (supply and demand) have steadied down and are in balance.

Error. The difference between the actual and the true value, often expressed as a percentage of either span or upper-range value.

Feedback. Information about the status of the controlled variable that may be compared with information that is desired in the interest of making them coincide.

Final control element. Component of a control system (such as a valve) which directly regulates the flow of energy or material to the process.

Frequency. Occurrence of a periodic function (with time as the independent variable), generally specified as a certain number of cycles per unit time.

Frequency corner. That frequency in the Bode diagram indicated by a
breakpoint—the intersection of a straight line drawn asymptotically to
the log-gain versus log-frequency curve and the unit log-gain abscissa.

Frequency-response analysis. A system of dynamic analysis that consists of
applying sinusoidal changes to the input and recording both the input and
output on the same time base using an oscillograph. By applying these
data to the Bode diagram, the dynamic characteristics can be graphically
determined.

Gain (magnitude ratio). The ratio of change in output divided by the change in
input that caused it. Both output and input must be expressed in the
same units, making gain a pure (dimensionless) number.

Gain, loop. The combined output/input magnitude ratios of all the individual
loop components multiplied to obtain the overall gain.

Gain, margin. The sinusoidal frequency at which the output/input magnitude
ratio equals unity and feedback achieves a phase angle of − 180 degrees.

Gain, static (zero-frequency gain). The output/input amplitude ratio of a
component or system as frequency approaches zero.

Handler. A small program that handles data flow to and from specific pieces
of hardware for use by the other software.

Hardware. Physical equipment; for example, mechanical, magnetic,
electrical, or electronic devices. Something that you can touch with your
finger.

Hunting. Oscillation or cycling that may be of appreciable amplitude caused
by the system's persistent effort to achieve a prescribed level of control.

Hysteresis. Difference between upscale and downscale results in instrument
response when subjected to the same input approached from opposite
directions.

Input. Incoming signal to measuring instrument, control units, or system.

Instrument. In process measurement and control; this term is used broadly to
describe any device that performs a measuring or controlling function.

Instrumentation. The application of instruments to an industrial process for
the purpose of measuring or controlling its activity. The term is also
applied to the instruments themselves.

Integral control action. Action in which the controller's output is proportional
to the time integral of the error input. When used in combination with
proportional action, it was previously called reset action.

Integral time. The calibrated time on the controller integral (reset) dial which
represents the time that will elapse while the open-loop controller repeats
proportional action.

Integral windup. The overcharging, in the presence of a continuous error, of
the integral capacitor (bellows, in a pneumatic controller) which must
discharge through a longtime constant discharge path and which prevents
a quick return to the desired control point.

Integrator. Often used with a flowmeter to totalize the area under the flow record; (for example, gallons per minute × minutes = total gallons. It produces a digital readout of total flow.

I/O. Input/Output: The interface between peripheral equipment and the digital system.

Lag. A delay in output change following a change in input.

Laplace transform. A transfer function that is the operational equivalent of a complex mathematical function permitting solution by simple arithmetic techniques.

Limiting. A boundary imposed on the upper or lower range of a variable (for example, the pressure in a steam boiler as limited by a safety valve).

Linearity. The nearness with which the plot of a signal or other variable plotted against a prescribed linear scale approximates a straight line.

Load. Level of material, force, torque, energy, power, or other variable applied to or removed from a process or other component in the system.

Log gain. Gain expressed on a logarithmic scale.

Loop. A signal path.

Manipulated variable. That which is altered by the automatic control equipment so as to change the variable under control and make it conform with the desired value.

Measuring element. An element that converts any system activity or condition into a form or language that the controller can understand.

Memory. Pertaining to that storage device in which programs and data are stored and easily obtained by the CPU for execution.

Nichols diagram (Nichols chart). A plot of magnitude and phase contours of return-transfer function referred to ordinates of logarithmic loop gain and abscissae of loop phase angle.

Noise. Unwanted signal components that obscure the genuine signal information that is being sought.

Off-line. (1) Pertaining to equipment or programs not under control of the computer. (2) Pertaining to a computer that is not actively monitoring or controlling a process.

Offset. The difference between what we get and what we want—the difference between the point at which the process stabilizes and the instruction introduced into the controller by the set point.

On-line. (1) Pertaining to equipment or programs under control of the computer. (2) Pertaining to a computer that is actively monitoring or controlling a process or operation.

Open loop. Control without feedback; for example, an automatic washer.

Optimum. The highest obtainable proficiency of control; for example, supply equals demand, and offset has been reduced to a minimum (hopefully zero).

Oscillograph recorder. A device that makes a high-speed record of electrical variations with respect to time; for example, an ordinary recorder might have a chart speed of ¾ inch per hour while an oscillograph could have a chart speed of ¾ inch per second or faster.

Output. The signal provided by an instrument; for example, the signal that the controller delivers to the valve operator is the controller output.

Overdamped. Damped so that overshoot cannot occur.

Overshoot. The persistent effort of the control system to reach the desired level, which frequently results in going beyond (overshooting) the mark.

Phase. The condition of a periodic function with respect to a reference time.

Phase difference. The time, usually expressed in electrical degrees, by which one wave leads or lags another.

Process. The equipment for which supply and demand must be balanced—the system under control, excluding the equipment that does the controlling.

Program. A series of instructions that logically solve given problems and manipulate data.

Proportional band. The reciprocal of gain expressed as a percentage. Refers to the percentage of the controller's span of measurement over which the full travel of the control valve occurs.

Proportional control. Control action in which there is a fixed gain or attenuation between output and input.

Ramp. An increase or decrease of the variable at a constant rate of change.

Rate action. That portion of controller output that is proportional to the rate of change of input. See *Derivative action.*

Reaction curve. In process control, a reaction curve is obtained by applying a step change (either in load or set point) and plotting the response of the controlled variable with respect to time.

Real-time clock. A device that automatically maintains time in conventional time units for use in program execution and event initiation.

Reproducibility. The exactness with which a measurement or other condition can be duplicated over time.

Reset action. See *Integral control action.*

Reset time. See *Integral time.*

Reset windup. See *Integral windup.*

Resistance. An opposition to flow that accounts for the dissipation of energy and limits flow. Flow from a water tap, for example, is limited to what the available pressure can push through the tap opening

$$\text{electrical resistance (ohms)} = \frac{\text{potential (volts)}}{\text{flow (amperes)}}$$

Response. Reaction to a forcing function applied to the input; the variation in measured variables that occurs as the result of step sinusoidal, ramp, or other kind of input.

Routine. A small program used by many other programs to perform a specific task.

Self-regulation. The ability of an open-loop process or other device to settle out (stabilize) at some new operating plant after a load change has taken place.

Servomechanism. An automatic control system that takes necessary corrective action as the result of feedback. The output may be mechanical position or something related to it as a function of time.

Servotechniques. The mathematical and graphical methods devised to analyze and optimize the behavior of control systems.

Set point. The instruction given the automatic controller determining the point at which the controlled variable hopefully will stabilize.

Signal. Information in the form of a pneumatic pressure, an electric current, or mechanical position that carries information from one control loop component to another.

Software. The collection of programs and routines associated with a computer.

Stability. That desirable condition in which input and output are in balance and will remain so unless subjected to an external stimulus.

Static behavior. Behavior which is either not a function of time or which takes place over a sufficient length of time that dynamic changes become of minor importance.

Steady state. A state in which static conditions prevail and all dynamic changes may be assumed completed.

Step change. A change from one level to another in supposedly zero time.

Summing point. A point at which several signals can be algebraically added.

System. Generally refers to all control components, including process, measurement, controller, operator, and valves, along with any other additional equipment that may contribute to its operation.

Terminal. A device for operator-machine interface; for example, CRT's, typewriters, teletypers with keyboard input, or telephone modems.

Thermocouple. A device constructed of two dissimilar metals that generates a small voltage as a function of temperature difference between a measuring and reference junction. This voltage can be measured and its magnitude used as a measure of the temperature in question.

Time constant. The product of resistance × capacitance ($T = RC$), which becomes the time required for a first-order system to reach 63.2 percent of a total change when forced by a step. In so-called high-order systems there is a time constant for each of the first-order components.

Transducer. A device that converts information of one physical form to another physical type in its output (e.g., a thermocouple converts temperature into millivoltage).

Transfer function. A mathematical description of the output divided by input relationship which a component or a complete system exhibits. It often refers to the Laplace transform of output over the Laplace transform of input with zero initial conditions.

Transportation lag. A delay caused by the time required for material to travel from one point to another; for example, water flowing in a pipe at 10 feet per second requires 10.0 seconds to travel 100 feet, and if this 100 feet exists between manipulation and measurement, it would constitute at 10-second lag.

Ultimate period. The time period of one cycle at the natural frequency of the system where it is allowed to oscillate without damping.

Value. The level of the signal being measured or controlled.

Variable. A level, quantity, or other condition that is subject to change. This may be regulated (the controlled variable) or simply measured (a barometer measuring atmospheric pressure).

Zero frequency gain. Static gain or change in output divided by the change in input which caused it, after sufficient time has elapsed to eliminate the dynamic behavior components.

Zero shift. Change resulting from an error that is the same throughout the scale.

Note: A more complete set of definitions may be found in the ISA publication, "Process Instrumentation Terminology," ANSI/ISA 551:1, 1979.

1-1. c	2-3. a	3-3. a	5-1. a
1-2. c	2-4. c	3-4. b	5-2. c
1-3. c	2-5. b	3-5. d	5-3. b
1-4. b	2-6. c	3-6. a	5-4. c
1-5. a	2-7. c	3-7. b	5-5. a
1-6. a	2-8. b	3-8. a	5-6. a
1-7. d	2-9. c	3-9. c	5-7. c
b	2-10. b	3-10. c	5-8. c
h	2-11. b		5-9. d
e	2-12. b	4-1. d	5-10. c
a	2-13. c	c	5-11. c
g	2-14. c	a	5-12. b
c	2-15. a	b	5-13. a
f	2-16. c	e	5-14. b
1-8. c	2-17. c	4-2. a	5-15. c
1-9. d	2-18. a	4-3. b	5-16. d
1-10. a	2-19. c	4-4. d	5-17. c
1-11. a	2-20. b	4-5. d	5-18. b
1-12. c	2-21. b	4-6. d	5-19. c
1-13. a	2-22. c	4-7. a	5-20. b
1-14. b	2-23. d	4-8. b	
1-15. c	2-24. d	4-9. b	
		4-10. a	6-1. a
2-1. b	3-1. c	4-11. d	6-2. b
2-2. c	3-2. c	4-12. b	6-3. a

488

6-4. a

6-5. a

6-6. b

6-7. c

6-8. a

6-9. b

6-10. c

6-11. c

6-12. a

6-13. c

6-14. a

6-15. a

7-1. b

7-2. c

7-3. a

7-4. c

7-5. b

7-6. c

7-7. a

7-8. b

7-9. b

7-10. b

7-11. 45 seconds

7-12. c

7-13. b

7-14. c

7-15. d

7-16. a

7-17. a

7-18. b

7-19. a

7-20. c

8-1. a

8-2. T

8-3. T

8-4. F

8-5. T

8-6. T

8-7. a. T

 b. T

 c. T

 d. F

 e. T

 f. T

8-8. d

8-9. c

8-10. b

8-11. a

8-12. b

8-13. d

8-14. c

8-15. c

8-16. d

8-17. c

8-18. c

8-19. d

8-20. b

9-1. b

9-2. c

9-3. c

9-4. a

9-5. a

9-6. a

9-7. a

9-8. c

9-9. b

9-10. b

9-11. a

9-12. c

9-13. b

9-14. d

9-15. d

10-1. b

10.2. a

10-3. b

10-4. b

10-5. b

10-6. c

10-7. b

10-8. a

10-9. b

10-10. c

11-1. C_r max = 4.57

 C_r min = 0.355

 ½ ball valve

11-2. C_r max = 300

 C_r min = 18.9

 6-inch, equal percentage, globe valve

11-3. C_r max = 41.3

 C_r min = 7.6

 2-inch, equal percentage, globe valve

11-4. C_r max = 1407

 C_r min = 56.3

 8-inch butterfly

11-5. C_r max = 3.43

 ½-inch globe or ball valve

11-6. b

11-7. b

11-8. c

11-9. a

12-1. c

12-2. d

12-3. a

12-4. b

12-5. d

13-1. d

13-2. a

13-3. b

13-4. b

13-5. a

13-6. d

13-7. a

13-8. b

13-9. d

13-10. d

13-11. b

13-12. d

14-1. d

14-2. c

14-3. a

14-4. a

14-5. b

14-6. c

14-7. b

14-8. d	16-1. a	17-10. c	19-3. a
14-9. b	16-2. d	17-11. b	19-4. a
14-10. d	16-3. a	17-12. c	19-5. d
	16-4. b		
15-1. a	16-5. d	18-1. b	20-1. c
15-2. b		18-2. b	20-2. d
15-3. a		18-3. a	20-3. a
15-4. a	17-1. b	18-4. c	20-4. a
15-5. a	17-2. a	18-5. b	20-5. d
15-6. a	17-3. c	18-6. d	20-6. c
15-7. a	17-4. a	18-7. d	20-7. b
15-8. b	17-5. d	18-8. b	20-8. b
15-9. b	17-6. a		20-9. c
15-10. a	17-7. c		20-10. d
15-11. a	17-8. b	19-1. b	
15-12. d	17-9. a	19-2. c	

Books

Lavigne, J. R. *An Introduction to Paper Industry Instrumentation.* San Francisco: Miller-Freeman Publications, 1972.

Lipták, B. G. *Instrument Engineers Handbook,* vols. I and II. Radnor, PA: Chilton Book Company, 1969, 1970.

Shinskey, F. G. *Process Control Systems.* New York: McGraw-Hill Book Company, 1979.

Shinskey, F. G. *pH and pION Control in Process and Waste Streams.* New York: John E. Wiley & Sons, 1973.

Shinskey, F. G. *Distillation Control for Productivity and Energy Conservation.* New York: McGraw-Hill Book Company, 1977.

Shinskey, F. G. *Energy Conservation Through Control.* New York: Academic Press, 1978.

Spink, L. K. *Principles and Practices of Flow Meter Engineering,* ed. 9. Foxboro, MA: The Foxboro Company, 1967.

Periodicals   .

Computerworld (W). Newton, MA: Computerworld, CW Communications, Inc.

Control Engineering. Barrington, ILL: Technical Publishing Company, Dun & Bradstreet.

Instrumentation Technology (M). Pittsburgh, PA: Instrument Society of America.

Instruments and Control Systems (M). Radnor, PA: Chilton Company.

Absolute pressure, 34, 37
Absolute pressure transmitter, 64–65
Acidity. *See* pH measurement
Actuators, 259–68
  electrical signal conversion and, 263–67
  electric motor, 267–68
  piston-and-cylinder, 263
  valve, 226–67, 259–61
Algebraic summing point, 179, 180
Aspirating relay, 208–9
Atmospheric pressure, 34
Automatic controllers. *See* Controllers
Auto-selector control, 345–48

Bar, 37
Batch controller, 231–34
Bell instrument, 44, 45
Bellows, metallic, 49–52
Bernoulli's theorem, 91–92, 270
Blocks, 397, 399, 400
Bode diagram, frequency response analysis and, 329–31
Bourdon tube, 46–48

Bubble tube method of level measurement, 73–74
Buoyancy level transmitters, 71–73

Calibrator, portable pneumatic, 59–60
Capacitance, 78
Capacitance measurement, 175
Capacity
  of control valves, 272–73
  of a process, 14, 18
Cascade control system, 247–52, 352–57
Cavitation, 274–76
Cell constant, 153
Central processing unit (CPU), 373–74
CGS units of pressure, 35, 37
Chromatographs, 174
Cippoletti weir, 110
Closed loop control system, 179–201. *See also* Controllers
  gain and, 180–83
  nonlinearities in, 189
  oscillation in, 184–87
  phase shifts in, 184
  pneumatic, 223–31
  stability in, 187–88

Computer hardware, 372–85
Computer input, 373
Computer interface equipment, 377–85
Computer memory, 373–77
Computer software and operation, 388–400
  analog flow loop, 399
  blocks, 397, 399, 400
  control software, 394–400
  Foxboro Control Package (FCP), 399–400
  Foxboro Process Basic (FPB) language, 396
  off-line system, 389–90
  on-line system, 388–89
  real-time clock and power fail/restart logic, 394
Computer system process control, 378–85
Computers, basic elements of, 372–74
Conductance, 152–54
Conductivity level sensors, 79
Conductivity measurements, 151–59
  applications of, 157
  calibration of instruments for, 154–57
  construction of cells for, 156–57
  electrodeless, 157–59
  polarization effects and, 156
Controllers, 12–29
  adjustment (tuning), 295–303
    closed-loop cycling method, 301–2
    proportional-only controller, 295–96
    proportional-plus-integral controller, 296–97
    proportional-plus-integral-plus-derivative controller, 297–301
    reaction curve used to determine, 320–24
    tuning maps, 298–99
  auto-selector or cutback, 345–48
  cascade, 352–57
  control modes of, 189–200
    proportional-plus-integral control, 196
    throttling control (proportional-only control), 191, 193–95
    two position control, 189–191
  direct and reverse actions of, 16
  duplex or split-range, 343–45
  elapsed-time, 402–11
  flow-ratio, 349–52

  pneumatic. *See* Pneumatic controllers
  programmed, 402–11
  responses of, 16–29
  selection of, 200–1
  selection of the action of, 16
  sequential, 405–6
Control valves, 270–92
  capacity of, 272–73
  pressure drop across, 273–76
  rangeability of, 276–79
  selecting, 278–79
  sequencing, 279–81
  sizing of, 272, 278–80
  viscosity corrections for, 281
Core memory, 374–75
Cutback control, 345–48

Dall tubes, 96
DDC (direct-digital control), 378
Dead time, 6–7, 17–18, 335
Dead weight testers, 42–44
  as gravity dependent, 38–39
  pneumatic, 43–44
Density measurement, 81–87
  hydrostatic head, 82–84
  radiation, 84
  temperature effects and, 84
  vibration, 84
Derivative action, 25–29. *See also*
    Proportional-plus-derivative control action; Proportional-plus-derivative-plus integral control
  frequency response analysis and, 335–37
  of pneumatic controllers, 215–16
  proportional control and, 199–200
  in SPEC 200 system, 244–45
DEWCEL, 147–48
Diaphragm box, 74–75
Differential-gap control, 191
Differential pressure, 37
Differential pressure transmitter
  density measurement with, 85–87
  electronic, 238–41
  for level measurement, 75–78
Direct action, 16
Disk, 375–76
Diskette, 384
Displacement level transmitters, 71–73
Dissociation, 159–61

Distillation, feedforward control of, 369–70

Drag body flowmeter, 98–99

Drums, computer, 375

Duplex controller, 343–45

Elapsed-time controller, 402–5

Elbow taps for flow measurement, 97–98

Electric motor actuators, 267–68

Electrolysis, 156

Electromotive force (emf) of thermocouples, 131–33

Electronic control systems, 237–255. *See also* Feedwater control system, electronic

pneumatic systems compared to, 254–55

Electronic process simulator, 414–20

Exponential-rise transient, 6

Feedback control loop, 3, 179–201. *See also* Closed-loop control system

basic characteristics of, 9, 12–14

Feedforward control, 360–70

advantages of, 360–61

definition of, 360

of distillation process, 369–70

of heat exchanger, 361–69

history of, 360

Feedwater control system, electronic, 237–54. *See also* SPEC 200 system

closed-loop operation of, 252–54

controllers in, 241–43

general description of, 247–52

principle of operation of, 243–47

transmitters in, 238–41

Filled thermal systems, 128–30

Final actuator, 13

Flapper-nozzle units, 205–14

Flashing, 274–76

Float-and-cable devices, 70

v-notch weir with, 116

Flow measurement, 90–124

constriction or differential head type devices, 91–92

elbow taps, 97–98

flow nozzle, 96–97

flow rates pressure relationship, 100–3

orifice plates, 94, 100

Pitot tube, 97

primary devices, 94–99

secondary devices, 99–100

target (drag force) device, 98–99

temperature and flow rate, 103, 105–8

variable area meters (rotameter), 109

Venturi tube, 95–96

displacement method of, 91

open-channel, 109–16

velocity flowmeters, 117–23

Flow nozzle, 96–97

Flow rate

differential pressure related to, 100–3

temperature and, 103, 105–8

Flow-ratio control, 349–52

Flumes, 110–16

Force, 36

Force-balance pneumatic pressure transmitter, 60–65

Frequency ratio, 334

Frequency response analysis, 328–40

and Bode diagram, 329–31

closed-loop response and, 339–40

control objectives and, 335

and derivative action, 335–37

and integral action, 335, 337–39

and proportional band, 335, 339

testing a system with, 331–35

Gain, 183

frequency response analysis and, 329–31

Gain margin, 335

Galvanometric motor, 265

Gauge pressure, 34, 37

Gravity

force and mass and, 36

pressure measurement and, 38–39

Hair element, 145–46

Head, 33–34, 39. *See also* Pressure

Head level, measurement and, 73–78

Heat exchanger, 2, 3, 12

Humidity, relative and absolute, 144

Humidity measurements, 144–48

Hydrogen ion activity (pH). *See* pH

Instrument Society of America (ISA)

symbols used by, 9, 10

Integral action (reset), 23–25, 28–29. *See also* Proportional-plus-derivative-plus-integral control; Proportional-plus-integral action
    frequency response analysis and, 335, 337–39
    of pneumatic controllers, 215–16
    proportional control and, 196–98
Integral time, 196–97
Integral windup, 198, 216
Interface equipment, computer, 337–85
Ionization, 159–61
Ion-selective measurement, 173–74

Level measurement, 69–81
    capacitance, 78
    conductance, 79
    displacement (buoyancy), 71–73
    float-and-cable, 70
    head or pressure, 73–74
    radiation, 79
    thermal, 81
    ultrasonic, 80–81
    weight, 79–80
Limp or slack diaphragm instrument, 44–46
Liquid density. *See* Density measurement
Liquid pressure measurement, 48
Lo-Loss tubes, 96

Magnetic flowmeter, 117–21
Manometers, 39–42
    mercury, 99
Manual control unit of pneumatic controller, 216–19
Mass, 36
Measurement, 7–9, 13
Measuring transmitters, 8–9
Memory, computer, 373–77
Meniscus correction, 41
Metering pumps, 91
mho (reciprocal ohm), 152–53
Microprocessor, 382
Multicapacity system, 313–15

Newton (N), 36–37
Nichols diagram, 339

Off-line system, 389–90
Offset, 22, 28, 194–95

On-line system, 388, 89
On/off control, 19–20, 189–91
    step-analysis method and, 316
Open-channel flow rate measurements, 109–16
Orifice plates, 94
    tap locations for, 100
Oscillation, 184–87
Output transducer, resistance pneumatic transmitter used with, 141, 142
Oxidation-reduction potential measurements, 172–73

Parshall flume, 112–15
Pascal, 35–37, 39
Percent-incomplete method, 310–13
pH (hydrogen ion activity), definition of, 161
pH measurement, 159–72
    control system, 170–72
    glass electrode system, 164–65
    ionization or dissociation and, 159–61
    reading the output of pH electrodes, 169–70
    reference electrodes and, 165–68
    temperature compensation and, 168–69
Phase margin, 335
Phase shift, frequency response analysis and, 329
Piston-and-cylinder actuators, 263
Pitot tube, 97
Pneumatic amplifier, 207–9
Pneumatic calibrator, portable, 59–60
Pneumatic controllers, 205–34
    batch controller, 231–34
    closed-loop control system of, 223–31
    derivative and integral action of, 215–16
    electronic systems compared to, 254–55
    flapper-nozzle units of, 205–14
    manual control unit of, 216–19
    requirements of, 211–12
    set-point mechanism of, 220–23, 225–26
    single-seat equal percentage valve of, 227
    transferring from automatic to manual, 217
    transferring from manual to automatic, 217–19
    valve actuators of, 226–27

Pneumatic indicators, 53–55
Pneumatic process simulator, 420–26
Pneumatic recorders, 53–55
Pneumatic relay, 62–63
Pneumatic transmitter, resistance, 140–42
Polarization, conductivity measurements
    and, 156
Portable pneumatic calibrator, 59–60
Positive displacement meters, 91
Potentiometric recorder, 137–38
Pounds per square inch (psi), 38, 39
Power fail/restart logic, 394
Pressure
    absolute, 34, 37
    definition of, 33–34
    differential, 37
    gauge, 34, 37
    seals, 48, 55, 57–59
    units of measurement for, 35–38
Pressure drop across control valve, 273–76
Pressure gauges, 34, 37, 46–48
Pressure level, measurement and, 73–78
Pressure-measuring instruments, 39–52
    bell instrument, 44, 45
    bellows, 49–52
    calibration standards for, 39–40
    calibration techniques for, 59–60
    deadener or damper for, 48–49
    dead weight testers, 42–44
    gauges, 46–48
    manometers, 39–42
    seals and purges for, 48
    slack or limp-diaphragm instrument,
        44–46
Pressure recorders and indicators, 53–55
Pressure transmitters, 52–59
    absolute, 64–65
    differential, 238–41
    force-balance pneumatic, 60–65
Process control computer system, 378–85
Processes
    controllability of, 17–18
    types of, 5–9
Programmed control systems, 402–11
Proportional action, pneumatic transmitter
    and, 209–11
Proportional band, 20–23, 193
    frequency response analysis and, 335,
        339
    ultimate, 334

Proportional control, 20–23, 28–29
    offset and, 194–95
    step-analysis method and, 316–18
Proportional-only control(ler), 191, 193–95
    adjusting, 295–96
    applications of, 195
Proportional-plus-derivative control
    step-analysis method and, 319–20
Proportional-plus-derivative-plus-integral
    control
    step-analysis method and, 320
Proportional-plus-integral action
    step-analysis method and, 318
Proportional-plus-integral control(ler),
    196–98
    adjusting, 296–97
Proportional-plus-integral-plus-derivative
    controller, 199–200
    adjustment procedure for, 297–301
Psychrometer, 145, 146
Pulsation dampener, 48–49
Purges, 48
Pyrometers, radiation, 144

Radiation, density measurement, 84
Radiation level measurement, 79
Radiation pyrometers, 144
Rangeability of control valve, 276–79
Ratio control system, 349–52
Reaction curve, controller adjustments de-
    termined by using, 320–24
Real processes, 427–32
Real-time clock, 394
Recorder
    potentiometric, 137–38
    wheatstone bridge, 142–43
Redox measurements, 172–73
Relay, pneumatic, 62–63
Resistance thermal detectors (RTDs),
    139–44
    output transducer used with, 141, 142
    wheatstone bridge recorder used with,
        142–43
Resistance-to-pneumatic convertors,
    140–42
Reverse action, 16
Reynolds number, 95
Rotameter (variable area meter), 109

Seal pressure system, 48, 55, 57–59
Semiconductor memory, 376–77

Sequencing-control valves, 279–81
Sequential controllers, 405–6
Signal transmission system for pressure,
    52–53
Simulated processes (simulators), 413–26
    electronic, 414–20
    pneumatic, 420–26
Single-seal equal percentage valve, 227
SI units of pressure, 35–37, 39
Slack or limp-diaphragm instrument,
    44–46
Sling psychrometer, 146
SPC (set-point control), 378
Specific gravity. *See* Density measurement
SPEC 200 system, 241–52
    automatic/manual switch in, 247
    controller adjustments in, 247
    derivative action in, 244–45
    deviation signal generation in, 244
    external integral (-R) option in, 246
    external summing (-S) option in, 246
    final summing and switching in, 245
    high and low limits in, 245–46
    increase/decrease switch in, 244
    power supply fault protection circuit in,
        246–47
    proportional band action in, 245
Split-range control, 343–45
Square root extractor, 250–52
Steam pressure measurement, 48
Step-analysis method of finding time con-
    stant. *See* Time constant, step-
    analysis method of finding
Swamping resistors for average tempera-
    ture measurement with ther-
    mocouples, 135–36
Symbols, 9–11

Target flowmeter, 98–99
Temperature
    flow rate and, 103, 105–8
    pH measurement and, 168–69
Temperature measurement, 126–44
    filled thermal systems, 128–30
    resistance thermal detectors, 139–44
    thermistors, 144
    thermocouples, 130–39
        average temperature measurement,
            135–36
        calibration curves, 137

    differential temperature measurement
        with, 136
    potentiometric recorders, 137–38
    reference junction compensation,
        133–34
    transmitter, thermocouple-to-current,
        138–39
Temperature transmitter, pneumatic,
    209–11
Thermal level measurement, 81
Thermistors, 144
Thermocouples, 130–39
    average temperature measurement,
        135–36
    calibration curves for, 137
    differential temperature measurement
        with, 136
    potentiometric recorders used with,
        137–38
    reference junction compensation and,
        133–34
Thermocouple-to-current transmitter,
    138–39
Throttling control, proportional-only con-
    trol, 191, 193–95
Time constant
    Bode diagram used for finding, 331
    step-analysis method of finding,
        304–24
        block diagrams, 304–7
        finding control modes, 315–20
        multicapacity system, 313–15
        on/off control action, 316
        percent-incomplete method, 310–13
        proportional control action, 316–18
        proportional-plus-derivative control
            action, 319–20
        proportional-plus-derivative-plus-
            integral control, 320
        proportional-plus-integral action, 318
        reaction curve used to determine con-
            troller adjustments, 320–24
        single-time-constant system, 308–9
        two-time-constant system, 309–10
Time-cycle control, 191
Transducer, output, 141, 142
Transmitters
    displacement (buoyancy), 71–73
    pH, 170, 171
    pneumatic, 140–42

Transmitters, (*cont'd.*).
  pressure, 52–59
  thermocouple-to-current, 138–39
Tuning maps, 298–99
Turbine flowmeter, 122–23

Ultrasonic level sensor, 80–81
Upsets, 17

Valve actuators, 226–27, 259–61
Valve positioners, 15, 261–63
Valve relay, 207–8

Valves
  control. *See* Control valves
  single-seat equal percentage, 227
Velocity flowmeters, 117–23
Venturi tube, 95–96, 208
Vibration, density measurement, 84
Viscosity corrections for control valves, 281
Vortex flowmeter, 121–22

Weight system, level measurement, 79–80
Weirs, 110, 111, 115–116
Wheatstone bridge recorder, 142–43